Sustainable Development Goals Series

The **Sustainable Development Goals Series** is Springer Nature's inaugural cross-imprint book series that addresses and supports the United Nations' seventeen Sustainable Development Goals. The series fosters comprehensive research focused on these global targets and endeavours to address some of society's greatest grand challenges. The SDGs are inherently multidisciplinary, and they bring people working across different fields together and working towards a common goal. In this spirit, the Sustainable Development Goals series is the first at Springer Nature to publish books under both the Springer and Palgrave Macmillan imprints, bringing the strengths of our imprints together.

The Sustainable Development Goals Series is organized into eighteen subseries: one subseries based around each of the seventeen respective Sustainable Development Goals, and an eighteenth subseries, "Connecting the Goals", which serves as a home for volumes addressing multiple goals or studying the SDGs as a whole. Each subseries is guided by an expert Subseries Advisor with years or decades of experience studying and addressing core components of their respective Goal.

The SDG Series has a remit as broad as the SDGs themselves, and contributions are welcome from scientists, academics, policymakers, and researchers working in fields related to any of the seventeen goals. If you are interested in contributing a monograph or curated volume to the series, please contact the Publishers: Zachary Romano [Springer; zachary.romano@springer.com] and Rachael Ballard [Palgrave Macmillan; rachael.ballard@palgrave.com].

Ken-Ichi Akao · Ayu Washizu
Editors

Climate Change Issues and Social Sciences

Towards a Carbon Neutral Society

Springer

Editors
Ken-Ichi Akao
Faculty of Social Sciences
Waseda University
Tokyo, Japan

Ayu Washizu
Faculty of Social Sciences
Waseda University
Tokyo, Japan

ISSN 2523-3084 ISSN 2523-3092 (electronic)
Sustainable Development Goals Series
ISBN 978-981-96-8950-7 ISBN 978-981-96-8951-4 (eBook)
https://doi.org/10.1007/978-981-96-8951-4

This Springer imprint is published by the registered company Springer Nature Singapore Pte Ltd.
The registered company address is: 152 Beach Road, #21-01/04 Gateway East, Singapore 189721, Singapore

Preface

Worldwide efforts are underway to achieve the long-term goal outlined in the Paris Agreement: to "achieve a balance between anthropogenic emissions from sources and removals by sinks of greenhouse gases in the second half of this century." Accordingly, many countries and regions have set targets to achieve carbon neutrality by 2050.

Achieving a carbon-neutral society requires not only unprecedented scientific and technological innovations (such as hydrogen technology) but also fundamental transformations in social systems. In recent years, research on social system innovation has gained momentum along with advancements in science and technology. Realizing a carbon-neutral society requires new social visions, business models, and innovative legal, political, and market reforms, making interdisciplinary knowledge in the social sciences essential. Furthermore, as this transition will inevitably involve shifts in cultural and normative frameworks, interdisciplinary insights from the humanities are equally crucial.

This book combines social science and humanities research, offering a comprehensive and interdisciplinary approach to achieving a carbon-neutral society. It aims to support researchers and practitioners in their efforts while fostering the development of innovative technologies essential for this transition.

The authors contributing to this book come from diverse fields of expertise, yet most share an affiliation with the Institute for Advanced Social Sciences at Waseda University. In response to its 2021 Carbon Neutral Declaration, the university has actively promoted research, education, and industry-government-academia collaboration to accelerate the transition to a decarbonized society. The Institute for Advanced Social Sciences established the Carbon Neutral Research Division in 2021, providing a robust research platform for developing methodologies to tackle the practical challenges of achieving a carbon-neutral society. This book represents one of the key research outcomes of these efforts.

In compiling this volume, we challenged the authors not only to present an ideal vision of a "carbon-neutral society" but also, by drawing on their respective fields of expertise, to propose concrete pathways and methodologies for realizing it. We hope this book serves readers as a comprehensive resource on the essential measures needed for societal transformation toward carbon neutrality. Several chapters have focused on sustainable development in Asia, a region where political, economic, and cultural diversity intersect in complex ways. Accordingly, this book can also be characterized as a comprehensive study of the humanities and social sciences aimed at advancing carbon neutrality in the Asian context.

The following section outlines the structure of this book and provides an overview of each chapter.

Book Structure and Chapter Summaries

Part 1: Culture and Norms (Chaps. 1–3) examines various discourses and issues in the humanities, welfare economics, and ethics.

Part 2: Society and Community (Chaps. 4–8) explores the consensus-building processes necessary for the transition, drawing on perspectives from environmental sociology, public opinion research, empirical economics, and environmental law.

Part 3: Corporate and Business (Chaps. 9–12) focuses on the role of businesses alongside citizens as key drivers of the transition. It examines new corporate norms for a carbon-neutral society and current corporate efforts to address environmental challenges using insights from corporate law, multinational companies, sustainability activities, and marketing.

Part 4: Development and Environment (Chaps. 13 and 14) highlights the challenges of carbon neutrality in the context of sustainable development, using case studies from Southeast Asia.

Chapter Highlights

Chapter 1 explores the historical symbolism of chrysanthemums and dandelions in Japan and illustrates how minimalist community narratives reflect harmony with nature. It argues that reviving these stories can inspire cultural shifts toward carbon neutrality by reshaping the environmental imagination.

Chapter 2 examines the emissions gap, focusing on the disparity between current climate policies and the reductions necessary to meet the Paris Agreement's 1.5-degree target. It emphasizes the role of intergenerational altruism and equity in bridging the gap, highlighting these ethical principles as key to fostering support for ambitious climate goals.

Chapter 3 explores individual moral responsibility for climate-related activities such as driving cars and meat consumption. It argues that a framework of non-agential shared responsibility effectively captures the systemic interdependence underlying climate change and advocates a global moral code to guide individual actions based on this framework.

Chapter 4 discusses the importance of urban policies for carbon neutrality and addresses institutional barriers to sustainable land use and transportation. Using historical institutionalism, this chapter examines Tokyo's development and reveals how past policy decisions have shaped suburban car dependency and inner-city transit reliance.

Chapter 5 envisions rural communities as drivers of carbon-neutral energy systems that utilize local renewable resources. It integrates Japan's Circular and Ecological Sphere concept with input-output analysis to model transitions to sustainable regional economies.

Chapter 6 examines the role of renewable energy in a low-carbon society and the justice issues surrounding its adoption, including local impacts, the Not-In-My-Backyard movement, zoning, and post-Fukushima challenges.

Chapter 7 explores nuclear damage compensation systems as critical yet often-overlooked mechanisms for carbon neutrality in the nuclear sector. It evaluates Japan's post-Fukushima risk-sharing framework and offers insights into responsible nuclear power management.

Chapter 8 argues that major policy changes supported by public opinion are essential for a carbon-neutral society. It calls for increased climate coverage in traditional media, investigative journalism, and systems promoting incidental exposure to climate content in online media.

Chapter 9 analyzes Japan's amended Cabinet Office Ordinance, which mandates climate-related disclosures in annual securities reports. It evaluates these requirements against international standards and concludes that the ordinance promotes sustainable business practices.

Chapter 10 examines how multinational companies advance the United Nations' Sustainable Development Goals through localized strategies and circular economy models, aligning global resources with local needs to foster regional revitalization.

Chapter 11 explores how corporate governance, executive roles, and ownership structures influence firm sustainability and emphasizes the need for cohesive governance systems that support carbon neutrality.

Chapter 12 focuses on green logistics in Japan's trucking sector and proposes a four-phase action plan—energy reform, digital transformation, structural reform, and automation—to reduce CO_2 emissions significantly.

Chapter 13 investigates Bali's traditional Subak irrigation system and the challenges posed by tourism growth, highlighting the balance between economic development and environmental sustainability.

Chapter 14 frames climate change as a human-rights issue, focusing on vulnerable populations in the Global South. Using Cambodia as a case study, it underscores the importance of inclusive policies for sustainable development.

Chapter 15 synthesizes discussions from previous chapters through the lens of transition methodology, presenting three parallel stages for advancing carbon neutrality. It aims to inspire readers with concrete ideas for action.

On behalf of the authors,

Ken-Ichi Akao
Ayu Washizu

Acknowledgments The realization of a carbon-neutral society is a central challenge in achieving the United Nations' Sustainable Development Goals. Therefore, this book has been included in Springer Nature's Sustainable Development Goals Series. The publication of this book in this important series was made possible by invaluable advice and support from Ms. Juno Kawakami of Springer Nature. We express our sincere gratitude for this opportunity.

Contents

Editors and Contributors

About the Editors

Dr. Ken-Ichi Akao is a Professor of environmental economics in the Faculty of Social Sciences, Waseda University. He received his BS, MS, and Ph.D. in Agricultural Sciences from Kyoto University. Dr. Akao joined Waseda University in 1995. His research interests include sustainable development, the tragedy of the commons, environmental policies, and international environmental agreements. He applies nonlinear dynamic theory to economic analysis and environmental issues. Selected publications include: "Tax Schemes in a Class of Differential Games" (*Economic Theory*, 2008); "Some Rationalizability Results for Dynamic Games" (with Tapan Mitra and Gerhard Sorgner, *International Journal of Economic Theory*, 2012); "A Tradable Permit System in an Intertemporal Economy" (with Shunsuke Managi, *Environmental and Resource Economics*, 2013); and "A Theory of Disasters and Long-Run Growth" (with Hiroaki Sakamoto, *Journal of Economic Dynamics and Control*, 2018). He serves as co-editor of the academic journal *Environmental Economics and Policy Studies* (Springer), as editor of *International Review of Environmental and Resource Economics* (now publishers), and as editorial board member of *Chinese Journal of Population, Resources, and Environment* (KeAi). He was the President of the Society for *Environmental Economics and Policy Studie*s from 2016 to 2018.

Dr. Ayu Washizu is a Professor at the School of Social Sciences, Director of the Institute for Economic Analysis of Next-generation Science and Technology, and Vice Director of the Research Institute for Environment Economics and Management at Waseda University in Tokyo. Her research interests include input-output analysis, renewable energy, and smart society utilizing information and communication technology. Dr. Washizu has published her work in academic journals such as *Cities, Renewable and Sustainable Energy Reviews*, the *Journal of Environmental Management*, *Telecommunications Policy*, *Environmental Economics and Policy Studies*, the *Journal of the Japan Institute of Energy*, and *Economic Systems Research*. Dr. Washizu holds a Ph.D. in Commerce, an MSc in Economics, and a BA in Economics from Keio University. She has served on several Japanese local government committees on public works, including the Public Works Evaluation Committee of the City of Yokohama and the Port Council of the City of Kawasaki. Dr. Washizu has also served on the editorial boards of academic journals such as *Energy and Resources*. In 2005, she was awarded the Best Presentation Paper Award at the 4th International Symposium on Environmentally Conscious Design and Inverse Manufacturing EcoDesign. In 2010, she was awarded the Outstanding Paper Award from the Japan Society of Material Cycles and Waste Management.

Contributors

Ken-Ichi Akao Waseda University, Tokyo, Japan

Masahisa Endo Waseda University, Tokyo, Japan

Shinji Hasegawa Waseda University, Tokyo, Japan

Izumi Kawashima Waseda University, Tokyo, Japan

Satoshi Kuroda Waseda University, Tokyo, Japan

Satoshi Kurokawa Waseda University, Tokyo, Japan

Yudi Liu Waseda University, Tokyo, Japan

Naori Miyazawa Waseda University, Tokyo, Japan

Takashi Nozu Waseda University, Tokyo, Japan

Tomoo Noguchi Waseda University, Tokyo, Japan

Yuko Shimazaki Tokai University, Tokyo, Japan

Kohei Tabei Waseda University, Tokyo, Japan

Kanako Takae Waseda University, Tokyo, Japan

Yi Cheong Eugene Tung Chinese University of Hong Kong, Sha Tin, Hong Kong, China

Ayu Washizu Waseda University, Tokyo, Japan

Akira Yoshida Waseda University, Tokyo, Japan

Toru Yoshikawa Waseda University, Tokyo, Japan

Dandelion and White Chrysanthemum Drops: Study on Environmental and Cultural History and Study on Imagination History

1

Satoshi Kuroda

Abstract

Through literary works such as Yoko Tawada's "The Emissary" and historical documents, I will decipher the historical images of chrysanthemums, white chrysanthemums, and dandelions in Japan. There is a mixture of the unique narratives of minimalist communities and maximalist myths and legends. The establishment of modern society and the modern state created the illusion of a uniform nation-state narrative. The minimalist regional imagination, which is gradually being lost as a result of this, represents the ideal relationship between nature and human beings in Japan in the past. By focusing on the power of stories and narratives, studying the history of imagination over the environment, and restoring the minimalist community narratives of the past, we will have the power to remake the environment for a new age.

Keywords

Environmental history · History of the imagination · Narrative of minimal communities (local communities)

S. Kuroda (✉)
Waseda University, Tokyo, Japan
e-mail: kurosato@waseda.jp

K. Akao and A. Washizu (eds.), *Climate Change Issues and Social Sciences*, Sustainable Development Goals Series, https://doi.org/10.1007/978-981-96-8951-4_1

1.1 Introduction

This chapter is an approach to humanities research for a carbon-neutral society. "The climate crisis is a crisis of culture, and therefore a crisis of the imagination." The greatest contribution that the humanities can make to the transition to a carbon-neutral society in the face of the environmental crisis is to restore the power of the imagination. We need to continue to promote the idea that the imagination can be a force for reshaping the environment.

How did the Japanese of the past imagine the environment surrounding human beings? The role of stories and the cultural codes of imagination stand on the accumulation of historical narratives passed down in the minimal communities (=local communities). Concentrating on the power of such stories and narratives, and researching the history of the mentality surrounding the environment and the imagination surrounding the environment, will be an urgent issue for environmental history in the future.

1.2 Study on the History of Imagination as Environmental History

Although the new epoch called the Anthropocene has been rejected as a geological epoch, it is becoming increasingly important as a term that describes the current state of the environment. Human beings can no longer live completely within the sphere of life that they have created for themselves, and we are now in a situation where, as a result of man's increased interference with nature, man has also become a part of nature. In other words, this is an era in which nature's retortion is becoming more and more visible.

Humanistic study focusing on the environment is advancing. In Japan, first, environmental folklore was advocated early on; second, in literary studies, environmental humanities are being attempted in response to American ecocriticism; third, in cultural anthropology, perspectives on the extra-human, such as multispecies anthropology and more-than-human, have come to be emphasized; and fourth, environmental history has also been proposed in historical studies since the 1990s, and climate restoration research, disaster history, famine theory, and the history of livelihoods have been promoted. Among these, the history of human mind focusing on the environment has been the subject of studies pioneered by Hojo, Katsutaka (Masuo et al., 2003).

Under the environmental crisis, humanities study is expected to focus on the power of stories and narratives. Amitav Ghosh tells us that "the crisis of climate is a crisis of culture, and therefore a crisis of imagination" (Ghosh, 2022). This coincides with Donald Hughes' indicating the study on human being's thinking and attitude focusing on the natural environment as the third topic of the environmental history (Hughes, 2018). Masami Yuki, who is advancing environmental humanities, proposes a "readjustment of human thought" in our era in which "normalcy has come to an end."

This overlaps with the study of the history of imagination proposed more than half a century ago by Katsumi Masuda, a scholar of Japanese literature (Masuda, 1968). In other words, this is to "examine the content of the folk imagination" by "assembling the various fragments that have remained in later generations," or to utilize the "inverted method of proof which intends to examine residual culture to infer a far earlier age" in order to "establish the invisible content of the human imagination."

How did Japanese people in the past imagine the environment surrounding humans? The study of the history of human mind and imagination focusing on the environment will be an urgent topic for the future environmental history. The most important contribution that the humanities can make to the environment is to continue to send out the message that the power of imagination is the power to change the environment. The role of storytelling and the cultural codes of imagination are based on the accumulation of historical narratives passed down in the minimalist communities (=local communities). Here, I will quote Yoko Tawada's novel *The Emissary* as an example.

1.3 The Controversy over Chrysanthemum Versus Dandelion

The Emissary is a medium-length novel by Yoko Tawada published in 2014 (Tawada, 2014). The setting is Japan in the near future, where the earth has been irrecoverably polluted, cities have been abandoned, the economy has collapsed, the government has been privatized and the country has been closed off. Due to the pollution, the elderly cannot die, and children have lost their biological functions and are as weak as baby birds. *The Emissary* is recognized as a dystopian novel that emerged in Japan in the aftermath of the Great East Japan Earthquake.

The novel introduces two anecdotes about dandelions and chrysanthemums: the controversy over chrysanthemum versus dandelion and the story of peeing on the chrysanthemums drawn on the wall of washroom.

<Quote of Chrysanthemum-Dandelion Controversy>

The petals of modern dandelions are about ten centimeters long. There was even a case in which a person entered a dandelion in the annual chrysanthemum show held at the civic center, and it was questioned whether it could be recognized as a chrysanthemum or not. The opponents argued that a large dandelion was not a chrysanthemum but merely a mutation of the dandelion, but the controversy was sparked by the counterargument that "mutation is a discriminatory term." In fact, the word "mutation" was rarely used in this context anymore, and instead the term "environmental assimilation" was in vogue. If they remained small, they would have been in the shade while most of the wildflowers were growing huge. The dandelion must have changed the size to survive in the current environment. (snip) Some opponents of the dandelion argued that chrysanthemum was a noble flower that was chosen as a family crest, while the dandelion was a weed and should not be mixed up with chrysanthemums. On the other hand, the pro-dandelion alliance, formed by the labor union of ramen noodle shops, quoted a wise saying of the imperial family, "There is no such thing as a weed," putting an end to the seven-month-long chrysanthemum-dandelion controversy.

The year after the publication of *The Emissary*, Yoko Tawada wrote a short essay titled *Dandelions Dancing in the Snow* in (Tawada, 2015) which she referred to Yasunari Kawabata's unfinished last work *Dandelion* (Kawabata, 1972) and wrote, *Only this book shines in yellow unlike other works as a weed dandelion that has been mixed in with the chrysanthemum show.* Tawada imagines a lone dandelion blooming forlornly in the garden of Kawabata's residence which had been almost abandoned in the midst of the award hubbub, just as the unfinished novel that Kawabata stopped writing after winning the Nobel Prize for Literature:

The novel *Dandelion* has a very unique position in Kawabata's literature, probably because of the plant "dandelion." *The dandelion is also a member of the chrysanthemum family, but it is not as valued as chrysanthemums. (snip) Peach branches are displayed on Peach Festival, bamboo on Tanabata Festival, cherry blossoms are enjoyed in Hanami (cherry-blossom viewing) season and Japanese pampas grass are displayed with the full moon on the fifteenth night, but dandelions do not have their place. The dandelion is one of our closest and most beautiful neighbor, but it is a flower that blooms outside of the culture of Japan.*

A flower that blooms outside of Japanese culture. Long before the common dandelion spread throughout Japan in the modern era, native species such as kanto dandelion existed. However, for some reason, there is not a single *waka* poem referring to the dandelion in Japan, and it was only in the early modern period that the dandelion came to be written in *haiku* poem.

The chrysanthemum-dandelion controversy contrasts two types of nature in Japanese culture:

Fig. 1.1 White chrysanthemums

"wild" and "tamed" nature. One is the chrysanthemum, an artificial flower with a single large petal which has been bred by human beings, and the other is the dandelion which grows as a weed in its native state, unnoticed and untouched by human beings, in the concrete under our feet, as if it were air.

There is a drama series on NHK (Japan Broadcasting Corporation) titled "*Ranman*," which was aired in the spring of 2023. Episode 105 is a scene from a chrysanthemum show. The main character, a botanist named Mantaro Makino, and his wife, Sueko, exhibit a single small *nojigiku* at a chrysanthemum show held by a wealthy merchant named Yanosuke Iwasaki (Fig. 1.1).

Sueko said, "It was a long time ago. Chrysanthemums were originally used to make medicine in the Tang Dynasty. When they came to Japan, Japanese people took great pains to create these very beautiful chrysanthemums. However, the original chrysanthemums also grew wild in this country. Its Japanese name is *nojigiku. Nojigiku* has been completely untouched by human hands for more than 1,000 years. It grows on sunny side rocks and cliffs by the sea and blooms in its original form. There is no superiority or inferiority of either chrysanthemum. However, if we have both *nojigiku* and these chrysanthemums, we can think of the continent and the sea, and the ingenuity of the Japanese people over centuries. And above all,"

Mantaro Makino (in flashback scene): "Above all, it is heartwarming to know that the people of this country love flowers so much, isn't it? If everyone had the love for flowers, there would be no conflict in the world."

Here, the contrast is not between chrysanthemums and dandelions, but between chrysanthemums which have been repeatedly bred and improved by man, and the wild *nojigiku* which has been blooming in Japan for more than 1,000 years, untouched by man. The "wild" and the "tamed" nature. The contrast is similar to that between the chrysanthemum and the dandelion in *The Emissary*.

1.4 Nature as the Other

In 2014, the same year that Yoko Tawada wrote *The Emissary*, she visited Fukushima. Let me quote again from *Dandelions Dancing in the Snow* (Tawada, 2015):

> *Another movie that comes to mind when I hear the word "dandelion" is Akira Kurosawa's Dream (1990). This film was made almost twenty years before the nuclear accident in Fukushima, but it is about a nuclear power plant in Japan that caused a huge explosion in an accident. Japan has already almost killed itself, and giant dandelions are growing there with their genes destroyed by radioactive materials. I was reminded of this movie when I visited Fukushima last year. When I was driving an area that had been contaminated by radioactive materials and was uninhabitable, I saw tall weeds growing up to cover the uninhabitable houses. There was also a row of magnificent cherry-blossom trees in the middle of an uninhabited town. The woman who showed us around told us that even before the nuclear accident, these cherry trees were famous, but now that the town was no longer inhabited, they were blooming many times more than ever before. (snip) This is not a fear from the nature itself. In an essay Cherry Blossoms in Full Bloom, Ango Sakaguchi writes that he found it odd that only the cherry blossoms were in full bloom while people were dying in the Great Tokyo Air Raids. Plants seem to bloom and wither reflecting people's joys and sorrows, but in fact they are others to us. Even if mankind destroys itself through war or nuclear power plants, cherry trees will still bloom profusely without shedding a tear.*

<Nature as the Other> This reminds me of the article *Thinking from the Standpoint of the Mountain*, published in 1944 by Aldo Leopold, a pioneer of environmental ethics (Leopold, 1986). This view of nature has often been depicted in Japanese novels. For example, Kazushi Hosaka's novel *Memories of the Seasons* is a story about the daily life of a father and son living in Kamakura (Hosaka, 1996). The protagonist, who has climbed a mountain near Kamakura, tells the following story:

> *No matter which direction you look from, a mountain is still a mountain and is the same mountain, and even if people don't see it, mountains and oceans exist, and they are, how to say, fully self-sufficient. Therefore, it really does not matter to the mountains and oceans that their beauty would not be known if humans had not existed. They exist in their own dynamism, or rather, they are constantly changing little by little.*
>
> *I said that the mountains were self-sufficient without being seen by humans, and the sea was self-sufficient in the sea, but this does not mean that they are isolated or unrelated to others. The nuts need to be brightly colored to stand out and be eaten by birds, which then drop them with their droppings in a different place than where they are growing now. The fact that the colors used to attract birds' attention are also perceived as beautiful by humans may indicate that human sensibilities are not completely separate from those of the animals that eat the berries.*

Human beings living in the Anthropocene have become rootless beings, disconnected from nature. However, *human sensitivity is not completely disconnected from the animals that eat the fruit of the trees*. The unique sensitivity of Japanese culture, in which nature is the other for humans, but not completely the other, seems to be represented by the dandelions that Yoko Tawada experienced in Fukushima and described in *The Emissary*.

1.5 Drops from *Shiragiku* (White Chrysanthemum)

In *The Emissary*, children pee on the chrysanthemum flowers drawn on the wall of the colorful and fun co-ed "washroom" at school (the foreign word "toilet" is forbidden because the country is in national seclusion).

Here, I would like to note that, in Japan's Kaetsuno region (Toyama, Ishikawa, and Fukui prefectures), the image of white chrysanthemums and holy water have been strongly associated in the local stories.

For example, (1) in Kyoka Izumi's long novel *Yukari no Onna* (woman with a relation) written in 1919, the main character, a man, is lured to a strange place called Shiragiku Valley deep in the mountains in the east of Kanazawa, Kaga, where he encounters a mysterious woman (Izumi, 1928). In the fictional place Shiragiku Valley, a sculpin is said to have a flower bloom on its head with a drop of chrysanthemum. It is also said that soaking one's body in a stream of fresh water spilled from these drops of chrysanthemum will heal one's illness. The image of chrysanthemums in Shiragiku Valley was strongly associated with this image of immortality in the Immortal Land.

In addition, (2) *Kikusake no Kojitsu* (legend concerning chrysanthemum liquor) in *Hokkoku Kidan Junjoki* Volume 1, issued in 1807 (Chosuidai, 1974), describes the origin of Kaga's sake called Kikusake, which is said to be made from medicinal water dripping from the chrysanthemums that bloom in the Asano River's water source, the Shiragiku Pool. Kyoka Izumi probably wrote a novel based on this strange story (Fig. 1.2).

Furthermore, (3) *Ishikawa-ken Nomi-gun shi* (history of Nomi county, Ishikawa prefecture) contains a story about a lumberjack named Rihei Ashitani, who wanders around in despair after his son is kidnapped by a *tengu* (Nomi-gun, 1923). At the end of the story, it says, *Rangiku growing under the cliffs are fragrant with the brocade woven by Princess Tatsuta, but when someone tries to pick them, you may suddenly hear the noise of drums.* In addition, (4) according to Bakusui Hori's *Sanshu Kidan* written in the latter half of the eighteenth century, a *shojo* (imaginary animal like an orangutan) appears in Idogiku Valley at the foot of the Etchu Tateyama mountain range, blowing wind and clouds (Hori, 1932). According to the story, "*The first time a person enters a place called Idogiku Valley, wind and cloud blows preventing human beings from entering. This often throws people away*" (5). Basho Matsuo, who visited Kaga Yamanaka hot spring at the end of the seventeenth century, is said to have composed the following poem: *In Yamanaka, refraining from picking chrysanthemums in the smell of the hot spring.*

Fig. 1.2 A *shojo* (imaginary animal like an orangutan) Wakan Sansai Zue, seventeenth century

The image of small white chrysanthemums blooming unnoticed, their petals swaying in the wind and releasing an aromatic fragrance, and the holy water of immortality that slipped down the chrysanthemums was widely spread in Hokuriku region. This should be due to the widespread legend of chrysanthemum water associated with Kikurihime-no-mikoto, the goddess of Mt. Hakusan, the holy mountain in the region (Muroyama, 2015; Tomita, 1823).

The image of holy water that slips down chrysanthemums has its origin in ancient China. A child named Jidou, who was loved by King Mu in ancient China's Zhou Dynasty, was banished to the deep mountains for the crime of straddling a pillow by mistake. It is said that Jidou drank water seeping from a chrysanthemum in the mountains and lived to be 800 years old. The image of chrysanthemums as elixir that grow in the Immortal Land spreads with the Chou-you-no-sekku festival (one of the five season-related festivals). In Japan, the story of the chrysanthemum and Jidou, which made hin called Kiku-Jidou has been told continuously since the end of the Kamakura period as a secret tale related to the Tendai accession law of the medieval emperors. The legend that the water produced with drops of chrysanthemum brought immortality and served as holy water or good medicine prevailed throughout Japan (Abe, 2020).

From ancient times, chrysanthemum was the flower of immortality that brought forth holy water deep in the mountains, with its origin in legends about Chinese emperors and Japanese emperors. In modern times, on the other hand, chrysanthemum became a popular term for anus. It is no coincidence, then, that chrysanthemums in the flower beds drawn on the wall in *The Emissary* are peed on, in contrast to the holy water that trickles through the chrysanthemums. Perhaps this dystopian novel was written as a device to highlight the two contradictory characteristics of chrysanthemums.

1.6 Imagination Held by Minimal Regions

In the Kaetsuno region, white chrysanthemums were flowers that wafted their fragrance in the wind in the Immortal Land deep in the mountains and brought forth holy water of immortality that healed illnesses. On the other hand, since the Middle Ages, the water created with drops of chrysanthemums, as told in the tale of Kiku-Jidou, was holy water that brought immortality and served as good medicine. As the tale of Kiku-Jidou was a tale of a medieval emperor, and the chrysanthemum crest was adopted as the crest of the emperor's family, chrysanthemum came to bear the ideology of the nation-state, especially in the modern age and thereafter.

White chrysanthemum and chrysanthemum, or the white chrysanthemum orally transmitted in the Kaetsuno region and the chrysanthemum that continued to live on in a mythology of medieval emperor as the tale of Kiku-Jidou are similar and different.

In pre-modern Japan, minimal communities, or local communities, produced stories that were unique to the respective communities. And smaller, more limited communities were differentiated from each other, and because of the locality, memories were segmented and each community created its own unique narrative. In the Kaetsuno region, *shiragiku* cures illnesses as refered to in (1) above, makes beautiful liquor (2), and also produces a famous hot spring (5), celebrating the blessing of water from Hakusan Gongen = Kikurihime-no-mikoto, the local deity. The area has spots with specific names such as "Shiragiku Valley" (1), "Shiragiku Pool" (2), and "Idogiku Valley" (4). In addition, inhuman creatures such as a mysterious woman (1), a *tengu* (3), and *shojo* (4) blow wind, cloud, and fragrance, preventing people from entering and leaving the area and issuing warnings to humans from nature. The history unique to the locality is vividly preserved and inscribed in the minimal community, and has been recounted over and over again.

The unique narratives of the minimal communities thus created have been developed through incorporating and mutually affecting with ancient and medieval myths, maximum narratives and omens. With the birth of modern society and the modern state, however, minimal narratives gradually lost their effect and were replaced by the illusion of a uniform nation-state. The disappearing minimal regional imagination represents the ideal relationship between nature and human beings in Japan in the past. It should have functioned as a resilience that was both tough and flexible in the age of the Anthropocene.

1.7 Concluding Remarks

In this chapter, I have examined the historical images of chrysanthemums, white chrysanthemums and dandelions in Japan, as seen in literary works such as Tawada Yoko's *The Emissary* and historical documents. There, I find a mixture of the unique narratives of minimal communities and the myths and legends of the maximum. The establishment of modern society and modern nation-states created the illusion of a standardized nation-state narrative. The minimal regional imagination that is gradually being lost represents the ideal relationship between nature and humans in Japan of the past. The restoration of imagination surrounding the environment, with the aim of achieving a carbon-neutral society, should become a force that reshapes the environment of the new era.

This article is based on a presentation given at the review session featuring "*Bungaku ha chikyu wo sozo suru*" (Literature Imagines the Earth) written by Masami Yuki organized by Association for the Study of Literature and Environment (ASLE-JAPAN) and held at Kanto Gakuin University on July 20, 2024. I would like to thank Masami Yuki and the participants of the session.

References

Abe, Y. (2020). *Chusei Nihon no oken Shinwa* (The myth of royal authority in medieval Japan). Nagoya: Nagoya daigaku shuppankai.

Chosuidai, H. (1974). *Hokkoku kidan junjoki.in Nihon zuihitsu taikei*, 2–18. Tokyo: Yoshikawa kobunkan.

Ghosh, A. (2022). *Oinaru Sakuran* (The Great Derangement). Tokyo: Ibunsha.

Hori, B. (1932). *Sanshu kidan.* Tshikawa ken toshokan kyokai.

Hosaka, K. (1996). *Kisetsu no kioku* (The memories of the seasons). Tokyo: Chuokoronsha.

Hughes, J. D. (2018). *Kankyoshi Nyumon* (What is environmental history?) Tokyo: Iwanami shote n.

Izumi, K. (1928). *Yukari no onna* (Woman with a relarion). Tokyo: Chikumashobo.

Kawabata, Y. (1972). *Tanpopo* (Dandelions). Tokyo: Shinchosya.

Leopold, A. (1986). *Yasei no uta ga kikoeru* (A Sand Country Akmanac and Skeches Here and There). Shinrinsyobo.

Masuda, K. (1968). *Kazan-retto no Shiso* (Ideology of the Volcanic Archipelago). Tokyo: Chikuma shobo.

Masuo, S., Kudo, K., & Hojo, K. (2003). *Shizen to kankyo no shinseishi* (History of the Mind of Nature and the Environment). Tokyo: Benseisha.

Muroyama, T. (2015). *Chusei Kaga Kikuzake no Syakaishi* (Social history of Kaga Kikusake in the Middle Ages). In Hokuriku shigaku, 63.

Nomi-gun. (1923). *Ishikawa-ken Nomi-gun shi* (History of Nomi-gun, Ishikawa prefecture). Tokyo: Rinsen Shoten.

Tawada, Y. (2014). *The Emissary*. Tokyo: Kodansha.

Tawada, Y. (2015). *Yuki no naka de odoru tanpopo* (Dandelions dancing in the snow). Bungaku, 16-3.

Tomita, K. (1823). *Kaga no kuni Kikuzake ko* (A study of Kikuzake in Kaga Province).

Yuki, M. (2022). *Seijo no shuen shiko no saichosei* (The end of normalcy, readjustment of thinking) in Shiso, 1183.

Yuki, M. (2023). *Bungaku ha chikyu wo sozosuru* (Literature imagines the earth). Tokyo: Iwanami shoten.

Intergenerational Altruism and Intergenerational Equity: The Source of Emissions Gap

2

Ken-Ichi Akao

Abstract

The emissions gap refers to the difference between the projected greenhouse gas emissions trajectory based on currently planned climate change mitigation policies and the efficient emissions path required to achieve the 1.5-degree (or 2-degree) target agreed upon in the Paris Agreement under the United Nations Framework Convention on Climate Change. This disparity reflects the gap between what we intend to do for future generations (intergenerational altruism) and what we believe should be done for them (intergenerational equity). Resolving the emissions gap is crucial for achieving a carbon-neutral society. Addressing this challenge requires reinforcing intergenerational altruism and garnering significant support for the ethical foundation of intergenerational equity underlying the 1.5-degree target. Based on these concerns, this study introduces and examines various discussions in the social sciences, particularly in economics, regarding intergenerational altruism and intergenerational equity.

K.-I. Akao (✉)
Waseda University, Tokyo, Japan
e-mail: akao@waseda.jp

Keywords

Emissions gap · Intergenerational altruism · Intergenerational equity · Preferences · Morality

2.1 Introduction

The emissions gap, shown in Fig. 2.1, refers to the difference between the greenhouse gas emission trajectory (hereinafter, NDC path) projected based on the current climate change mitigation policies (Nationally Determined Contributions, NDCs) of countries and the cost-effective greenhouse gas emission trajectory (hereinafter, 1.5-degree emissions path) required to achieve the 1.5-degree (or 2-degree) target agreed upon in the Paris Agreement under the United Nations Framework Convention on Climate Change (UNFCCC).

If the 1.5-degree emissions path is implemented, CO_2 emissions will be nearly zero by the middle of this century (United Nations Environment Programme, 2023). Therefore, achieving a carbon-neutral society equates to the world resolving the emissions gap and aligning with the 1.5-degree emissions path. For this reason, the United Nations has called on each country to ambitiously review its mitigation policies. As a result, many countries announced plans

K. Akao and A. Washizu (eds.), *Climate Change Issues and Social Sciences*,
Sustainable Development Goals Series, https://doi.org/10.1007/978-981-96-8951-4_2

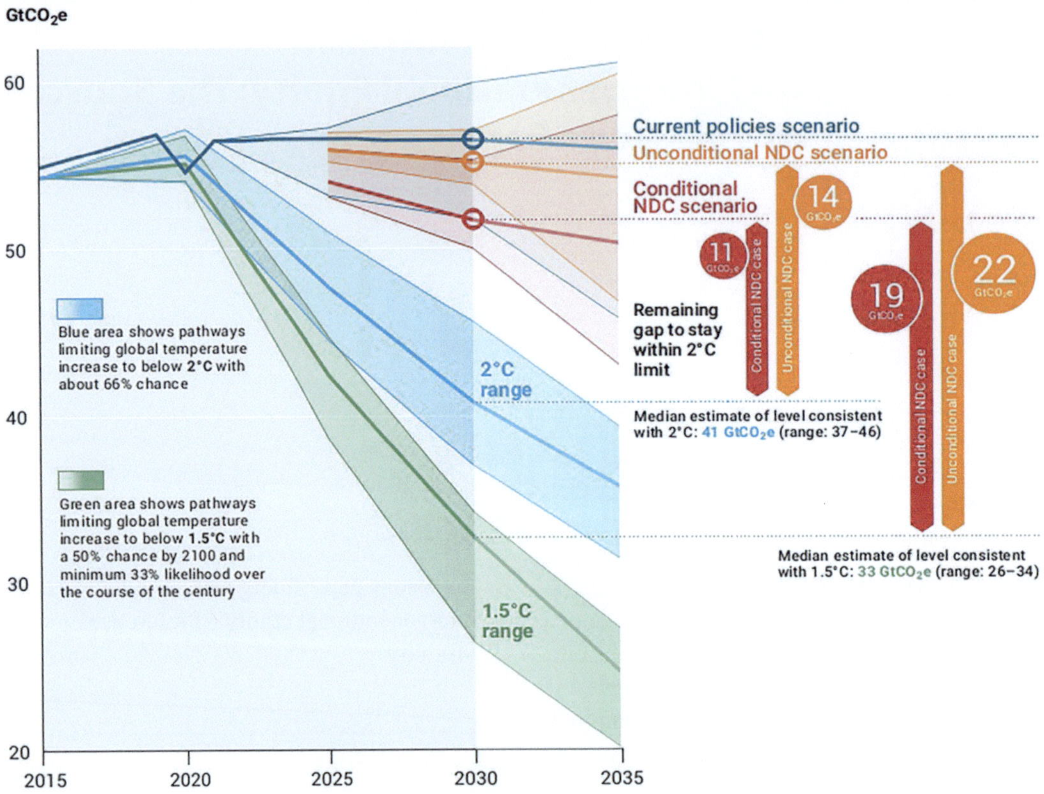

Fig. 2.1 Emissions gap. *Source* UNEP (2023, p. XXI, Fig. ES.4)

to strengthen their emission reduction targets at the 2021 UNFCCC COP26, marking a groundbreaking event. However, there is still no clear prospect of closing the gap. At COP28 in 2023, the first evaluation of the global stocktake, which evaluates the progress of the world's climate change measures every five years based on the Paris Agreement, expressed the following concern:

> [S]ignificantly greater emission reductions are required to align with global greenhouse gas emission paths in line with the temperature goal of the Paris Agreement and recognizes the urgent need to address this gap. (United Nations, 2023)

Effective mitigation policies require massive investments to close the emissions gap. For example, the Intergovernmental Panel on Climate Change (AR6, WG 3, Chap. 15) estimates that the annual funds required for mitigation policies range from USD 2.3 trillion to USD 4.5 trillion and states that an increase in investment by approximately 3 to 6 times compared to current levels is necessary (Kreibiehl et al., 2022). Can the current generation implement such bold policies and substantial investments?

The extraordinarily long-term nature of climate change complicates this issue. Climate change takes decades or even longer to become evident; however, once it begins, it continues for thousands of years. Due to its ultra-long-term nature, even though greenhouse gases have accumulated in the atmosphere since the onset of industrialization in the eighteenth century, anthropogenic climate change has only been recognized very recently. This is a major reason behind the delay in climate action today. The ultra-long-term nature of climate change also implies that it will take a considerable time for current measures to produce noticeable effects. A sobering prediction is that even if anthropogenic greenhouse gas emissions were immediately reduced to zero, sea levels would continue to rise for thousands of years (Intergovernmental Panel on Climate Change, 2021).

An important implication of the ultra-long-term nature of climate change is that future generations, whom we will never meet and whose lifestyles and preferences we cannot predict, will benefit more from the current mitigation policies than us. However, serious damage due to climate change is already occurring worldwide; therefore, policies for climate change adaptation for the current generation are a pressing concern. We must allocate limited budgets and resources between mitigation policies to save future generations and adaptation policies to save the current generation. Furthermore, the current generation is facing several serious social issues. These include poverty, conflict, unemployment, economic stagnation, aging, population decline, and an aging infrastructure. Many resources are required to solve these problems that directly affect our well-being.

How much of our limited resources will we allocate to close this emissions gap? These resources could also be used by us, our families, and others in our generation who face difficult situations. How willing are we to use these resources for future generations, to forgo using them for people living today, and especially to prioritize climate change mitigation measures over other pressing social issues? The ultra-long-term nature of climate change raises fundamental questions about the allocation of resources between current and future generations across thousands of years. This is a challenge that humanity has rarely faced.[1]

A succinct answer to this intergenerational resource allocation problem is the NDC path, which reflects our actual approach to future generations based on our preferences. But what does the 1.5-degree emissions pathway represent? It illustrates the moral standard of how future generations should be treated. Thus, the NDC pathway expresses what we *intend to* do for future generations, while the 1.5-degree pathway represents what we *should* do for them. The emissions gap, therefore, highlights the divergence between our preferences and our morality toward future generations. This challenge can be reframed as the task of reconciling our preferences for future generations—referred to here as *intergenerational altruism*—with our morality toward them, termed *intergenerational equity*.[2]

To address this challenge, it is essential to understand what intergenerational altruism is, what intergenerational equity is, and how they are related. This study provides an exploratory examination of these issues. The remainder of this paper is organized as follows: The next section introduces a divergence problem in economics that parallels the emissions gap: the difference between the socially optimal greenhouse gas emissions path derived from standard economic models and the emissions path many economists consider morally correct. The former is comparable to the NDC path, suggesting that the NDC path possesses economic rationality. However, like the NDC path, this optimal path lacks sustainability, casting doubt on its economic rationality. By contrast, although the latter resembles the 1.5-degree emissions path, the two are grounded in different moral principles. This points to the fact that the morality of intergenerational equity is not unique. The question raised here is what

[1] The threat of nuclear war serves as a precedent for ultra-long-term issues because it imperils humanity's survival. What sets climate change apart from the threat of nuclear war is that its cause is created by the myriad lives and economic activities of people worldwide. Controlling the problem is extremely difficult because it involves the numerous decisions made by people worldwide. A comparable challenge to climate change is the biodiversity loss. This loss shares an ultra-long-term nature because it is irreversible. Moreover, habitat destruction and pollution stem from the widespread human activities worldwide. While the impacts of biodiversity loss have not received as much attention as the problem of climate change, the issue is often likened to the screws of an airplane: a few missing screws are not a problem, but if the number increases, the plane will crash. The current efforts to address climate change are also a touchstone for solving the problem of biodiversity loss.

[2] Preferences can be understood in various ways (Hausman, 2012). In this study, choices based on pleasure and pain are referred to as preferences. This definition is narrower than that of Hausman (2012). A subtle example is a moral behavior that, while not unrelated to economic interests, is nevertheless pleasant to perform. According to our definition of preferences, this behavior (choice) is considered to be preference-based. In this study, morality refers to what an individual believes they should do, regardless of preferences.

ethical framework can be supported to ensure intergenerational equity.

The two sections following Sect. 2.2 discuss these two issues in detail. Section 2.3 focuses on intergenerational altruism. It shows that what has traditionally been considered socially optimal in economics should be reconsidered when intergenerational altruism is explicitly incorporated into economic models and when tackling ultra-long-term issues such as climate change. Section 2.4 discusses intergenerational equity and highlights issues with its central concept of egalitarianism. This is followed by an examination of strong sustainability, which I argue is the moral foundation of intergenerational equity behind the 1.5-degree target. I also attempt to rationalize its validity through Rawls' veil of ignorance. Section 2.5 examines a practical research challenge: enhancing intergenerational altruism and intergenerational equity to bridge the emissions gap. The final section reviews this chapter and identifies issues that require further research.

2.2 Emissions Gap in the Economics of Climate Change

In economics, the divergence between intergenerational altruism and intergenerational equity was recognized as early as the beginning of the twentieth century. Pigou (1932), the founder of welfare economics, pointed out that we value the well-being of future generations less than that of the current generation, and argued that this is ethically unacceptable (Roemer & Suzumura, 2007, p. xiv). Ramsey (1928), who laid the foundation for today's economic dynamic analysis, argued that for normative analysis to find an optimal path for society, the utility (a number representing well-being) of each generation should not be discounted in the calculation of utilitarian social welfare but should be simply added up without discounting.[3]

Despite such warnings from predecessors, mainstream economics has been using a discounted utilitarian social welfare function not only in empirical analysis but also in normative analysis. Its use in the empirical analysis is explained by the fact that we are discounting the utility that we will obtain in the future. In addition, it is useful for economic forecasting and policy evaluation because it can reproduce human behavior well. However, the reason for its use in normative analysis remains unclear. The rationale is based on the following two-step procedure. Well-being is based on our preferences, which are manifested in our actions. However, the second step remains questionable, as discussed in Sect. 2.3.

Many economic plans and public policies are projects of approximately 30 years at most, so the choice between the two types of utilitarianism does not make a significant difference. However, this choice is decisive in the case of climate change issues with ultra-long-term characteristics. For example, a very small discount rate of 1% per year gives a discount factor $\delta = 1/1.01 \approx 0.99$,, which is very close to the value of the discount factor when not discounted ($\delta = 1$. See footnote 3). However, this discounts the utility of people 500 years from now to one ten-thousandth of that of the current generation. Consequently, even if people 500 years from now lose 10,000 units of utility to gain 1 unit of utility now, discounted utilitarianism sees this as socially desirable. Critics of discounted utilitarianism criticize those who do not realize this as lacking imagination (Ramsey, 1928).

Figure 2.2 shows the greenhouse gas emissions paths of Nordhaus (2008). The path labeled as "optimal" represents the optimal path of the Dynamic Integrated Climate and Economy (DICE) model, which maximizes the discounted utilitarian social welfare function based on traditional economic models. Conversely, the path labeled as "Stern" is the optimal path estimated

[3] The utilitarian social welfare function refers to the following function: $\sum_{t=1}^{\infty} \delta^{t-1} u_t$, where t stands for time, u_t is the one-period social welfare at time t, and $\delta \in (0, 1]$ is the discount factor. $\delta = 1$ is the case without discount. If $\delta < 1$, the function represents discounted utilitarianism.

by Stern (2007) based on the undiscounted utilitarian social welfare function that treats all generations equally. As illustrated in the figure, there is a significant divergence between these two optimal paths, similar to the emissions gap shown in Fig. 2.1. The discounted utilitarian social welfare function in the DICE model is empirically supported, while the undiscounted utilitarian social welfare function in the Stern model is morally supported. Therefore, the divergence of the two optimal paths signifies a gap between intergenerational altruism and intergenerational equity.

The optimal path of the DICE model is similar to that of the NDC path shown in Fig. 2.1. The International Energy Agency (2015) warned that the NDC path at that time would lead to a temperature rise of 2.7 degrees by the end of the century. Similarly, the optimal path of the DICE model predicts a temperature rise of 2.61 degrees.[4] The DICE model is based on people's preferences, predicting actual human behavior, and simultaneously informing us of the mitigation policies that we altruistically wish to implement for future generations. However, the anticipated temperature rise of more than 2 degrees contradicts the Paris Agreement that the world has agreed upon.[5]

As the DICE model is extremely useful for considering climate change policies, research has been conducted on how to modify the model's optimal path to eliminate the emissions gap while maintaining its discounted utilitarian framework. Such studies aim to revise the model to estimate greater future damage, which leads to a stronger optimal emission reduction. One method is to reduce the discount rate (increase the discount factor) for future generations. Additional preference-related parameters such as the intertemporal elasticity of substitution should also be adjusted simultaneously to preserve the model's reproducibility of past paths. This improvement in the model has already been incorporated into the optimal path of the DICE model shown in Fig. 2.2. Another method is to update the model with the latest research results on the mechanisms of climate change, such as those of Dietz et al. (2021), to revise the future damage estimates.

Looking next at morally supported paths, it can be seen that Stern's path is similar to the 1.5-degree emissions path. However, these two intergenerational equity paths are based on different ethics. More precisely, Stern's path is based on utilitarian ethics that evaluate social desirability by the sum of utilities over generations, while the 1.5-degree emissions path is based on so-called strong sustainability that demands a sustainable global environment for future generations.[6] This indicates that the moral basis of intergenerational equity is not unique.

The international community has already agreed on the 1.5-degree target. However, the likelihood of the emissions gap not being eliminated by the current planning period (2030) is very high. When it becomes clear that the realization of the 1.5-degree emissions path is difficult, we will be forced to reconsider the target itself and, therefore, the rationale underlying it. Then, the question arises as to what kind of intergenerational equity we support. As we will see later, several proposals have been made regarding intergenerational equity, and negative views are known.

[4] While the optimal path is literally optimal, the NDC path is an aggregation of each country's individual optimal path reflecting their strategic considerations, and its theoretical counterpart is a Nash equilibrium path of the differential game (for example, see Akao et al., 2021). As a theoretical prediction, a Nash equilibrium path is generally inefficient, and more greenhouse gases are emitted than the socially optimal path. The difference in temperature rise at the end of the century between the two paths may correspond to the difference between the optimal path and the equilibrium path.

[5] Nordhaus probably does not think that a path leading to a 2.6-degree temperature increase is acceptable for society. This is because, in a paper written more than a decade before climate change became a societal concern, Nordhaus pointed out that a temperature rise of 2 or 3 degrees is something we have not experienced in hundreds of thousands of years, and warned that we cannot know from the past what will happen to the global environment with a temperature rise of more than 2 degrees (Nordhaus, 1975).

[6] Section 2.4 discusses the notion of strong sustainability.

Can we find rational basis to morally justify the 1.5-degree target?

2.3 Intergenerational Altruism

Not only has the time-separable and exponential discounted function ($\sum_{t=1}^{\infty} \delta^{t-1}u_t$, shown in footnote 3) been used as a social welfare function, as discussed in the previous section, but it has also been commonly used in economics to represent individual preferences.[7] However, its functional form assumes extremely specific intergenerational altruism, where each agent derives utility solely from the well-being of the next generation and not from the well-being of subsequent generations. The utility of subsequent generations is only indirectly reflected.[8] For example, the well-being of the generation after next is reflected in the utility of the next generation, and then indirectly reflected in the utility of the current generation through the utility of the next generation. Galperti and Strulovici (2017) refer to this as indirect altruism.

It is natural for us to have altruistic feelings toward generations beyond our children, both directly and explicitly. However, if we depart from the restrictive assumption of indirect altruism, we face time inconsistency (Galperti & Strulovici, 2017). This leads to situations where what is considered desirable may be undesirable in the future. In the context of climate change, this refers to a scenario in which future generations may not adopt climate change mitigation policies as actively as the present generation desires.

Time inconsistency poses a significant challenge in normative analysis, particularly when defining social optimality. If the greenhouse gas emissions path that maximizes the utility of the current generation differs from the one that maximizes the utility of future generations, what should be considered a socially optimal path? This problem arises as long as the optimal path for the current generation is time-inconsistent. One proposed solution is to disregard each generation's social preferences, including intergenerational altruism, and focus solely on the utility derived from their pleasure and pain. The goal is to choose a path of this "selfish" utility that is both Pareto efficient and time-consistent.[9] For a specific altruistic utility function, such an optimal

[7] There exists a relationship between the social welfare function and the individual utility function. Let's assume that each person in the current generation i has a utility function expressed as the sum of discounted utilities over time: $\sum_{t=1}^{\infty} \delta^{t-1}u_t^i$, where the discount factor δ is common for all individuals (although the one-period utility function u_t^i may vary between individuals). According to the standard assumptions of economics, the equilibrium state of the economy maximizes the following social welfare function, represented as the weighted sum of individual utilities: $\sum_{t=1}^{\infty} \delta^{t-1}u_t, u_t := \sum_i (w_i u_t^i)$ (Negishi, 1960). Here, the weighted utility function u_t refers to the utility function of a representative individual. The weight $w_i, i > 0$ determined by the distribution of endowments among individuals. By adjusting the distribution of endowments through income redistribution policies, a social welfare function that assigns equal weight to each person can be achieved: $\sum_i U_i,\ U_i := \sum_{t=1}^{\infty} \delta^{t-1} u_t^i$. This social welfare function aligns with standard utilitarianism. Note that the social welfare function derived in this manner is not discounted utilitarian, as discounting is not a social procedure but rather a component of individual preference.

[8] We denote the sequence of individual utilities (excluding intergenerational altruism) as $u_1, u_2, u_3 \ldots$. Let U_2 represent the utility including altruism of the next generation, and let δ denote the discount factor. Then, the utility including altruism of the current generation is expressed as $U_1 = u_1 + \delta U_2$. Similarly, for any generation t, $U_t = u_t + \delta U_{t+1}$ holds. Substituting this into the utility of the current generation, we obtain:

$$\begin{aligned} U_1 &= u_1 + \delta U_2 = u_1 + \delta u_2 + \delta^2 U_3 \\ &= u_0 + \delta u_1 + \delta^2 u_2 + \delta^3 u_3 + \ldots \\ &= \sum_{t=1}^{\infty} \delta^{t-1} u_t \end{aligned}$$

[9] This proposal suggests, using the notation of footnote 8, considering the path of u_t instead of U_t and regarding a path of $\{u_t\}$ that is Pareto efficient and time consistent as an optimal path. Pareto efficiency refers to a state where no individual's utility can be increased without decreasing the utility of another individual. If it is not Pareto efficient, it is possible to increase someone's utility without decreasing the utility of others. This value judgement, that increasing an individual's utility without decreasing the utility of others is considered desirable for society, is known as the Pareto principle. A Pareto efficient state is the optimal state where no further improvement is possible in the sense of the Pareto principle.

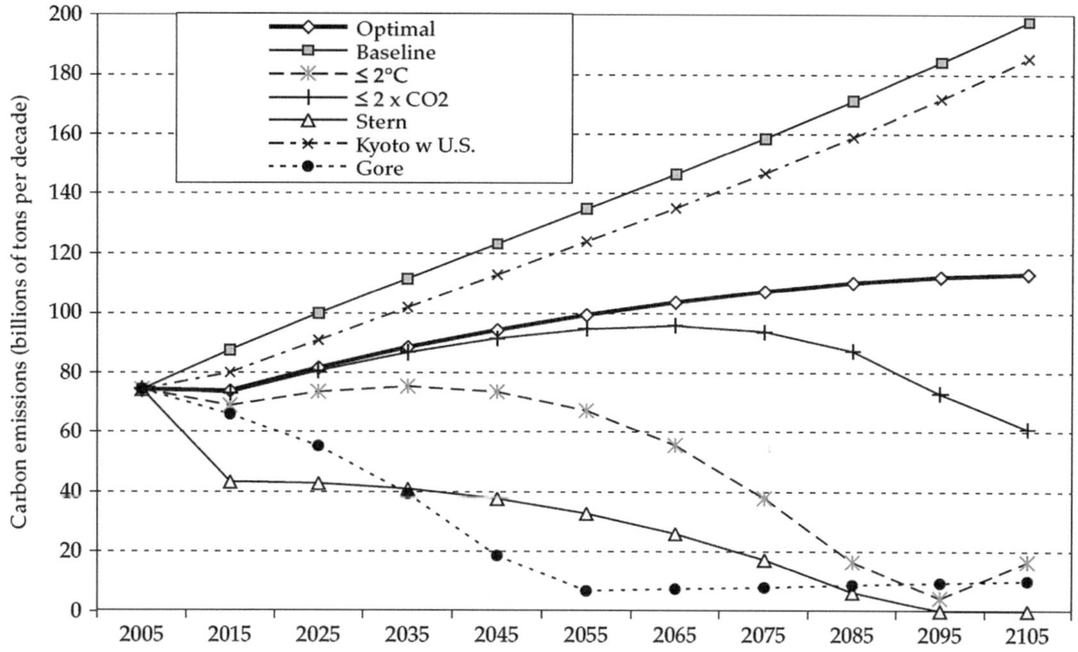

Fig. 2.2 Emissions gap in economics. *Source* Nordhaus (2008, p. 102, Figs. 5–6)

path exists, which is a solution to the maximization problem of the time-separable and exponential discounted function ($\sum_{t=1}^{\infty} \delta^{t-1} u_t$) (Galperti & Strulovici, 2017).

Although the functional form is the same, the social welfare function discussed in the previous section is derived from a completely different context. The functional form has been used solely because it can be well-defined as the objective function in an infinite planning horizon problem and ensures that the solution is time-consistent. In other words, it is not based on any functional form of individual preference. On the contrary, individual preferences are formulated so as not to conflict with such a social welfare function—namely, the time-separable and exponentially discounted utility function, which requires a restrictive assumption of indirect altruism for its rationalization.[10]

Based on the above discussion, it might seem surprising that standard macroeconomic models with time-separable and exponentially discounted utility functions, including the DICE model, can effectively describe reality and predict the future. However, although it may come as another surprise, there is a mathematical result showing the following: if the trajectory of the economy can be understood as a stationary Markov process and interpreted as a competitive equilibrium path, then the economy can be represented as a solution to an optimization problem. The optimization problem follows standard economic assumptions, including a representative individual's utility function that is time-additive and exponentially discounted.[11]

[10] Furthermore, many economic models, including Stern (2007), specify the utility function as the logarithmic function for the sake of analytical simplicity.

[11] Refer to Mitra and Sorger (1999). Akao et al. (2012) achieve similar outcomes in a dynamic game. A stationary Markov process refers to a dynamic process where the state in the next period is determined by the current state, regardless of the specific time point. Standard economic assumptions stipulate that the feasible set of inputs and outputs is represented by a convex set, and the one-period utility function u_t of the representative individual is concave.

This representative individual's utility function has traditionally been considered a social welfare function representing the desirability of a society. However, it is difficult to expect that a utility function that has been specifically defined without any basis from the perspective of intergenerational altruism represents our well-being, which includes altruism toward future generations. In other words, the optimality of the optimal path in traditional economic models is not as obvious.

Economics has traditionally assumed that behavior reveals preferences and that preferences represent pleasure and pain. However, the preceding discussion demonstrates that behavior does not always accurately reflect preferences. When intergenerational preferences exhibit time inconsistency, the present generation faces a dilemma in decision-making: whether to adhere to their own preferences or make decisions that account for the likelihood that future generations will not behave as expected. Consequently, people's behavior is influenced by this conflict. Evaluating well-being requires extracting preferences from individuals' actions, while disregarding elements associated with conflict. This assessment is complicated by intergenerational altruism.

2.4 Intergenerational Equity

When discussing intergenerational equity, one problem must first be mentioned. This is the non-identity problem in Parfit's discussion of intergenerational issues (Parfit, 1984, Ch. 17). This problem is caused by the obvious fact that current choices change the future; therefore, people born under a certain choice of the current generation may not be born under a different choice. Our interest in intergenerational issues lies in how the well-being of future generations can be affected by our choices. However, if our choices change, the people who were the subject of the original choice will no longer exist, making it impossible to consider how their well-being would be affected. In other words, we cannot identify the future generations we aim to care for. As a result, arguments that apply concepts like rights and obligations to intergenerational issues, appealing to our duties toward future people and those yet to be born, encounter difficulties, akin to dealing with Schrödinger's cat.

This problem arises when we treat future generations as external, objectively knowable entities and ask what they would want from this transcendent perspective. However, this approach is fundamentally flawed because future generations are endogenous—they are shaped by our choices. Therefore, to avoid the non-identity problem, intergenerational equity must be approached from a different perspective.[12]

Rawls' thought experiment with the veil of ignorance (Rawls, 1971) is one such route. Rawls assumes a hypothetical state (called the original position) covered by a veil of ignorance, where one does not know what circumstances they will be born into, what preferences they will have, what talents they will possess, and, in the context of intergenerational equity, what generation they will be born into, and asks what kind of society one would want in such a state. The people in the original position are homogeneous; therefore, their choices will be desirable for everyone. Therefore, people will agree on having such a society. This thought experiment does not involve future generations as external entities; therefore, it is not troubled by Parfit's non-identity problem.

According to Rawls, people choose equal freedom. However, the only accepted inequality is that which benefits the least advantaged members of society—a concept Rawls calls the difference principle. The ethic of equity supported by

[12] The standard approach that avoids the non-identity problem treats future generations as anonymous individuals and represents the well-being of each generation in numerical terms. This approach is adopted in social choice theory, a field of economics, as well as in egalitarian philosophy (e.g., Asheim, 2010; Hirose, 2015). Such an approach is subject to criticisms similar to those directed at the utilitarian and maximin approaches, which will be introduced below. Specifically, these criticisms question why well-being should be expressed in numerical terms and whether the proposed criteria for intergenerational equity can offer compelling solutions in all circumstances. However, as Arrow's impossibility theorem suggests, it is challenging to find criteria for intergenerational equity that fully address these concerns.

this difference principle is the maximin criterion, which states that social welfare increases only when the utility of the person with the lowest utility is raised. In economics, numerous analyses have been conducted based on this criterion in the context of intergenerational equity (see Pezzey & Toman, 2002). These studies were inspired by the Club of Rome's *Limits to Growth*, which predicted societal decline due to resource depletion and environmental destruction (Meadows et al., 1972).

However, there are three problems associated with the maximin criterion. First, it expresses the desirability of society as a whole through a social welfare function that considers the utility of each generation as a variable. Therefore, like utilitarianism, it is a form of consequentialism and can be considered a type of utilitarianism in a broad sense, as one of the Bergson-Samuelson social welfare functions, which generalizes the utilitarian social welfare function. In other words, the maximin criterion applies a utilitarian interpretation to Rawls' theory of justice.

In response to this criticism, one might argue that the functional form reflects one aspect of Rawls' concept of justice. However, it advocates a notion of justice that is difficult to accept in the context of intergenerational issues. Specifically, the maximin criterion favors the current generation as the least fortunate, when it is possible for future generations to be better off than the current generation. This suggests that savings—and, therefore, investment in future generations—are unnecessary. In the context of climate change, this implies that we can emit more carbon. This is the second issue with the maximin criterion, specific to the application of the difference principle to intergenerational issues.

Rawls' solution to this problem was to consider that our utility includes intergenerational altruism. This appears to be a promising solution, as a person in the original position is concerned only with their own interests, whereas savings are intended for future generations. However, Arrow (1973) and Dasgupta (1974) demonstrate that introducing altruism causes the maximin optimal path to oscillate and become time-inconsistent. It is paradoxical that the utility of each generation could fluctuate on the egalitarian path. An even more serious issue is time inconsistency. This poses a critical problem in normative analysis because the socially optimal path should remain optimal for *all* generations. However, if the path is time-inconsistent, what is desirable for society at one moment may not be so at a later point, which contradicts the concept of an optimal path. Therefore, the principle of intergenerational equity must be time-consistent.[13]

Third, the social contract chosen under the veil of ignorance is not limited to the difference principle. Before Rawls, Harsanyi (1953) used the idea of the veil of ignorance. Harsanyi assumes that a person is completely ignorant of the position they will take (called Harsanyi's impartial observer). Harsanyi interprets this veil of ignorance as a risk and infers that in that situation, people would want a society that increases the expected value of utility.[14] Furthermore, if we consider the equal possibility of being born into any generation, expected utility becomes equivalent to the (undiscounted) utilitarian social welfare function. Consequently, utilitarianism is selected under the veil of ignorance. It is through the procedure of justice devised by Rawls—who critiques utilitarianism—that utilitarianism is justified.

The Rawlsian veil of ignorance is a type of uncertainty (referred to as ignorance), and the maximin rule corresponds to the Wald criterion, which is one of the decision-making rules

[13] Rawls refers to the agreement on savings chosen in the original position, due to the introduction of intergenerational altruism, as the "savings principle" (Rawls, 1971, Sect. 44). In the revised edition of *A Theory of Justice* in 1999 and in Rawls (2001, Sect. 49), he asserts that the difference principle applies within generations, while the savings principle applies to intergenerational equity. By discarding the difference principle in intergenerational issues, he avoids the time inconsistency associated with it. Additionally, he argues that the savings principle should satisfy the property interpreted as time consistency. However, he does not clarify what the savings principle that meets this time consistency entails. Please refer to footnote 16 for related discussions.

[14] It is based on the expected utility hypothesis. Such preferences were axiomatically formalized by von Neumann and Morgenstern (1945) as foundational elements of game theory, and since then, expected utility has been commonly used in economics as the preferences of agents facing risk.

under uncertainty. Harsanyi's concept of utilitarianism corresponds to the Laplace criterion. Other decision-making rules in situations of ignorance include the minimum maximum regret (Savage criterion) and the Hurwicz criterion (Milnor, 1954). Therefore, the justice agreed upon under the veil of ignorance is not unique. Furthermore, this non-uniqueness is not limited to Rawls' approach. Hirose (2015) introduces moralities of egalitarianism beyond Rawlsian egalitarianism, such as luck egalitarianism, Telic egalitarianism, Prioritarianism, and Sufficientarianism. Throughout history, various moralities have emerged in response to the circumstances of each era. As argued by MacIntyre (1981), the moralities of past ages survive and coexist, even if they lose their validity, and the coexistence of incompatible moralities often causes serious conflicts in society. The emissions gap can be considered one of these conflicts (Akao, 2023). A pragmatic solution is to identify the most supportive intergenerational equity ethics and examine whether a policy can be justified by these ethics.

In the context of climate change, the current policy goal is to achieve the 1.5-degree target cost-effectively; that is, to implement an emissions path that realizes the target at minimum cost. There is likely widespread agreement regarding the validity of cost-effectiveness as a value criterion, provided that the concept of costs is not limited to mere monetary expenses but is understood as encompassing all factors that lead to a reduction in people's utility. Therefore, the question arises as to which ethical principles underpin the 1.5-degree target. Clearly, this is neither discounted utilitarianism (as seen in Nordhaus' optimal path), nor undiscounted utilitarianism (as observed in Stern's optimal path), nor Rawls' egalitarianism (the maximin path). Rather, it is the ethics of sustainability—specifically, *strong sustainability*—that informs the 1.5-degree target.

Strong sustainability refers to the concept that a society is sustainable only if critical natural capitals, which are environmental assets that have a significant impact on societal sustainability and cannot be replaced by human intervention, do not diminish.[15] Although the global climate system can be locally modified by human intervention, it cannot be replaced as a whole. Recognizing that the entire Earth system is critical natural capital, countries agreed on the ambitious 1.5-degree target to ensure its preservation—in other words, to meet the standard of strong sustainability.

How can strong sustainability be justified as an ethic of intergenerational equity? Bourban (2022) examines the communitarian, Rawlsian, and capability approaches, arguing that the capability approach extended to non-humans is most suitable for this purpose. However, it seems possible to justify strong sustainability as procedural justice without relying on the concept of capability or extending the capability approach to non-humans. Below, I outline this concept.

When using the veil of ignorance for intergenerational issues, there are differences from intragenerational ones. When considering intragenerational issues, we consider justice of distribution in one world. However, the world has changed over time regarding intergenerational issues. Therefore, people not only consider how they want to be treated at each point in time but also how they want the man-made and natural infrastructure of the socio-economy to develop over time.[16] Under the veil of ignorance, people recognize that while our lives have become more comfortable due to advances in science, technology, and capital accumulation, natural resources and the environment have continued to deteriorate, resulting in global issues that now threaten our lives, such as climate change. Consequently, people would seek a path that avoids further degradation of the critical natural capital represented by natural resources

[15] The antithesis to strong sustainability is weak sustainability. It posits that society is sustainable if the utility sequence of generations does not decline. For further discussion on these two sustainability concepts, refer to Akao (2023).

[16] Such intertemporal concern is introduced by Rawls as the savings principle. However, in economics, savings are a result generated by fundamentals comprising preferences, technology, and resource endowment, along with the societal system. It is more fundamental to consider the developmental path of fundamentals, excluding preferences that are included in the veil of ignorance, as agreed upon intertemporally.

and the environment. Additionally, people would desire a path that directs scientific and technological advancements toward conserving natural resources and reducing environmental burdens. Both the environment and technology are public goods that endure over time. The desire for public goods remains constant, regardless of when people are born or the circumstances they face. Therefore, strong sustainability is chosen under the veil of ignorance as a principle of justice in distribution across generations.

2.5 Strengthening Intergenerational Altruism and Intergenerational Equity

To close the emissions gap, it is essential to strengthen intergenerational altruism and increase support for the ethics of intergenerational equity. Which policies can be considered for these purposes? This section introduces the research related to this question.

Saijo (2022) proposes a social system and its practice, referred to as future design, to prioritize the interests of future generations. He argues that while humans possess a natural inclination—referred to by Saijo as futurability—to prioritize the interests of future generations over their own, current democratic and market systems lack mechanisms to consider the well-being of future generations. This deficiency leads to myopic decision-making and undermines our futurability. He concludes that it is necessary to construct and implement a system that allows individuals to express their concerns for future generations. One idea proposed and investigated by Saijo involves placing individuals in a society where they experience representing future generations. Saijo's research group aims to elucidate, through laboratory and field experiments, the impact of the presence of such individuals—referred to as imaginary future persons—on the decision-making of the current generation, as well as the experiences that qualify a person to represent future generations (for details, see Saijo, 2022).

Maalouly et al. (2023 and 2024) approach the issue of strengthening intergenerational altruism as follows: First, they perceive the characteristics of intergenerational altruism as altruism toward a subject they will never meet. Based on the empathy-altruism hypothesis in social psychology (Batson, 2011), they argue that empathic concern (feeling emotions that align with the emotions of others who need help) brings about altruism. Therefore, to strengthen intergenerational altruism, the current generation needs to recognize that future generations are in a situation where they need help. However, future generations do not appear; therefore, there is a need to discuss the future situation virtually. In Maalouly et al. (2023), a future AI speaks about the future situation of people through a telenoid. In Maalouly et al. (2024), in addition to the telenoid, a human representing future generations and a voice-only medium speak about the situation of future people. In all experiments, communication with future generations has been found to strengthen altruism.

As mentioned above, although the number is limited, and efforts are very recent, research has begun on how to strengthen intergenerational altruism and increase support for intergenerational equity ethics.

2.6 Conclusion

In this study, we discussed the emissions gap as a divergence between our intergenerational altruism and the morals of intergenerational equity. The assumption behind this dichotomy is that people can think morally about the common good independently of their own preferences, an idea attributed to Binmore (2005). As evidence of this duality of the human mind, I would like to point out that while Rawls' (1971) theory of justice has been controversial, the veil of ignorance—the thought experiment used to derive his theory—has been widely accepted without objection. This acceptance is highly unusual and astonishing, as I will demonstrate below: the veil of ignorance demands a highly artificial state of ignorance detached from specific circumstances and history. Nevertheless, readers of *A Theory of Justice* accept this unimaginable situation, so far removed from our everyday lives, and view the inference about the social contract that people behind the veil of ignorance would agree upon as something that is considered publicly supported. This is a remarkable fact. I believe that the reason for this acceptance lies precisely in the inherent human capacity to contemplate the common good independently of individual pleasure and pain.

This duality between preference and morality creates a conflict over whether we follow our own preferences or adhere to moral principles.[17] The emissions gap is precisely one of these conflicts, extending beyond an internal struggle within individuals to a global scale, with serious consequences for the well-being of future generations worldwide.

Conflicts may also arise due to time-inconsistent preferences driven by intergenerational altruism. As discussed in Sect. 2.3, one of the problems this conflict creates is that people's words and actions do not necessarily align with their true preferences. Extracting preferences from observable data, such as people's words and actions, and estimating well-being are central issues in social science. However, if our words and actions do not reliably reveal our preferences—and thus our pleasure and pain—how should we estimate well-being? Without clarity on this, it becomes difficult to determine the optimal path based on preferences, namely, the path that maximizes well-being. Estimating well-being is crucial to address climate change issues and, at the same time, is a challenging task. It remains a pioneering area of research.

As Boehm (2012) argues, morality evolved among our distant ancestors who lived in groups of approximately 30 individuals, including non-kin. They faced situations where pursuing individual interests diminished the group's competitiveness against other groups, making cooperation essential. To ensure a group's survival during competition and an individual's success in leaving offspring within the group, a delicate balance between self-interest and group interest was necessary. This provides further evidence of the duality between preferences and morality from an evolutionary perspective.[18]

Since our ancestors acquired language, morality was transmitted as a concept of good and evil. Children learn these concepts alongside language. As the size of a group increased and society became more complex, each community developed its own concepts of good and evil. As MacIntyre (1981) analyzed, since the Enlightenment, such morality required a rational explanation. This necessity arose from the expansion of society beyond individual communities, the need for coordination and integration of the moral frameworks of each community, and the replacement of God with reason, which produced advancements in science and technology that enriched our lives. Consequently, morality was rationalized, acquired legal force, and underpinned policies in modern society.

I believe that this moral evolution process also applies to the morality of climate change. However, it is morality about a new problem that has come to be recognized by science, and it has not gone through a biological or cultural evolutionary process. In other words, it is not supported at the level of genes or memes. Therefore, its rationalization is crucial for this morality to gain support. In Sect. 2.4, I attempted to rationalize strong sustainability, which is the basis for the 1.5-degree target, using the veil of ignorance. However, there is no doubt that further exploration of persuasive intergenerational equity ethics is necessary.

[17] Here, conflict is considered to arise when human behavior is not solely determined by preferences. Apart from morality, conflicts arise when facing true uncertainty (distinct from risk, where probability distributions are not given, and in situations where subjective identification of probability distributions is not possible), and when time inconsistency exists (not only intergenerationally, as seen in Sect. 2.3, but also intrapersonally with future selves. See Ainslie (2001).

[18] Our distant ancestors are believed to have encountered situations akin to the cooperative and prisoner's dilemma games in game theory. Moreover, Boehm (2012) suggests that egalitarianism was a common moral framework in primitive societies facing such situations.

We are constantly exposed to news about climate change and the need for CO_2 reduction. Through this, the morality of climate change permeates our daily lives. Understanding how this exposure influences the formation of new moral perspectives that support the 1.5-degree target and intergenerational equity should be an important area of study.

Finally, in this study, we have treated preferences and morals as two independent concepts, but this is not the case in reality. Our social preferences encompass not only intergenerational altruism but also the pleasure of adhering to moral principles, the joy of contributing to the creation of effective public systems, and the shame associated with acting against the interests of society (Boehm, 2012). Therefore, as a matter of intergenerational equity, if strong sustainability gains widespread support or is sufficiently rationalized and established as an ethical standard, adherence to strong sustainability will become a source of satisfaction. A positive feedback loop may form between preferences and morals, potentially leading to increased support for such ethical frameworks. Research into the dynamic relationship between preferences and morals is important, as it may offer valuable insights into addressing the emissions gap and contribute to the realization of a carbon-neutral society.

Acknowledgements This work was supported by JSPS KAKENHI Grant Numbers 19K21706 and 23K20604.

References

Ainslie, G. (2001). *Breakdown of will*. Cambridge University Press.

Akao, K. I. (2023). Sustainable development: Controversies and theoretical results in economics. In S. Urata, K. I. Akao, & A. Washizu (Eds.), *Sustainable development disciplines for society*. Sustainable Development Goals Series. Springer. https://doi.org/10.1007/978-981-19-5145-9_8

Akao, K. I., Mitra, T., & Sorger, G. (2012). Some rationalizability results for dynamic games. *International Journal of Economic Theory, 8*(4), 361–379.

Akao, K. I., Sakamoto, H., Uchida, K., & Wasa, Y. (2021). Payoff dominant equilibrium in a differential game model. Paper presented at Society for the Advancement of Economic Theory 2021.

Asheim, G. B. (2010). Intergenerational equity. *Annual Review of Economics, 2*, 197–222.

Arrow, K. J. (1973). Rawls' principle of just saving. *The Swedish Journal of Economics, 75*, 323–335.

Batson, C. D. (2011). *Altruism in humans*. Oxford University Press.

Binmore, K. (2005). *Natural justice*. Oxford University Press.

Boehm, C. (2012). *Moral origins: The evolution of virtue, altruism, and shame*. Basic Books.

Bourban, M. (2022). Strong sustainability ethics. *Environmental Ethics, 43*, 291–314.

Dasgupta, P. (1974). On some alternative criteria for justice between generations. *Journal of Public Economics, 3*, 405–423.

Dietz, S., Rezai, A., van der Ploeg, F., & Venmans, F. (2021). Are economists getting climate dynamics right and does it matter? *Journal of the Association of Environmental and Resource Economists, 8*(5), 895–921.

Galperti, S., & Strulovici, B. (2017). A theory of intergenerational altruism. *Econometrica, 85*(4), 1175–1218. http://www.jstor.org/stable/44955156

Harsanyi, J. C. (1953). Cardinal utility in welfare economics and in the theory of risk-taking. *Journal of Political Economy, 61*, 434–435.

Harsanyi, J. C. (1955). Cardinal welfare, individualistic ethics, and interpersonal comparisons of utility. *Journal of Political Economy, 63*(4), 309–321. http://www.jstor.org/stable/1827128

Hausman, D. M. (2012). *Preference, value, choice, and welfare*. Cambridge University Press.

Hirose, I. (2015). *Egalitarianism*. Routledge.

International Energy Agency. (2015). *Energy and Climate Change: World Energy Outlook Special Briefing for COP21* (11 pp). https://www.aben.com.br/Arquivos/359/359.pdf

Intergovernmental Panel on Climate Change. (2021). Summary for policymakers. In: V. Masson-Delmotte, P. Zhai, A. Pirani, S. L. Connors, C. Péan, S. Berger, N. Caud, Y. Chen, L. Goldfarb, M. I. Gomis, M. Huang, K. Leitzell, E. Lonnoy, J. B. R. Matthews, K. TMaycock, T. Waterfield, O. Yelekçi, R. Yu, & B. Zhou (Eds.), *Climate Change 2021: The Physical Science Basis. Contribution of Working Group I to the Sixth Assessment Report of the Intergovernmental Panel on Climate Change* (pp. 3–32). Cambridge University Press. https://doi.org/10.1017/9781009157896.001

Kreibiehl, S., Jung, T. Y., Battiston, S., Carvajal, P. E., Clapp, C., Dasgupta, D., Dube, N., Jachnik, R., Morita, K., Samargandi, N., & Williams, M. (2022). Investment and finance. In P. R. Shukla, J. Skea, R. Slade, A. Al Khourdajie, R. van Diemen, D. McCollum, M. Pathak, S. Some, P. Vyas, R. Fradera, M. Belkacemi, A. Hasija, G. Lisboa, S. Luz, & J. Malley (Eds.), *IPCC, 2022: Climate Change 2022: Mitigation of Climate Change. Contribution of Working Group III to the Sixth Assessment Report of the Intergovernmental Panel on Climate Change*. Cambridge University Press. https://doi.org/10.1017/9781009157926.017

Maalouly, E., Yamazaki, R., Nishio, S., Nørskov, M., Kamaga, K., Komai, S., Chiba, K., Atsumi, K., & Akao, K. I. (2023). Assessing the effect of dialogue on altruism toward future generations: A preliminary study. *Frontiers in Computer Science, 5*, 1–12. https://doi.org/10.3389/fcomp.2023.1129340

Maalouly, E., Yamazaki, R., Nishio, S., Nørskov, M., Kamaga, K., Komai, S., Chiba, K., Atsumi, K., & Akao, K. I. (2024). The effect of conversation on altruism: A comparative study with different media and generations. *PLOS One, 19*(6), e0301769. https://doi.org/10.1371/journal.pone.0013636

MacIntyre, A. C. (1981). *After virtue. A study in moral theory* (286 pp). University of Notre Dame Press.

Meadows, D. H., Meadows, D. L., Randers, J., & Behrens III, W. (1972). *The limits to growth: A report for the club of Rome's project on the predicament of mankind*. Universe Books.

Milnor, J. W. (1954). Games against Nature. In C. H. Coombs, R. L. Davis, & R. M. Thrall (Eds.), *Decision processes* (pp. 49–60). Wiley.

Mitra, T., & Sorger, G. (1999). Rationalizing policy functions by dynamic optimization. *Econometrica, 67*(2), 375–392. http://www.jstor.org/stable/2999589

Negishi, T. (1960). Welfare economics and existence of an equilibrium for a competitive economy. *Metroeconomica, 12*, 92–97. https://doi.org/10.1111/j.1467-999X.1960.tb00275.x

Nordhaus, W. D. (1975). *Can we control carbon dioxide?* (IIAS Working Paper 75-63, 49 pp.).

Nordhaus, W. D. (2008). *A question of balance: Weighing the options on global warming policies*. Yale University Press.

Parfit, D. (1984). *Reasons and Persons*. Oxford University Press.

Pezzey, J. C. V., & Toman, M. A. (2002). *The economics of sustainability* (1st ed.). Routledge. https://doi.org/10.4324/9781315240084

Pigou, A. C. (1932). *The economics of welfare* (4th ed.). Macmillan.

Ramsey, F. P. (1928). A mathematical theory of saving. *The Economic Journal, 38*, 543–559.

Rawls, J. (1971). *A theory of justice*. Harvard University Press.

Rawls, J. (2001). *A Theory of Justice* (rev. ed.). Harvard University Press.

Roemer, J., & Suzumura, K. (2007). *Intergenerational equity and sustainability*. Palgrave Macmillan.

Saijo, T. (2022). Future design for sustainable nature and societies. In: W. Leal Filho, A. M. Azul, F. Doni, & A. L. Salvia (Eds.), *Handbook of sustainability science in the future*. Springer. https://doi.org/10.1007/978-3-030-68074-9_145-1

Shrader-Frechette, K. S. (1991). *Environmental ethics* (2nd ed.). Boxwood Press.

Stern, N. H. (2007). *The economics of climate change: The stern review*. Cambridge University Press. https://doi.org/10.1017/CBO9780511817434

United Nations. (2023). *First global stocktake*. Proposal by the President. Draft decision -/CMA.5 Outcome of the first global stocktake, Para 21. https://unfccc.int/sites/default/files/resource/cma2023_L17_adv.pdf

United Nations Environment Programme. (2023). *Emissions Gap Report 2023: Broken Record—Temperatures hit new highs, yet world fails to cut emissions (again)*. Nairobi. https://doi.org/10.59117/20.500.11822/43922

von Neumann, J., & Morgenstern, O. (1944). *Theory of games and economic behavior*. Princeton University Press.

3 Shared Responsibility in Climate Change: Reassessing Individual Moral Responsibility Within Collective Harm

Kanako Takae

Abstract

This chapter examines individual moral responsibility for collective harm, specifically in the context of climate change, arguing that individuals bear moral responsibility for contributing to such harm, implying that we should refrain from activities such as driving cars and consuming meat. The argument is developed in two parts: by demonstrating that the moral wrongness of actions contributing to collective harm increases in proportion to the scale and severity of that harm, and by exploring the type of responsibility we can assign to individuals in response to this wrongness. Through this exploration, I critically examine the limitations of an individual-based framework and argue that it fails to capture the interconnected nature of actions leading to collective harm. Subsequently, I emphasize the systemic interdependence of contributors, illustrating that we are part of a shared system, in which our actions collectively contribute to harm. From this perspective, I introduce the concept of non-agential shared responsibility, providing a framework for understanding individual responsibility within the broader context of collective harm such as climate change. Lastly, I argue that, to fully support and enrich this responsibility, we must revise our daily moral codes to align with the notion of global citizenship, recognizing that our interconnectedness and systemic interdependence extend beyond individual levels to a global scale, and across generations.

Keywords

Collective harm · Individual moral responsibility · Systemic interdependence · Shared responsibility · Climate-change ethics

3.1 Introduction

When assessing the moral wrongness of an action, its frequency and scale can substantially influence the judgment. For instance, although stealing a snack from a local store by a single child is undoubtedly wrong, it is not typically perceived as a serious threat. However, if numerous children repeatedly steal from the store, eventually driving it into bankruptcy, the cumulative impact amplifies the moral significance of theft. This broader harm, resulting from the collective actions of many individuals, is commonly referred to as "collective harm."

K. Takae (✉)
Waseda University, Tokyo, Japan
e-mail: kanako.takae@gmail.com

K. Akao and A. Washizu (eds.), *Climate Change Issues and Social Sciences*, Sustainable Development Goals Series, https://doi.org/10.1007/978-981-96-8951-4_3

Climate change is an example of collective harm. Although individual actions, such as excessive car use or high meat consumption, may seem insignificant when considered in isolation, their cumulative effect, when multiplied by millions of people, leads to substantial global harm. This raises a pressing ethical question: How should we evaluate the moral implications of everyday actions that contribute to climate change, especially within the broader context of pervasive human behavior and overpopulation?

This chapter explores the argument that individuals bear moral responsibility for contributing to collective harm, particularly focusing on climate change. It first examines how the moral wrongness of an action can be assessed within its framework of collective harm. Upon establishing that the moral wrongness of an action can be evaluated in proportion to the severity of the collective harm caused, the discussion shifts toward how individual moral responsibility can be assigned in response to such wrongness. This requires a critical re-examination of the conventional understanding of how individual moral responsibility is attributed. Building on this critique, this chapter explores how to reconceptualize individual moral responsibility in the face of global challenges such as climate change.

Reassessing the current understanding of individual moral responsibility is crucial for effectively addressing collective harm. This chapter aims to show that society can manage climate change and move toward a carbon-neutral future only by updating this framework.

The rest of this chapter is arranged as follows. Section 3.2 addresses skeptical arguments that challenge the attribution of individual moral responsibility to climate change. After refuting these skeptical views, Sect. 3.3 introduces two major approaches, namely, consequentialism and non-consequentialism, for evaluating the moral wrongness of an action. I argue that consequentialism provides a more suitable framework for assessing moral wrongness in the context of collective harm, as it allows for the evaluation of the degree of wrongness relative to the scale and severity of the harm caused. Section 3.4 discusses how individual moral responsibility can be grounded in this consequentialist understanding of moral wrongness. I argue that the conventional consequentialist framework, which ties individual moral responsibility to causal responsibility, is inadequate for assigning moral responsibility to individuals in cases of collective harm, as it results in the dilution of responsibility. To resolve this issue, in Sect. 3.5, I propose an alternative concept of responsibility, namely, non-agential shared responsibility.

3.2 Skepticism About Individual Moral Responsibility for Climate Change

Some scholars are skeptical about attributing moral responsibility to individuals to address climate change. They argue that because climate change is a collective harm, it should primarily be managed at the institutional and corporate levels, rather than focusing on individual actions. Some even claim that, while individuals should support public policies aimed at mitigating climate change, they are not necessarily required to reduce their own emissions (Johnson, 2003; Kagan, 2011; Maltais, 2013; Sinnott-Armstrong, 2005).

However, the claim that individuals have no moral responsibility while simultaneously having a duty to support climate-change policies seems morally puzzling. For instance, consider a person who advocates for policies to reduce carbon emissions but continues to drive a high-emission car when lower-emission alternatives are readily available. Similarly, imagine an individual that supports the abolition of factory farming while consuming beef steaks every weekend. In these cases, their behaviors appear inconsistent with the policies they endorse.

Indeed, it is reasonable to argue that individual behaviors should align with climate-change policies if one supports them. In other words, there is an ethical requirement that individuals ensure that their actions do not contradict the policies they advocate. Therefore, a responsibility to support policies aimed at mitigating climate change logically extends to our individual actions.

Other scholars are skeptical about the effectiveness of focusing on individual actions, and argue that they may overlook broader systemic issues. For instance, Michael Mann, a climatologist and geophysicist at the University of Pennsylvania, contends that "focusing on individual choices around air travel and beef consumption heightens the risk of losing sight of the gorilla in the room" (Mann, 2019). Mann advocates for systematic changes to reduce everyone's carbon footprint without relying on their individual commitments. He suggests measures such as pricing carbon emissions, incentivizing renewable energy, and improving existing technologies, rather than enforcing bans that could disrupt people's current lifestyles. He also notes that focusing on individual actions may inadvertently support climate-change deniers who portray climate advocates as oppressive, "freedom-hating totalitarians" (Mann, 2019). Consequently, Mann argues that climate change should be addressed as a political issue requiring systematic solutions, rather than an ethical issue focused on individual lifestyles.

While it is undeniable that climate change requires collective solutions, and a political approach is essential for achieving them, too much focus on systematic changes and treating climate change primarily as a political issue can obscure ethical questions concerning our daily choices and actions. Approaches emphasizing systematic change often assume that collective harm arises from the social structure or arrangement of actions; hence, it is considered resolvable through systematic changes, implying that the individual actions do not require change. For instance, Mann asserts: "We don't need to ban cars; we need to electrify them (and we need that electricity to come from clean energy). We don't need to ban burgers; we need climate-friendly beef" (Mann, 2019). Accordingly, Mann presupposes that climate change is a systematic problem related to how we manage our lifestyles without considering the possibility that they may be ethically problematic.

Neglecting these ethical dimensions risks accepting technological solutions uncritically, and overlooks how systematic changes, such as political arrangements, can impact individual lives. Contrary to Mann's claim, decisions made at higher levels often require corresponding changes at individual levels. For instance, during the COVID-19 pandemic, institutional policies such as mobility restrictions, mask mandates, and vaccine requirements directly affected individual behaviors. These restrictions illustrate another example of collective harm. While a single person's failure to comply may not cause significant damage, widespread noncompliance exacerbates the pandemic, and worsens the situation for everyone. This situation underscores that solving collective harm frequently requires changes in individual actions.

Climate change is no exception. Although the impact of such changes is not as immediately apparent as in the case of a pandemic, effective mitigation efforts at the institutional level often depend on widespread changes in individual behavior. The European Commission (EC), a politically independent executive arm, states that civil society "will need to continue increasing its role of creating awareness among citizens about long-term decarbonisation, including actions that can be taken at individual level and lifestyle choices that each citizen can make" (European Commission, 2018, p. 290). Similarly, the Japanese government, consistent with the Paris Agreement,[1] emphasizes the significance of lifestyle innovation, stating that, "...shifting the way of life of each citizen towards sustainability, provides a major impact directly and indirectly on climate change through consumer behaviors and the use of fuels, energy, and resources," highlighting the expansion of *ethical* consumption as one example (Japan Ministry of the Environment, 2021, p. 92).

These perspectives suggest that individual lifestyles cannot remain unaffected. While Mann argues that focusing on individual ethical issues may empower climate-change deniers to

[1] The Paris Agreement is a legally binding international treaty on climate change, adopted by 196 countries at the UN Climate Change Conference (COP21) in December 2015. It came into effect in November 2016. The agreement is primarily aimed at limiting global warming to well below 2 °C above pre-industrial levels, while aiming to restrict the temperature rise to 1.5 °C.

portray climate advocates as oppressive, thereby hindering political progress, neglecting individual moral responsibility in the context of climate change poses its own risks. Failure to recognize individual accountability may make people more vulnerable to the impacts of collective harm caused by climate change.

Therefore, exploring a theoretical framework for individual moral responsibility in relation to collective harm is crucial. Such a framework is essential to not only justify necessary restrictions and changes, but also critically examine potentially unjustified ones. The following discussion explores the concept of individual moral responsibility in response to collective harm, particularly focusing on climate change. Specifically, I analyze the type of responsibility that underpins the argument that individuals should prefer public transportation to cars, or adopt vegetarian or vegan diets instead of consuming meat.

This exploration proceeds in two steps, namely, illustrating how the moral wrongness of an action can increase with the scale of collective harm, as opposed to when the action is considered in isolation, and examining how this amplified moral wrongness allows us to argue for individual moral responsibility for collective harm.

Given the scope of this chapter, I focus on the nature of such responsibility while leaving the question of its extent open for further discussion. In other words, while acknowledging such a responsibility exists, the degree of its demandingness remains open to debate.

3.3 Moral Wrongness of Actions in Collective Harm

This section introduces two primary approaches for assessing the moral wrongness of an action, and examines the more appropriate approach for evaluating moral wrongness in the context of collective harm.

The moral wrongness of an action is generally assessed through two major approaches, namely, consequentialism and non-consequentialism. Consequentialism evaluates the moral value of an action based on its consequences, asserting that the rightness or wrongness of an action is determined by its outcome. For example, from a consequentialist perspective, stealing is considered morally wrong due to the negative consequences it causes, such as harm inflicted on the victim.

In contrast, non-consequentialism assesses the moral value of an action independent of its consequences. It focuses on the inherent nature of an action, such as whether it adheres to certain moral principles. For instance, non-consequentialism condemns stealing based on its violation of fundamental moral principles, such as the right to property, regardless of the actual harm caused or the consequences that follow.

Consequentialism incorporates numerical factors and quantifiable impacts into its assessments, whereas non-consequentialism focuses on the intrinsic nature of actions. A well-known example illustrating this difference is the trolley problem, a thought experiment in which one must decide whether to pull a lever to divert a trolley from running over five people stuck on one track to another where only one person is stuck.

In this scenario, consequentialism evaluates the action based on the number of lives saved versus those lost, arguing that pulling the lever to save five lives at the expense of one is morally preferable. In other words, the consequentialist approach emphasizes the overall impact, and seeks to maximize positive outcomes, making the tradeoff of sacrificing one person for a greater number morally justifiable. Non-consequentialism, in contrast, argues that sacrificing one person, even to save five, is morally wrong. This perspective holds that individuals should not be used merely as a means to an end, and that each person's inherent worth should be respected, regardless of the overall outcome.

With this distinction in mind, let us now examine how each approach assesses the moral wrongness of an action that contributes to collective harm, such as driving a car or eating meat. I begin with non-consequentialism, and discuss its challenges when arguing that these actions are morally wrong.

From a non-consequentialist perspective, two key arguments can be made: the intrinsic value of individuals and the condition of universalizability. Non-consequentialism emphasizes that individuals possess inherent worth, and that actions disregarding this intrinsic value are considered morally wrong, as illustrated by the trolley problem. Universalizability examines whether an action can be applied consistently as a moral principle. For example, breaking a promise violates universalizability, because if everyone were to break promises, the concept of promising would become meaningless. In other words, if an action cannot be universally applied without leading to contradiction or undermining the very practice it supports, it cannot be considered morally principled, and is therefore deemed morally wrong.

However, applying these arguments to actions, such as driving a car or eating meat, presents difficulties. First, although these actions contribute to collective harm, they do not directly violate the inherent worth of individuals in the same immediate and explicit manner as sacrificing a person in the trolley problem. For instance, driving a car does not directly harm specific individuals, and eating meat, while causing significant suffering to non-human animals, does not typically attribute intrinsic value to these animals in non-consequentialist theories, as it does to humans.

Second, universalizability also faces challenges. Unlike breaking a promise, where the intent to break promises is inherent in the action, driving a car or eating meat usually does not involve a direct intention to cause collective harm, such as climate change or animal suffering. Instead, the intent behind these actions is often benign, such as enjoying a weekend trip or sharing a meal with friends.

Thus, non-consequentialism struggles to argue that actions such as driving a car or eating meat are morally wrong, because they neither directly violate the intrinsic value of individuals in an immediate sense, nor clearly breach the condition of universalizability.

Next, we explore how consequentialism addresses the moral wrongness of such actions, and examine whether it provides a more robust argument in the context of collective harm.

Non-consequentialist judgements are rigid as they focus on the inherent moral wrongness of an action, independent of the consequences. In contrast to non-consequentialism, consequentialism provides a flexible framework for moral evaluation by assessing the rightness or wrongness of an action based on its outcome. For example, in the trolley problem, non-consequentialism deems the act of diverting the trolley morally impermissible, regardless of the number of lives saved. Conversely, consequentialist judgements vary depending on the outcome of an action. Therefore, the act of diverting the trolley can be considered right or wrong depending on the specific situation, including the number of lives affected. This flexibility allows consequentialism to evaluate actions based on their impacts, emphasizing the importance of achieving the best overall outcomes.

This adaptability makes consequentialism particularly useful for evaluating the actions that contribute to collective harm. While driving a car or eating meat might not appear morally problematic in isolated instances, when aggregated, these actions can lead to collective harm such as climate change, environmental damage, and animal suffering. Consequentialism captures these cumulative effects on the broader system, and assesses the moral wrongness of actions based on these impacts.

Indeed, understanding collective harm requires recognizing the interrelatedness of individual actions, as such harm arises from their interconnected nature and combined impact. For instance, a store's bankruptcy cannot be explained without considering the network of individual actions that led to it. Evaluating actions in isolation fails to account for the existence of collective harm, which highlights the limitations of non-consequentialism. Because non-consequentialism focuses solely on the inherent nature of an action, it does not distinguish between actions considered individually or as part of a larger aggregate, making no difference in its moral judgment. Consequently, it overlooks the broader consequences that emerge from interconnectedness of individual actions.

By acknowledging the interconnectedness of actions within broader networks, consequentialism is better equipped to assess the overall impact and moral wrongness of actions that contribute to collective harm. From this perspective, the degree of moral wrongness of an action is not fixed but varies in proportion to the scale of the collective harm it contributes to. For example, eating meat in the current era—characterized by overpopulation, industrial farming, and significant contributions to greenhouse gas emissions—carries significantly greater moral weight than it would have in less interconnected times, such as in pre-industrial societies where meat consumption had minimal environmental impact. In other words, as the collective impact of an action such as driving a car or consuming meat intensifies harm in relation to climate change, the moral wrongness of each individual contributing action likewise increases.

Therefore, consequentialism effectively captures how the growing severity of collective harm amplifies individuals' moral responsibility for their contributions to that harm. By linking moral wrongness to the cumulative effects of individual behaviors, consequentialism offers a dynamic framework that adjusts moral evaluations based on the broader consequences of collective harm.

3.4 Individual Moral Responsibility Based on Causal Connection

Consequentialism typically assigns moral responsibility to individuals based on the contributory role that each action plays in generating harm, specifically through causal connections. This section argues that while consequentialism provides a framework for understanding the moral wrongness of an action in relation to collective harm, this conventional understanding of moral responsibility does not adequately support the claim that individuals are morally responsible for such wrongness in the context of collective harm.

From a consequentialist perspective, as previously shown, the moral wrongness of an action increases in proportion to the scale of collective harm to which it contributes. As individual actions collectively lead to such harm, the most straightforward way to understand this is through the causal connection between each action and the resulting collective harm. Consequently, the moral responsibility assigned to each individual engaging in these contributing actions can be interpreted as causal responsibility.

However, despite its simplicity, the notion of causal responsibility for collective harm faces significant theoretical challenges. Two main issues arise in this regard. First, the impact of each individual action is miniscule, leading to a dilution of responsibility, which makes each person's contribution seem too insignificant to effectively influence or standardize behavior. Essentially, no single act, such as driving a car or eating meat, directly causes climate change. This issue can be termed the "responsibility gap," referring to the disconnect between morally problematic outcomes and the responsibility assigned to individual contributors.

Second, individual actions alone often lack the causal power to effect significant change in collective harm. Even if a person stops driving a car or eating meat, their efforts alone cannot halt climate change or mitigate its adverse effects. As long as the majority of people continue their current lifestyles, individual efforts remain largely ineffective. This issue can be described as the "problem of inefficacy."

Both the "responsibility gap" and the "problem of inefficacy" present significant challenges to arguments for individual moral responsibility in addressing climate change, especially when such responsibility demands changes to daily lives. Therefore, while consequentialism offers a way to understand the moral wrongness of an action that contributes to collective harm, it does not provide a satisfactory basis for holding individuals morally responsible for those actions in the context of large-scale harm such as climate change.

What underlies both the "responsibility gap" and the "problem of inefficacy" is the assumption that assigning responsibility requires the causal connection between one's actions and their outcomes. For example, if my action causes you to get hurt, I am responsible for your injury. Conversely, if you injure yourself by tripping, I am not responsible for that.

This kind of responsibility, often referred to as "causal responsibility," ties accountability to a traceable link between an individual act and a specific outcome.

The challenge in applying causal responsibility to collective harm is that, as the number of people involved increases, the causal connection between each person's actions and the resulting harm becomes weaker. Consequently, even if the collective harm is severe and the moral wrongness of each contributing action increases, the individual sense of responsibility diminishes as their contribution becomes a smaller part of the larger group effort.

This dilution of responsibility is morally problematic as it leads to the following consequence. Although we are each part of the causes of climate change—and, in this sense, each contributing action carries moral wrongness—no individual is deemed responsible for it. This disconnect between individual actions and collective harm creates a moral paradox. While everyone is involved in contributing to the harm, the structure of causal responsibility makes it difficult to hold any individual accountable for the overall damage. This, consequently, undermines efforts to address climate change at the individual level, and calls into question the adequacy of causal responsibility as a framework for understanding moral responsibility in the context of collective harm.

3.5 Systemic Interdependence and Non-Agential Shared Responsibility

The failure of causal responsibility lies in evaluating actions in isolation. As highlighted in Sect. 3.3, collective harm cannot be understood fully without acknowledging that each contributing action is interconnected. Actions do not occur in a vacuum; they are part of a broader network of behaviors. Therefore, when discussing individual moral responsibility, it is essential to recognize our actions as integral parts of a social system.

Depicting actions in this interconnected manner makes it clear that those engaged in these actions are deeply intertwined. This interconnectedness is crucial for assigning moral responsibility to each person. Each individual action is linked to others, creating a web of interdependence, where the cumulative effect of these actions results in collective harm. This perspective helps prevent the dilution of responsibility that arises from the perspective of causal responsibility as it shifts the focus away from the direct causal connection between each action and collective harm. Instead, this approach frames moral accountability as "shared responsibility," emphasizing the collective nature of responsibility, rather than isolating individual contributions. Consequently, the "responsibility gap" and the "problem of inefficacy" do not emerge. Understanding moral responsibility in the context of collective harm, therefore, requires seeing ourselves as participants in a shared system, where our actions are not merely isolated incidents, but part of a larger, collective dynamic.

Notably, the sense of shared responsibility in this context differs from agential responsibility, which often implies shared intentions or goals. A clear case of this type of shared responsibility is presented in Christie's novel, *Murder on the Orient Express*.[2] In the story, 12 passengers participate in the stabbing and murder of a man who is later revealed to be a gangster responsible for kidnapping and murdering a child. Passengers, all of whom have personal ties with the child or family, collectively decide to seek justice by murdering the gangster. This is a case of collective harm, as each passenger inflicts a wound, but it is the accumulation of their actions that results in the man's death. Their actions are united by a shared intention to punish the man for his past crimes, making them collectively responsible for

[2] This example was mentioned by Conly (2022) in her work "Overpopulation and Individual Responsibility," which is included in *The Oxford Handbook of Population Ethics*.

his death. The passengers do not act independently; their coordinated actions create a unified outcome that cannot be achieved alone.

In contrast, we do not share the intention to produce collective harm in the case of climate change. As seen in Sect. 3.3, we neither intend to engage in contributing actions together, nor jointly aim to produce collective harm. Each action that contributes to climate change, such as driving a car or consuming meat, is typically performed independently, and without a collective goal. This indicates that shared responsibility in this context is non-agential.

Non-agential shared responsibility arises from our systemic interdependency; we are part of a shared system that coordinates our actions, and can collectively contribute to harm. A notable example is our responsibility for societal inequalities addressed by affirmative actions. Affirmative action encompasses policies and practices designed to rectify the historical and ongoing inequalities faced by marginalized groups such as racial minorities, women, and the disabled. For instance, given the under-representation of women in academia, some positions are designated exclusively for women as affirmative action measures. This reflects non-agential responsibility because, while no individual's actions are specifically intended to discourage women from entering academia, systemic injustices have produced this outcome.

Some argue that affirmative action introduces intergenerational unfairness, as male researchers in the current job market may face disadvantages despite not being solely responsible for systemic gender inequality. To address this concern, the concept of shared responsibility offers the following response: These inequalities have accumulated over generations, yet the current generation needs to make sacrifices to rectify them, as our shared responsibility for systemic injustice transcends generations. In other words, even though previous generations have contributed to systemic injustice, the current generation still bears the moral responsibility to rectify it.

Similarly, the current generation is not solely responsible for climate change or the harm it causes. Therefore, it may seem unfair for this generation, along with future generations, to bear the burden of mitigating this harm. However, shared responsibility suggests that our obligation to address this harm is not confined to those who originally caused it; it is passed down through generations. As part of a system that perpetuates this harm, we all share the responsibility for mitigating it.

This interconnectedness can be understood as a form of membership in a global society, or global citizenship. Being a member of society entails shared responsibilities that extend beyond individual actions and the agential responsibilities that stem from them. Indeed, from the moment of birth, we are embedded in society without actively choosing to be, yet we are nonetheless expected to fulfill social obligations. This implies that each person has an individual moral obligation to uphold shared responsibilities, even in the absence of direct agential involvement. This perspective allows us to argue that our participation in the causes of climate change makes each contributing action, such as driving cars or consuming meat, more morally wrong than it would be if considered in isolation, thereby attributing moral responsibility to individuals to reduce or avoid such actions.

3.6 Concluding Remarks

This chapter shows that an individual-based framework is insufficient for appropriately understanding individual moral responsibility in the context of climate change, which exemplifies collective harm. Instead, it should be understood as a non-agential shared responsibility. The chapter is divided into two parts; first, it explores how to assess the moral wrongness of an action contributing to collective harm, and second, it examines how to argue for individual responsibility in response to such wrongness.

Initially, I critically examined the skeptical arguments that question the attribution of individual responsibility for large-scale harm. Subsequently, I introduced two primary approaches for assessing the moral wrongness of actions, consequentialism and non-consequentialism, and argue that consequentialism provides a more appropriate framework for evaluating the moral wrongness of actions that contribute to collective harm.

Despite its merits, however, I argued that the consequentialist understanding of moral responsibility, namely, causal responsibility, fails to capture the collective nature of our actions. I further proposed an alternative view: shared responsibility, particularly non-agential shared responsibility. This approach better captures the interconnectedness of our actions within broader systems, and recognizes the moral obligations we hold as members of a global society.

Considering this re-examination of individual moral responsibility in the context of collective harm, it is safe to conclude that, to shift toward a carbon-neutral society, we should enrich the concept of global citizenship. This entails recognizing that our interconnectedness and systemic interdependence extend beyond the individual level to a global scale and across generations. Consequently, there is need to update individual moral responsibility accordingly, and integrate our actions and daily lives within this framework of non-agential shared responsibility. This calls for a reassessment of the moral wrongness of our individual actions, such as driving a car and eating meat, in light of their broader impact.

Ultimately, by reconceptualizing moral responsibility in this way, we can address collective harm such as climate change more effectively.

References

Conly, S. (2022). Overpopulation and individual responsibility. In G. Arrhenius, K. Bykvist, T.Campbell, & E. Finneron-Burns (Eds.), *The oxford handbook of population ethics* (pp. 430–443). Oxford University Press.

European Commission. (2018). *In-depth analysis in support of the commission communication com (2018) 773: A clean planet for all—A European long-term strategic vision for a prosperous, modern, competitive and climate neutral economy.*

Japan Ministry of the Environment. (2021). *The long-term strategy under the Paris agreement*. Government of Japan.

Johnson, B. (2003). Ethical obligations in a tragedy of the commons. *Environmental Values, 12*(3), 271–287.

Kagan, S. (2011). Do I make a difference? *Philosophy & Public Affairs, 39*(2), 105–141.

Maltais, A. (2013). Radically non-ideal climate politics and the obligation to at least vote green. *Environmental Values, 22*(5), 589–608.

Mann, M. (2019, September 12). *Lifestyle changes aren't enough to save the planet. Here's what could.* Retrieved September 30, 2024, from https://time.com/5669071/lifestyle-changes-climate-change/

Sinnott-Armstrong, W. (2005). It's not my fault: Global warming and individual moral obligations. In W. Sinnott-Armstrong & R. B. Howarth (Eds.), *Perspectives on climate change: Science, economics, politics and ethics* (pp. 285–307). Elsevier.

4 Explaining and Overcoming Barriers to Urban Carbon Neutrality: Historical Institutionalism in Transport-Land Use Coordination

Yudi Liu

Abstract

Urban and regional policies are key to achieving carbon neutrality as the majority of the global population is now living in urban built environments. However, progressive national agendas have been emerging, but institutional barriers often persist in constraining policymakers from guiding sustainable land use-transport coordination. This chapter argues that the research tradition of historical institutionalism (HI) offers insights into understanding and overcoming these barriers, helping explain current policies by identifying path dependence and power contestation rooted in contingent decisions over time. It considers the case of Tokyo to show that past policy choices shape recent contestations between carbon-emissive automobiles and low-carbon electrified railways. In Tokyo, the contingent absence of private automobiles owing to industrial policies until the mid-twentieth century delayed the progress of sustainable transport planning. This allowed pro-development coalitions to accelerate highway construction in the late century and led to Tokyo's puzzling metropolitan structure today: a transit-served inner city and automobile-reliant outer suburbs. These sequential events suggest that reformers should deliberate path dependence on layered development agendas set by powerful actors to advance sustainable transport and land use planning and overcome institutional barriers toward achieving carbon neutrality.

Keywords

Carbon neutrality · Historical institutionalism · Land use planning · Sustainable transport · Transit-oriented development

4.1 Introduction

Human beings increasingly live in urban built environments, leading to an urban age (Gleeson, 2012). The urban age proposition interacts well with human society's enormous impact on climate change, which is embedded in the Anthropocene proposal (Lewis & Maslin, 2015). This interplay highlights a fact: the urban space is key to carbon neutrality, while urban policymaking and planning are indispensable paths to it. However, cities are complex social constructs and are proposing varied policy agendas toward the mutably conceptualized carbon neutrality, emphasizing the tech-

Y. Liu (✉)
Waseda University, Tokyo, Japan
e-mail: yudiliu@aoni.waseda.jp

K. Akao and A. Washizu (eds.), *Climate Change Issues and Social Sciences*,
Sustainable Development Goals Series, https://doi.org/10.1007/978-981-96-8951-4_4

nical transition but overlooking political contestations (Tozer & Klenk, 2019).

Transport and land use underpin the urban built-up environment, but the carbon neutrality scholarship has focused more on energy transition over transport-land use coordination until recently (Li et al., 2023; Zhang & Hanaoka, 2022). Recent findings resonate with the seminal urban planning assumption that a transit-served and compact built environment contributes to carbon reduction, whereas a sparse and automobile-dependent one undermines it (Newman & Kenworthy, 1999; Suzuki et al., 2013). Practical concepts, such as transit-oriented development (TOD), facilitate the transition in land use-transport coordination. However, cities use TOD to produce varied physical forms and serve different political interests (Shen & Wu, 2020; Suzuki et al., 2015).

Tokyo is an atypical global metropolis that challenges the dichotomy between compact transit cities and sparse automobile suburbs. Specifically, it incorporates some of the best rail transit systems in the world, coupled with deeply problematic private car-dependent urban sprawl conditions (Hebbert & Nakai, 1988b; Sorensen, 2011). Antecedent studies have identified that the conditions are remarkably distinct between the populous inner cities and sparse outer suburbs in Tokyo (Liu et al., 2022), as its transit services and nearby land developments follow market dynamics (Cervero, 1998). Nevertheless, the policy agenda and political contestations behind the condition remain puzzling.

This chapter examines the coexistence of a robust rail transit capacity and an automobilization trend in Tokyo to exemplify that land use-transport coordination does not always come from a progressive policy agenda. Rather, it can rise in contingent policy choices in the past, thereby casting path dependence to influence decision-making today. Specifically, this study proposes historical institutionalism (HI), a new institutionalist branch, as an explanatory approach for understanding the hidden pitfalls and opportunities behind the current strategy. In doing so, the analysis explains why current policies fostering both rail transit and automobile-oriented land uses coexist in Tokyo, and how path dependence from contingent policy choices in the past aids or impedes a better transition toward carbon neutrality.

4.2 Path-Dependent Transport and Land Use

Transport has been saliently shaping urban land use and city forms since the Industrial Revolution, transforming traditional cities for pedestrians into industrialized ones for rail transit and, subsequently, private automobiles (Newman & Kenworthy, 1999). Automobiles mostly consumed highly carbon-emissive gasoline until the recent emergence of electric vehicles (EV), while much of the rail transit worldwide has shifted to electric power since the early twentieth century. This contrast in carbon emissions is one of the main reasons for changing the automobile-dependent suburbanization trends in industrialized countries (Ewing, 1997; Ewing & Cervero, 2010).

Since the 1980s, the new urbanism movement in the United States has facilitated concepts such as TOD, smart growth, and neotraditional neighborhoods to reverse the course of automobile suburbanization toward transit-oriented compact development (Carlton, 2009; Ellis, 2002). While European cities have had a higher density than their North American counterparts, they are still encouraging further transit-served compact city forms (Breheny, 1992; P. Hall, 2013). These reformist concepts sometimes take high-density Asian cities as practical exemplars (Curtis et al., 2009), while influencing the planning practices shaping cities in the Global South (Wood, 2022).

Land use-transport coordination is often path-dependent and locked-in. Path dependence means that a formal or informal institution will become increasingly difficult to change once established. A classic example is the QWERTY keyboard's ubiquitousness in digital devices, despite its suboptimal design (also see Sect. 4.3) (Pierson, 2004; Sorensen, 2015). Automobile dependence in the United States exemplifies the path dependence in land use-transport coordination, as the reformist planning concepts have so far changed very few automobile cities. In California, transit ridership has hardly increased despite continuous investments (Taylor et al., 2020).

In Western Europe, in contrast, planners have encouraged rail transit together with other transport modes considered more sustainable than private cars, including walking, cycling, and rapid bus transit (BRT), to reduce carbon dioxide emissions from vehicles, thereby framing the sustainable transport planning practice (Hickman & Banister, 2014). For instance, rail transit and buses in London often run services along the same route, with an increasing number of planning projects improving the walking and cycling environment (Hickman et al., 2010). Planning policies thus have contributed to differentiating the two regions' transport and land use (Hebbert, 1998).

Japan is in the middle between carbon-emissive automobile suburbia and carbon-reductive sustainable transport towns. Its most populous metropolitan areas rely on heavy rail transit, while smaller cities are converting into automobile suburbs (Ohashi & Phelps, 2021; Sorensen, 2011). Japanese cities also appeal to global observers with compact street networks shaped by traditional urban forms (Shelton, 2012). However, in the capital Tokyo or across the country, BRT and contemporary light rail transit (LRT) have been recent policy innovations (Kato et al., 2008), with exclusive bike lanes remaining rare compared to European cities.

Tokyo is the densest and wealthiest part of the market-led Japanese transportation system, with the highest level of rail transit reliance nationwide. Until the mid-twentieth century, the city relied on rail transit and had poor road conditions among developed cities and lower automobile ownership than Singapore and Kuala Lumpur, which had lower income levels (Sorensen, 2002). Since the 1980s, existing transit-served central areas held steady. However, cars and highways have simultaneously surged in numbers, inducing a United States-like suburbia with the deregulation of large retail stores (Upham, 1994). Consequently, its metropolitan outer suburbs are becoming increasingly sparser and automobile-dependent (Liu et al., 2022).

This condition reveals that technical differences in aspects such as population density and economic progress do not certainly determine the relationship between sparse suburbs dependent on private cars and compact cities reliant on low-carbon transport. The land use-transport coordination is not simply a technical goal but is socially constructed by actors across urban policymaking sectors in multifold power relationships (Kębłowski et al., 2019). The two entangled forms of land use-transport coordination intervene in the debate on achieving urban carbon neutrality, and a reinterpretation of the power relationships is of the utmost importance for offering such insights.

4.3 Historical Institutionalism: Path Dependence and Contingency

HI is suitable for exploring why and how a formal or informal institution emerges, stabilizes, and changes (Hall & Taylor, 1996). As one of the three main branches of new institutionalism, together with rational-choice and sociological institutionalism, HI is neither a theory nor a method but a research approach structured by an established conceptual toolkit to understand institutional stability and change (Steinmo, 2008). Path dependence is arguably the most well-known concept in the toolkit, suggesting that institutions become sticky and locked-in as time passes as reformers increasingly need more resources, know-how, coordination, or positive expectations to introduce a new institution (Pierson, 2004).

Institutions can change abruptly at exogenously triggered critical junctures (Blyth et al., 2016) or incrementally by endogenous forces (Mahoney & Thelen, 2009). Institutional changes vary in the relationship between the preexisting and new institutions, differentiated by the institutional defenders' veto power and level of discretion in interpretation/enforcement (Sorensen, 2015, p. 30). Under a low level of discretion, reforms may displace the old institution with a new one when the defenders hold weak veto power but layer them together if the power is strong. Meanwhile, a high level of discretion may drift the reform when defenders hold strong veto power or lead to the adaptation of the new institution's content even if they hold bare veto power.

Contingency is another concept guiding HI analyses. It refers to the availability and choice among multiple institutional options during reforms (Sorensen, 2015). Contingency does not align with coincidence but emphasizes that "relatively small events, if occurring at the right moment, can have large and enduring consequences" (Pierson, 2004, p. 44). Therefore, the timing and sequencing of historical events matter for an institution to change in a certain direction instead of some other direction. They constitute and alter the conjuncture of institutional arrangements, or "the linking of discrete elements or dimensions of politics in the passage of time" (p. 55).

Political power shaped by the most influential actors is central to HI. It tends to define power as material advantages and resources (Pierson, 2016). However, the primarily materialist and structuralist approach does not ignore ideational influences and agency (Blyth et al., 2016; Hay & Wincott, 1998). For example, Pierson (2016) argues that HI should consider Lukes' (1974) ideational power thesis, whereby the winning political coalition institutionalizes its advantage by inculcating ideas to make others unconscious of the contestation. In doing power, HI commonly compares institutional arrangements across contexts to generate and structure insights (Sorensen, 2015).

Lockwood et al.'s (2017) research agenda on adopting HI in sustainable energy transition exemplifies the carbon neutrality scholarship's interest in borrowing the approach's capacity to explain the transition as an institutional change and understand its contextual differences. Similarly, Sorensen's (2010, 2023) continuous work shows how HI bridges past events and presents decisions to reinterpret land use and urban property institutions related to urban sustainability. The next section succeeds in these efforts to explain the coexistence of rail transit and automobile-shaped built environments in Tokyo and offer insights into the transition toward carbon neutrality.

4.4 Parallel Lines: Rail Transit and Automobile Institutions in Tokyo

Contrary to the contestation between rail transit (and other sustainable transport modes) and private cars in the Anglosphere since the nineteenth century (Carlton, 2009; Newman & Kenworthy, 1999), the two transport sectors and relevant land uses have been parallel lines until recently in Japan. After the Meiji Restoration in 1868, railways grew rapidly to dominate ground transport across the country, carrying more passengers and goods at faster speeds than any transport means back then. In contrast, automobiles slowly emerged to serve the upper class in the late nineteenth century and then increased in numbers through the import of US-made cars in the early twentieth century (Yamamoto, 1993).

During the early twentieth century, many influential automakers emerged in the West and challenged the operation of mass-rail transit systems across regions (Haywood, 1997; Wachs, 1984). This situation was much less the case in Tokyo before World War II, as Japanese automobile manufacturing appeared moderately later. Meanwhile, industrialists developed their suburban railway enterprises as private conglomerates engaging with real estate, retailing, and many other businesses and holding direct political power to influence state affairs. During WWII, policymakers excluded US-made cars from the market and allowed railway owners to take over bus businesses (Nakanishi, 2009).

The scarcity of automobiles partially explained the poor road conditions across Japan compared to other industrialized countries immediately after WWII (Black & Rimmer, 1982). From the 1940s to the 1960s, elites across sectors aspired to shift to the automobile era, as evidenced by railway owners' contemplations about the displacement of railways by cars on highways (Kobayashi, 1962) and leading planners' road visions (Ishikawa, 1948; Yamada, 1966). They aspired to transform the rail transit-served and compact Tokyo into a modern metropolis with magnificent boulevards as long as 40–100 m, although the boulevards were not planned as highways exclusive to cars, as illustrated in planner Hideaki Ishikawa's (1948) vision of Tokyo in 1968.

The vision was not realized immediately as the economic conditions after the war did not allow so. More importantly, private cars could barely increase due to the infant industry policy for fostering domestic automobile manufacturers (Hook, 1996). During the three postwar decades, Japan gradually became the second wealthiest nation, but automobiles and highways grew slowly because of import quotas and tariffs for foreign cars (Liu et al., 2024). Tokyo's postwar rail transit-led development thus startlingly occurred when elites favored further development toward an auto age and did not consider imposing the strict traffic regulations seen in other developed countries, such as Britain (Hickman & Banister, 2014).

During this period, suburban railway companies took the opportunity to expand their influence, operating rail transit services as arteries to transport passengers from central Tokyo to the suburbs, and then using bus services as capillaries to carry them from suburban stations to homes (Yajima & Ieda, 2014). Notwithstanding, in 1981, the state removed all import quotas and tariffs, as Japanese automakers emerged in the domestic and global markets. This situation then revived policymakers' expectation on road construction. As Fig. 4.1 presents, the public investment ratio in road and railway was 1.45 times in 1965. Afterward, it surged from 3.25 times in 1975 to 26.67 times in 1990, and since then, the ratio has remained more than ten times in most cases despite austerity in general.

Consequently, automobiles began to carry more passengers than railways across Japan, with the highway system growing from 2,022 km in 1977 to 3,721 km in 1986 and, subsequently, to more than 10,000 km within three decades (MLIT, 2010). Notably, this number grew when the national economy stagnated. Meanwhile, railways nationwide did not grow much during the same period, and major mega infrastructure projects, such as the Tokyo Bay Aqualine connecting central Tokyo and Boso Peninsula in Chiba Prefecture, often only incorporated roads. The policy-stimulated highway boom realized policymakers' delayed postwar vision aspirations for roads (Feldhoff, 2007). During the recent two decades, policymakers introduced new taxes and tolls to fund highway construction while expecting railways to be financially self-sufficient in principle (Kurosaki, 2020).

Moreover, the rise of trucking gradually marginalized railways in the freight market in Tokyo (MoT, 1976), resulting in the disposal of railway yards owned by the Japanese National Railway (JNR) after its privatization and separation in 1987. These former public land parcels have now become high-density complexes in the city center and sometimes sparse automobile-shaped spaces in the suburbs, as Fig. 4.2 shows the conversion of the JNR Musashino classification yard (*sōshajō*) into a suburban district incorporating shopping malls run by IKEA, Costco, and Mitsui LaLaport. These malls proliferated during the deregulation of large stores influenced by diplomatic pressure under the United States-Japanese trade friction since the 1990s (Upham, 1994).

Finally, rail transit for sustainable development incrementally appeared on the national policy agenda in the twenty-first century. Considering the 1989 *Railway housing law* tailored for the Tsukuba Express project as a forerunner, the 2007 *Law on revitalization and rehabilitation of local public transport systems*, 2013 *Basic law on transport policy*, and 2014 revision of the *Special law on urban renaissance* emerged to assist the funding and operations of rail transit systems for networked compact cities (Utsunomiya, 2018, p. 555). During this period, innovative transit systems for sustainable development, such as LRT, BRT, and public-private partnership-funded systems, emerged, although this progress happened later than many Western counterparts (Utsunomiya, 2004).

4.5 Layering the Land Use-Transport Coordination in Tokyo

As antecedent research shows, Tokyo has experienced a reinforcing path dependence on the institution favoring private railways since the early

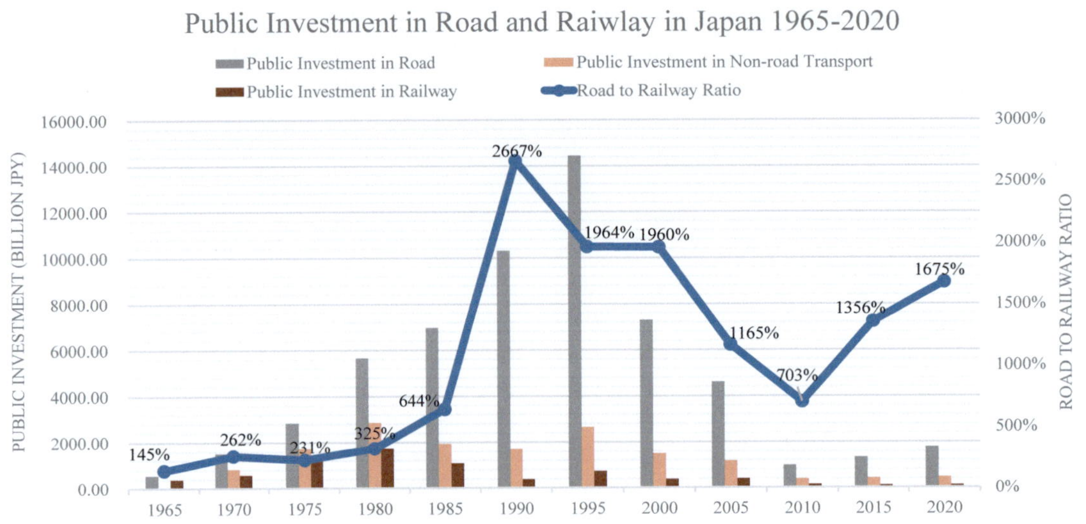

Fig. 4.1 Public Investment in Road and Railway in Japan. *Data Source* White Book of Transport and White Book of Ministry of Land, Infrastructure, Transport, and Tourism (MLIT)

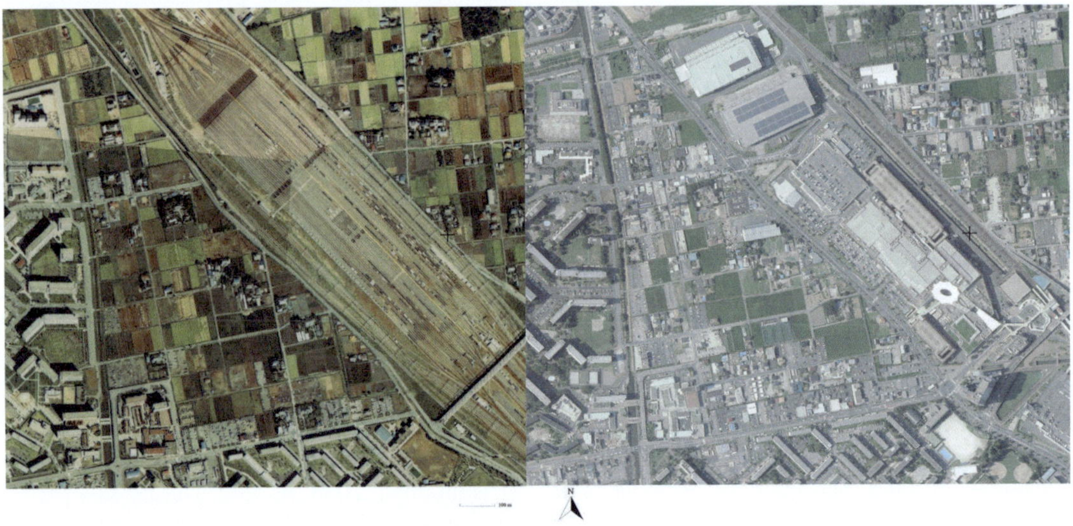

Fig. 4.2 Satellite Image of Musashino Classification Yard Conversion. *Source* The Geospatial Information Authority of Japan satellite image 1979–1983 and 2007—with usage allowance (https://mapps.gsi.go.jp/maplibSearch.do#1)

twentieth century (Liu et al., 2024). However, this historical institution from policy innovations that assume rail transit's importance as a path toward low-carbon sustainable development in the twenty-first century. In other words, Tokyo encounters two contexts of TOD: a forerunning TOD practice for profit under private railway conglomerates and a rather delayed one for the social good under recent policy incentives. Pro-road construction policies ascended in spaces that were less appealing to private operators when policy-incentivized rail transit had not appeared.

Automobiles had not threatened Tokyo and the central parts of its surrounding prefectures by bringing a sprawling land use pattern similar to that of American metropolises before the

1980s despite enduring challenges in infrastructure provision, which is regarded as a local form of sprawl (Okata & Murayama, 2011). Their late emergence contingently reinforces the path dependence on railways, especially lucrative private railways. However, it also weakens policy responsiveness to the pitfalls of automobilization. Car use might be constrained by limited parking spaces, difficulty in obtaining driving licenses, and employers' commuting subsidies in Tokyo (Jones et al., 1984), but contrary to British traffic control, these restrictions were not intentionally imposed and did not hinder automobiles' continuous increase after the 1980s.

Figure 4.3 summarizes the history and contrasts it to the contesting automobile and sustainable transport-land use institutions in the United States, indicating the institutions backing private railways since the 1910s, automobiles since the 70s, and innovative transit systems since the 2000s in Japan incrementally layer on each other. New institutions have surfaced without displacing the old due to the hardly interrupted conservative rule, favoring private corporations, economic growth, and infrastructure construction. Private railway and automobile industrialists run varied transport modes but commonly ally with the incumbent party for favorable policies (Hebbert & Nakai, 1988a), but notably, innovative transit systems do not benefit from the mechanism.

Clearly, urban policies toward carbon neutrality should be more responsive to the power relationship shaped by the industrialist-conservative coalition behind the land use-transport coordination in Tokyo. Private railways and private cars are two sides of the same coin from this perspective, while the institutional foundation of innovative transit systems in sparse regions is weak and requires reinforcement. Reversals to the path dependence on privately funded transport may be difficult, but leveraging its financial strength to reform the current institutions is desirable. For example, the private railways' operational experience has become a foundation for operating sustainable local transit systems in Japan (Taniguchi, 2018), which is not the case in other countries.

Furthermore, closing the public investment gap between railways and roads is a worthwhile and practical step toward achieving carbon neutrality. This analysis emphasizes that Japanese institutions might not have sufficiently prioritized sustainable transport and land use behind Tokyo's reputation as an exemplar of TOD (Cervero, 1998). To achieve carbon neutrality, policymakers should further consider and invest in electrified railways as a low-carbon means of transport against carbon-emissive automobiles. In Japan, local cities, rural areas, and metropolitan peripheries of Tokyo require further attention because private railway companies can hardly profit from these places, especially under the continuing depopulation (Kurosaki, 2020).

At its core, Tokyo exemplifies a socio-political dimension of urban land use and transport through an HI lens. Technical differences in population density, economic productivity, and technological expertise do not always differentiate transport-shaped land uses from sparse automobile suburbs to compact transit-served cities. Instead, contingent policy choices in sequential events reinforce path dependence and sometimes lead to unexpected institutional outcomes (Pierson, 2004). In this case of Tokyo, powerful institutional defenders from the industrialist-conservative coalition incrementally preserved policies favoring their interests and layered new institutions with the old, as shown by the rail transit-automobile coexistence.

4.6 Conclusion

This chapter proposes HI as an approach to institutionally understanding land use-transport coordination beyond the technical dichotomy of low-carbon transit-served cities and carbon-emissive automobile suburbs. Through concepts such as path dependence, contingency, incremental change, and critical juncture from a comparative perspective, this discussion draws on previous insights into the path dependence on the institution favoring private railways (Liu et al., 2024) to argue that the weaker presence of cars and roads in the early and mid-twentieth century contingently

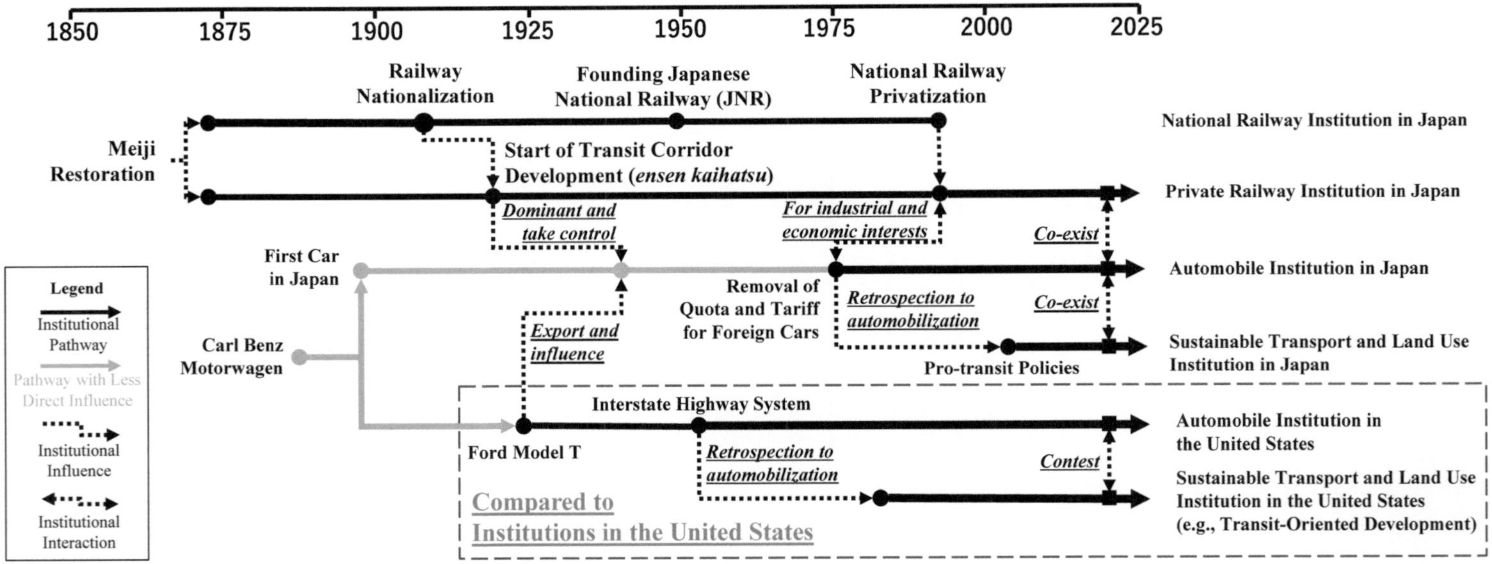

Fig. 4.3 Transport-Land Use Institutions in Japan and the United States. *Source* Elaborated by Author

undermined the current policy responsiveness to pitfalls in automobilization. This led to persistent public investment in highways over railways despite recent innovations in encouraging local transit.

Subsequently, Tokyo relied on privatized rail transit in the center and automobiles in peripheries, endorsed by the continuous rule under the industrialist-conservative coalition, despite these transportation modes' contrasting land uses and impacts on carbon neutrality. Although its foundation is still weak, sustainable local transit has emerged to address this issue recently. This illustrated case demonstrates that HI enables the observation of the socio-political dimension of transport and land use. Moreover, it reveals that power relationships and temporal processes, rather than technical differences in density, wealth, or expertise, shape the transport and land use patterns. This chapter suggests that policymakers should consider reforms for alternative policy options and leverage path-dependent institutions to invest more in low-carbon transport and land use toward achieving carbon neutrality.

References

Black, J. A., & Rimmer, P. J. (1982). Japanese highway planning: A Western interpretation. *Transportation, 11*(1), 29–49. https://doi.org/10.1007/BF00165593

Blyth, M., Helgadottir, O., & Kring, W. (2016). Ideas and historical institutionalism. In O. Fioretos, T. G. Falleti, & A. Sheingate (Eds.), *The Oxford handbook of historical institutionalism*. Oxford University Press. https://doi.org/10.1093/oxfordhb/9780199662814.013.8

Breheny, M. (1992). The compact city: An introduction. *Built Environment, 18*(4), 241–246.

Carlton, I. (2009). *Histories of transit-oriented development: Perspectives on the development of the TOD concept* (Working Paper No. 2009,02). Working Paper. https://www.econstor.eu/handle/10419/59412

Cervero, R. (1998). *The transit metropolis: A global inquiry* (Illustrated ed.). Island Press.

Curtis, C., Renne, J. L., & Bertolini, L. (2009). *Transit oriented development: Making it happen*. Ashgate Publishing Ltd.

Ellis, C. (2002). The new urbanism: Critiques and rebuttals. *Journal of Urban Design, 7*(3), 261–291.

Ewing, R. (1997). Is Los Angeles-Style Sprawl desirable? *Journal of the American Planning Association, 63*(1), 107–126. https://doi.org/10.1080/01944369708975728

Ewing, R., & Cervero, R. (2010). Travel and the built environment: A meta-analysis. *Journal of the American Planning Association, 76*(3), 265–294.

Feldhoff, T. (2007). Japan's construction lobby and the privatization of highway-related public corporations. In *Living Cities in Japan*. Routledge.

Gleeson, B. (2012). Critical commentary. The urban age: Paradox and prospect. *Urban Studies, 49*(5), 931–943. https://doi.org/10.1177/0042098011435846

Hall, P. (2013). *Good cities, better lives: How Europe discovered the lost art of urbanism*. Routledge.

Hall, P. A., & Taylor, R. C. (1996). Political science and the three new institutionalisms. *Political Studies, 44*(5), 936–957.

Hay, C., & Wincott, D. (1998). Structure, agency and historical institutionalism. *Political Studies, 46*(5), 951–957. https://doi.org/10.1111/1467-9248.00177

Haywood, R. (1997). Railways, urban form and town planning in London: 1900–1947. *Planning Perspectives, 12*(1), 37–69.

Hebbert, M. (1998). *London: More by fortune than design*. Academy Press.

Hebbert, M., & Nakai, N. (1988a). Deregulation of Japanese planning in the Nakasone Era. *The Town Planning Review, 59*(4), 383–395.

Hebbert, M., & Nakai, N. (1988b). *How Tokyo grows: Land development and planning on the metropolitan fringe*. Suntory-Toyota International Centre for Economics and Related Disciplines: London School of Economics and Political Science.

Hickman, R., Ashiru, O., & Banister, D. (2010). Transport and climate change: Simulating the options for carbon reduction in London. *Transport Policy, 17*(2), 110–125. https://doi.org/10.1016/j.tranpol.2009.12.002

Hickman, R., & Banister, D. (2014). *Transport, climate change and the city*. Routledge.

Hook, W. B. (1996). *The competitive city: The impact of transport and land policy on Japan's economic growth and development* [PhD Dissertation]. Columbia University.

Ishikawa, H. (1948, March). Nijunengo no tōkyō (Tokyo in 20 years). *Kagaku no tomo (Friend of Science), 4*(3). https://doi.org/10.11501/8099163

Jones, M., White, P., Toda, T., Sugie, Y., & Sugiki, T. (1984). Passenger transport planning, management and policy in Japan. In *Proceedings of Conference Held at Wolfson College*, Oxford, Working Paper 75.

Kato, H., Shiroyama, H., & Fukayama, T. (2008). Process management in public transit planning: Case study of introduction project of Light Rail Transit in Toyama Japan. *Transportation Research Record, 2042*(1), 33–40. https://doi.org/10.3141/2042-04

Kębłowski, W., Van Criekingen, M., & Bassens, D. (2019). Moving past the sustainable perspectives on transport: An attempt to mobilise critical urban transport studies with the right to the city. *Transport Policy, 81*, 24–34. https://doi.org/10.1016/j.tranpol.2019.05.012

Kobayashi I. (1962). Chikatetsudō no jidai ha sudeni sareri (The era of the subway has already gone). In *Kobayashi Ichizo Zenshū (Kobayashi Ichizo Collection)* (pp. 243–247). https://doi.org/10.11501/2941029

Kurosaki, F. (2020). Railways in Japan. In *Handbook on railway regulation.* https://doi.org/10.4337/9781789901788.00016

Lewis, S. L., & Maslin, M. A. (2015). Defining the Anthropocene. *Nature, 519*(7542), 171–180. https://doi.org/10.1038/nature14258

Li, L., Huang, X., & Yang, H. (2023). Optimizing land use patterns to improve the contribution of land use planning to carbon neutrality target. *Land Use Policy, 135*, Article 106959. https://doi.org/10.1016/j.landusepol.2023.106959

Liu, Y., Manabe, R., Nitanai, R., & Murayama, A. (2024). Historical institutionalism in action: Incremental prevalence of Transit-Oriented Development in Tokyo 1945–1982. *Progress in Planning*, 100850. https://doi.org/10.1016/j.progress.2024.100850

Liu, Y., Nath, N., Murayama, A., & Manabe, R. (2022). Transit-oriented development with urban sprawl? Four phases of urban growth and policy intervention in Tokyo. *Land Use Policy, 112*, Article 105854. https://doi.org/10.1016/j.landusepol.2021.105854

Lockwood, M., Kuzemko, C., Mitchell, C., & Hoggett, R. (2017). Historical institutionalism and the politics of sustainable energy transitions: A research agenda. *Environment and Planning C: Politics and Space, 35*(2), 312–333. https://doi.org/10.1177/0263774X16660561

Lukes, S. (1974). *Power: A radical view.* British Sociological Association.

Mahoney, J., & Thelen, K. (2009). *Explaining institutional change: Ambiguity, agency, and power.* Cambridge University Press.

Ministry of Land, Infrastructure, Transport and Tourism. (2010). *Nettowa-ku no kei to genjo* (History and condition of the network). https://www.mlit.go.jp/common/000996075.pdf

Ministry of Transport. (1976). *1976 unyū hakusho (White paper of transport).* https://www.mlit.go.jp/hakusyo/transport/shouwa51/ind020303/004.html

Nakanishi, K. (2009). *Japan private railway history research—Urban transport development and their structure* (Supplementary ed.). Minerva Shobo.

Newman, P., & Kenworthy, J. (1999). *Sustainability and cities: Overcoming automobile dependence* (Illustrated ed.). Island Pr.

Ohashi, H., & Phelps, N. A. (2021). Suburban (mis) fortunes: Outer suburban shrinkage in Tokyo Metropolis. *Urban Studies, 58*(14), 3029–3049.

Okata, J., & Murayama, A. (2011). Tokyo's urban growth, urban form and sustainability. In A. Sorensen & J. Okata (Eds.), *Megacities: Urban form, governance, and sustainability* (pp. 15–41). Springer Japan. https://doi.org/10.1007/978-4-431-99267-7_2

Pierson, P. (2004). *Politics in time: History, institutions, and social analysis* (1st ed.). Princeton University Press.

Pierson, P. (2016). Power in historical institutionalism. In O. Fioretos, T. G. Falleti, & A. Sheingate (Eds.), *The Oxford handbook of historical institutionalism.* Oxford University Press. https://doi.org/10.1093/oxfordhb/9780199662814.013.7

Shelton, B. (2012). *Learning from the Japanese city: Looking east in urban design.* Taylor and Francis.

Shen, J., & Wu, F. (2020). Paving the way to growth: Transit-oriented development as a financing instrument for Shanghai's post-suburbanization. *Urban Geography, 41*(7), 1010–1032. https://doi.org/10.1080/02723638.2019.1630209

Sorensen, A. (2002). *The making of urban Japan: Cities and planning from Edo to the twenty first century.* Routledge. https://doi.org/10.4324/9780203993927

Sorensen, A. (2010). Urban sustainability and compact cities ideas in Japan: The diffusion, transformation and deployment of planning concepts. *Crossing Borders: International Exchange and Planning Practices*, 117–140.

Sorensen, A. (2011). Post-suburban Tokyo? Urbanization, suburbanization, reurbanization. In N. A. Phelps & F. Wu (Eds.), *International perspectives on suburbanization: A post-suburban world?* (pp. 210–224). Palgrave Macmillan UK. https://doi.org/10.1057/9780230308626_12

Sorensen, A. (2015). Taking path dependence seriously: An historical institutionalist research agenda in planning history. *Planning Perspectives, 30*(1), 17–38. https://doi.org/10.1080/02665433.2013.874299

Sorensen, A. (2023). Urban intensification and land value capture in Toronto: Conjunctural analysis, critical junctures, and developmental pathways in urban planning. *Environment and Planning A: Economy and Space.*

Steinmo, S. (2008). Historical institutionalism. *Approaches and Methodologies in the Social Sciences: A Pluralist Perspective*, 118–138.

Suzuki, H., Cervero, R., & Iuchi, K. (2013). *Transforming cities with transit: Transit and land-use integration for sustainable urban development.* World Bank Publications.

Suzuki, H., Murakami, J., Hong, Y.-H., & Tamayose, B. (2015). *Financing transit-oriented development with land values: Adapting land value capture in developing countries.* World Bank Publications.

Taniguchi, M. (2018). Ensen Kaihatsu (railway area developments) in Japan: A comparison with transit-oriented development (TOD). In *Routledge handbook of transport in Asia* (pp. 287–295). Routledge.

Taylor, B. D., Blumenberg, E., Wasserman, J. L., Garrett, M., Schouten, A., King, H., Paul, J., & Ruvolo, M. (2020). *Transit blues in the golden state: Analyzing recent california ridership trends.* https://escholarship.org/uc/item/32j5j0hb

Tozer, L., & Klenk, N. (2019). Urban configurations of carbon neutrality: Insights from the Carbon Neutral Cities Alliance. *Environment and Planning C: Politics and Space, 37*(3), 539–557. https://doi.org/10.1177/2399654418784949

Upham, F. K. (1994). Speculations on legal informality: On Winn's "relational practices and the marginalization of law": Comment. *Law and Society Review, 28*(2), 233–241.

Utsunomiya, K. (2004). When will Japan choose light rail transit? *Japan Railway and Transport Review, 38*, 10–16.

Utsunomiya, K. (2018). The value of local railways: An approach using the contingent valuation method. *Research in Transportation Economics, 69*, 554–559. https://doi.org/10.1016/j.retrec.2018.05.003

Wachs, M. (1984). Autos, transit, and the Sprawl of Los Angeles: The 1920s. *Journal of the American Planning Association, 50*(3), 297–310. https://doi.org/10.1080/01944368408976597

Wood, A. (2022). Problematising concepts of transit-oriented development in South African cities. *Urban Studies, 59*(12), 2451–2467.

Yajima, T., & Ieda, H. (2014). *Tokyo: A global city created by railways*. The Institute of Behavioral Sciences.

Yamada, M. (1966). Approach to the 21St century: A development policy for the Tokyo Metropolitan Region. *Ekistics, 22*(129), 114–122.

Yamamoto, H. (1993). *Technological innovation and the development of transportation in Japan*. United Nations University Press Tokyo.

Zhang, R., & Hanaoka, T. (2022). Cross-cutting scenarios and strategies for designing decarbonization pathways in the transport sector toward carbon neutrality. *Nature Communications, 13*(1), 3629. https://doi.org/10.1038/s41467-022-31354-9

5 Sustainability Transition to a Low-Carbon Society: Focusing on Rural Areas

Ayu Washizu and Takashi Nozu

Abstract

To achieve a carbon-neutral and sustainable society, communities must be built that efficiently use local renewable energy. In this chapter, we first present a vision of the carbon-neutral energy system that we should strive for. We argue that to realize this vision, smart energy systems should be built in rural areas rich in renewable energy resources, as well as in the agriculture and forestry industries that support them. In this chapter, we will explain the Japanese government's circular and ecological sphere concept as a method for revitalizing rural villages, followed by a methodology for transitioning to carbon-neutral rural villages based on the theoretical framework of "sustainability transition." According to this methodology, to create a carbon-neutral society, promoters should (1) play a central role in establishing a vision for the region's future and (2) proactively implement projects that improve the environment, society, and economy in accordance with the vision while involving various stakeholders. Furthermore, (3) the promoter must continue to expand its business by iteratively refining the concept based on feedback from the results. This chapter employs this methodological framework to assess best practice processes while clarifying future measures' requirements. Finally, we argue that to achieve this goal, agriculture and forestry must be integrated into the economic cycle mechanism and discuss the ideal future economic structure. Specifically, we introduce an analysis method known as input–output analysis and use it to explain the new input–output structure that results from the constructing rural village using renewable energy.

Keywords

Circular and ecological sphere · Sustainability transition · Renewable energy · Input–output analysis · Industrial structure

5.1 Introduction

Achieving a carbon-neutral and sustainable society necessitates research on social innovations that contribute to the development of smart rural areas harnessing renewable energy (RE). Figure 5.1 presents a blueprint for Japan's future carbon-neutral energy system. Metropolitan areas along the coasts of Japan have large populations and concentrations of economic activity.

A. Washizu (✉) · T. Nozu
Waseda University, Tokyo, Japan
e-mail: washizu@waseda.jp

K. Akao and A. Washizu (eds.), *Climate Change Issues and Social Sciences*,
Sustainable Development Goals Series, https://doi.org/10.1007/978-981-96-8951-4_5

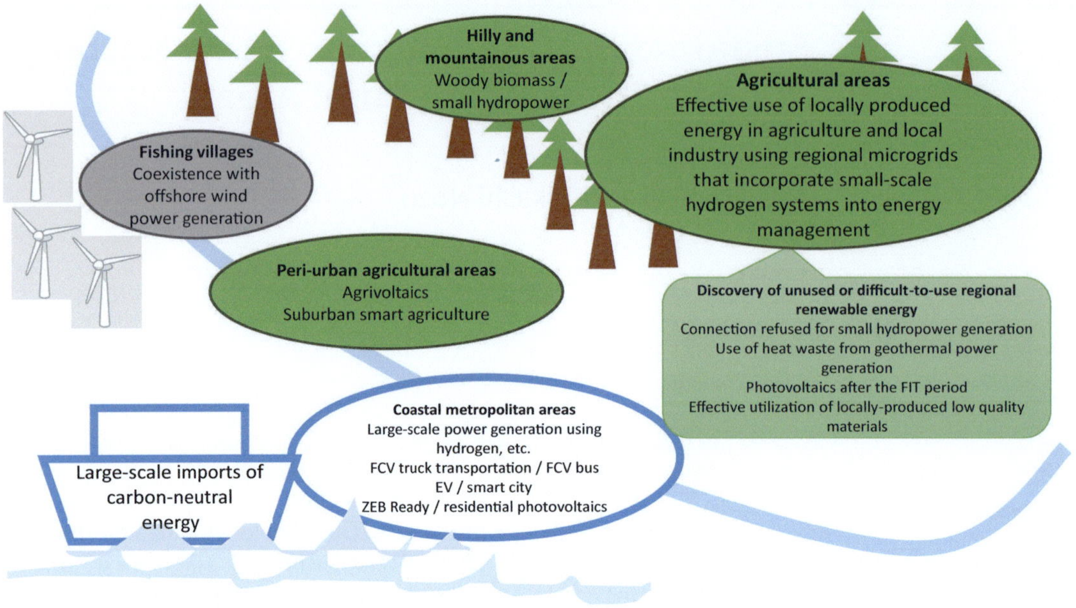

Fig. 5.1 Carbon-neutral energy system

Coastal areas are home to many heavy chemical industries, including the steel and petrochemical industries, which play a vital role in society but pose technical challenges with regard to achieving decarbonization. Achieving complete carbon neutrality in coastal metropolitan areas is difficult even if the renewable energy available in these areas (mainly photovoltaics) is fully utilized. To achieve carbon neutrality, it is crucial for these areas to undergo fundamental shifts in energy and industrial technologies, such as developing a large-scale imported hydrogen utilization system and exploring innovations in areas such as hydrogen reduction steelmaking and circular carbon chemistry. However, the realization of these technological transformations necessitates prior completion of engineering innovations.

Although rural areas have abundant RE, their demand for energy is limited. This can be attributed to the small population size in rural areas as well as the low coefficient of energy consumption per unit of production value in agriculture, forestry, and fisheries. Consequently, achieving carbon neutrality in these areas is technically possible. Therefore, it is desirable to develop local RE production and consumption-type rural areas that do not depend on external resources; social science and social innovation play a vital role in this process (rather than engineering and technological innovation). Furthermore, due to their large spatial area, rural areas in Japan are expected to contribute to the decarbonization efforts through the establishment of carbon neutrality.

However, in practice, there has been no progress in the utilization of RE in the agriculture, forestry, and fisheries industries. Aging and population decline is a serious issue in rural areas, and survival of the agriculture, forestry, and fisheries industries is at risk. There are also sporadic issues such as conflicts between power generation companies and local residents over photovoltaics and wind power generation. From a technical perspective, smart regional energy management using information and communication technology (ICT) is essential given the variable, small-scale, and distributed nature of RE, and the seasonal variability of energy demand in rural areas. The urgent issue of achieving a carbon-neutral society necessitates the implementation of a bottom-up approach to developing smart rural areas. This approach involves the comprehensive use of RE and circular utilization of local resources while simultaneously addressing local problems. It is believed that this approach will ultimately result in the maximum utilization of RE. The Financial Stability Board (FSB) established the Task Force on Climate-related Financial Disclosures (TCFD) and the Task Force on Nature-related Financial Disclosures (TNFD) in response to a request from the G20; therefore, large companies have been increasingly willing to engage in capital investments in areas related to the environment and natural capital, and this momentum should be readily utilized to address issues.

However, there is no concrete vision for developing rural areas that can utilize RE and simultaneously address local issues, and no solution has been provided for achieving a carbon-neutral and sustainable society. Regional development methodology is required for developing smart rural areas that utilize RE. Therefore, in this chapter, we explore the systems that support energy autonomy initiatives in rural areas, and organize and discuss the process of developing rural areas that utilize RE based on good practices. Furthermore, we focus on regional development methodologies for building smart rural areas that utilize RE, and the analysis of carbon-neutral industrial structures to discuss the effects of transitioning to a new society.

5.2 Process of Building Rural Areas That Utilize RE

5.2.1 Formation of Circular and Ecological Sphere[1]

The circular and ecological sphere is a concept set forth in the 5th Basic Environmental Plan in 2018 as the target sustainable society of Japan. The circular and ecological sphere is a state in which regions that can continuously create projects to improve the environment, economy, and society by utilizing local resources are interconnected through the circulation of people, goods, funds, and information. It is essential that projects implemented in each region are profitable and economically sustainable. The conditions that the circular and ecological sphere must meet in such cases are (1) regional independence, (2) partnerships with diverse stakeholders, both inside and outside the regions, and (3) simultaneous resolution of environmental, social, and economic issues. The "environmental" term referenced here includes the three elements of regional decarbonization, circular economy, and co-existence with nature. Initiatives for utilizing RE in rural areas are also important themes for the circular and ecological sphere. Figure 5.2 presents the process of circular and ecological sphere formation used by the Ministry of the Environment of Japan. The Ministry of the Environment is collaborating with the Global Environment Outreach Center (GEOC) and eight intermediary organizations known as Environment Partnership Offices (EPO) across the country to promote the creation of circular and ecological spheres according to this process. The network of blue arrows (directed upward; Fig. 5.2) depicts how the project expands, involving various stakeholders. The circular arrow (shown in orange) in the center of the diagram denotes the continuous process wherein the stakeholders constantly discuss the feedback on results and refine the concept. As of October 2023, circular and ecological sphere creation is being practiced in 127 regions across the country, and the logic formation of experiential knowledge based on cumulative cases is desirable for further project development.

5.2.2 Sustainability Transition

"Sustainability transition" involves incorporating accumulated knowledge from research related to individual socio-technical innovations and creating a new sustainable social system while emphasizing the need for policies to promote the transition to a sustainable system (i.e., "sustainability transition") (Chen et al., 2022; Tasaki et al., 2023). In terms of the sustainability transition, it is also desirable to propose measures needed to form

[1] Please refer to MOE (2023a).

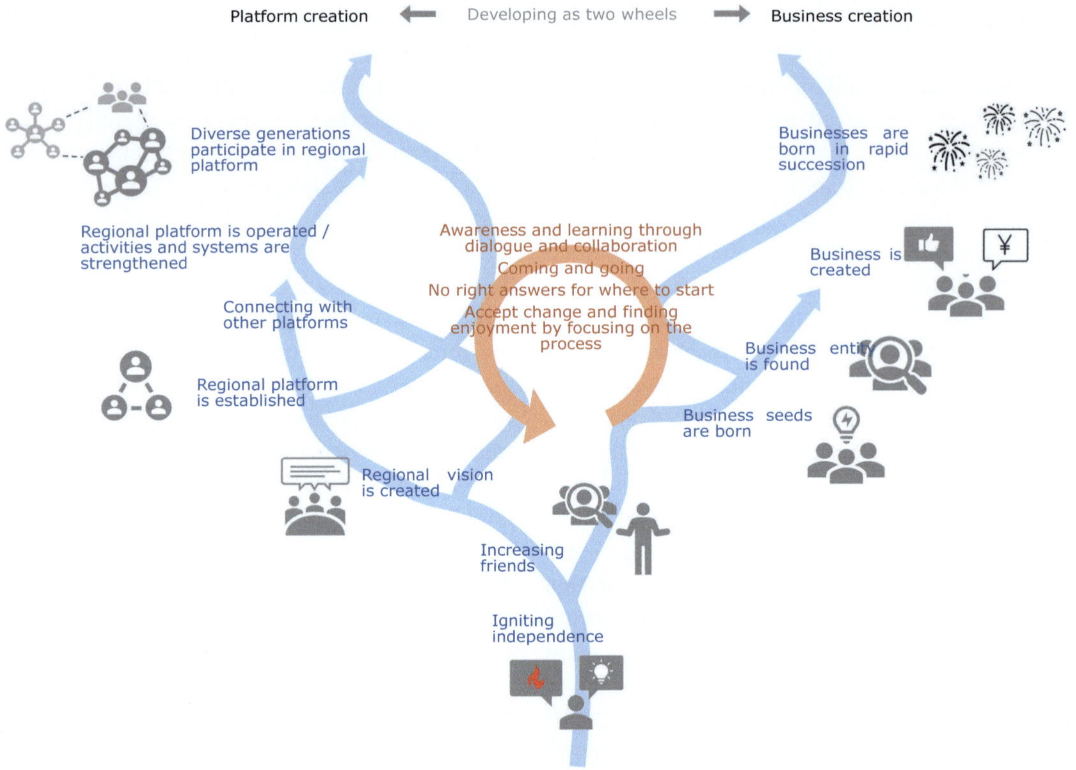

Fig. 5.2 Process of circular and ecological sphere formation. *Source* "Guide to creating a circular and ecological sphere (in Japanese)" (Ministry of the Environment) (https://chiikijunkan.env.go.jp/assets/pdf/manabu/tebiki.pdf) (accessed on 2025/8/15)

a circular and ecological sphere to overcome the many economic, social, and environmental challenges faced by a region while utilizing digital technology and RE.

Figure 5.3 shows the concept of a multilevel perspective (MLP), which is a premise of the sustainability transition (Geels, 2019). Microlevel innovative technologies, such as renewable energy technologies, are adopted within the context of a macrolevel landscape marked by an enhanced recognition of the imperative for carbon cycling. These shifts are accompanied by changes in mesolevel regimes, encompassing social systems, culture, and business practices, ultimately leading to the integration of these technologies into mainstream practices.

The steps for triggering a sustainability transition based on MLP are as follows (Loorbach, 2010; Matsuura, 2017): (1) an executive committee consisting of diverse members is formed, and (2) under the leadership of an elected leader, (3) tasks are set, and (4) a future vision for adapting to changes in macrolevel exogenous conditions is examined. Additionally, (5) the gap between the future vision and the current regime is examined, and first, (6) the regime is maximally utilized to obtain subsidies for transition experiments (niche innovations), and eventually, (7) the niche experiments gain mainstream acceptance through the collective agreement of numerous stakeholders, ultimately leading to a successful transition.

However, the transition does not occur swiftly or sequentially through a single project but rather involves a gradual trial-and-error process, encompassing repeated projects or a combination of multiple projects. In such cases, the trial-and-error process requires an intermediate evaluation of each project as well as a mechanism for providing feedback on the evaluation results to the field.

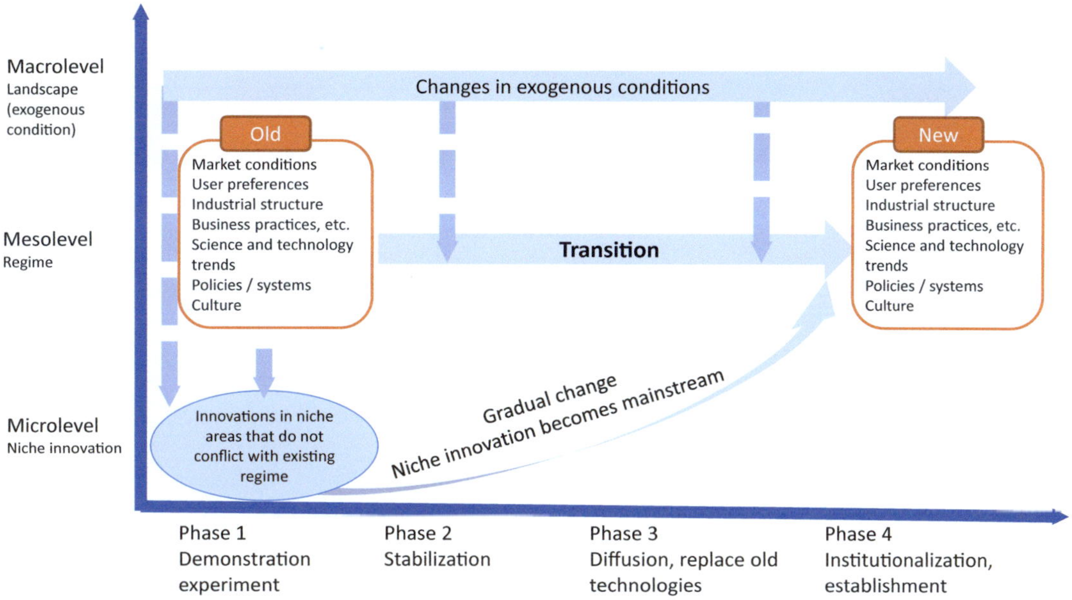

Fig. 5.3 Multilevel perspective (MLP) concept

A project is evaluated based on its appropriate implemented in accordance with the basic principles of transition management. The basic principles are as follows; these are also the perspectives for evaluating projects. The nine basic principles (or evaluation criteria) are as follows: (1) accurate understanding of the issues, (2) long-term perspective, (3) flexible system, (4) system that enables timely interventions in the event of a problem, (5) awareness of change triggers, (6) creation of niche innovation, (7) system that permits internal steering, (8) emphasis on learning diverse perspectives and options, and (9) interactions between stakeholders.

The feedback and spiral-up arrows in Fig. 5.2 indicate that a sustainability transition will be realized by continuously improving the initial project through intermediate evaluation and feedback mechanisms, and expanding the scope of the project.

5.2.3 Strategic Niche Management (SNM)

The impetus for transitions arises from the initiation of niche innovation. As niche innovations do not align with current regimes; they require strategic development in experimental fields that are isolated from the existing regime. The experiment implementation process of "(1) identification of concept → (2) experiment planning → (3) experiment implementation → (4) expansion of experiment into niches → (5) review of experiment protection" has an established methodology for strategic niche management (SNM) of new technologies, as shown in Table 5.1 (Weber et al., 1999).

5.2.4 Summary

In summary, to form a carbon-neutral society, the promoting entity must (first stage) take the lead in establishing a future vision for a region; (second stage) work toward goals and implementing projects that improve the environment, society, and economy while involving various stakeholders; and (third stage) continue project expansion by providing feedback on results and refining the concept. This is illustrated in Fig. 5.4.

A leader knowledgeable in how to proceed with this type of process for the promoting entity would be efficient in the first stage. Therefore,

Table 5.1 Strategic niche management (SNM)(Weber et al., 1999)

Experimental procedure	Main content
(1) Identification of new technology/ concept	Establish the technology that should be used to solve the problem
(2) Experiment planning	Prepare the experimental environment
(3) Experiment implementation	Conduct experiments in collaboration with a wide range of partners other than the experimental participants; learn from the experiments (primary learning, secondary learning)
(4) Expansion of experiment into niches	Disseminate information, expand stakeholder networks, involve competitors, horizontally expand to other regions, and change regulatory and political frameworks
(5) Review of experiment protection	Gradually release protections

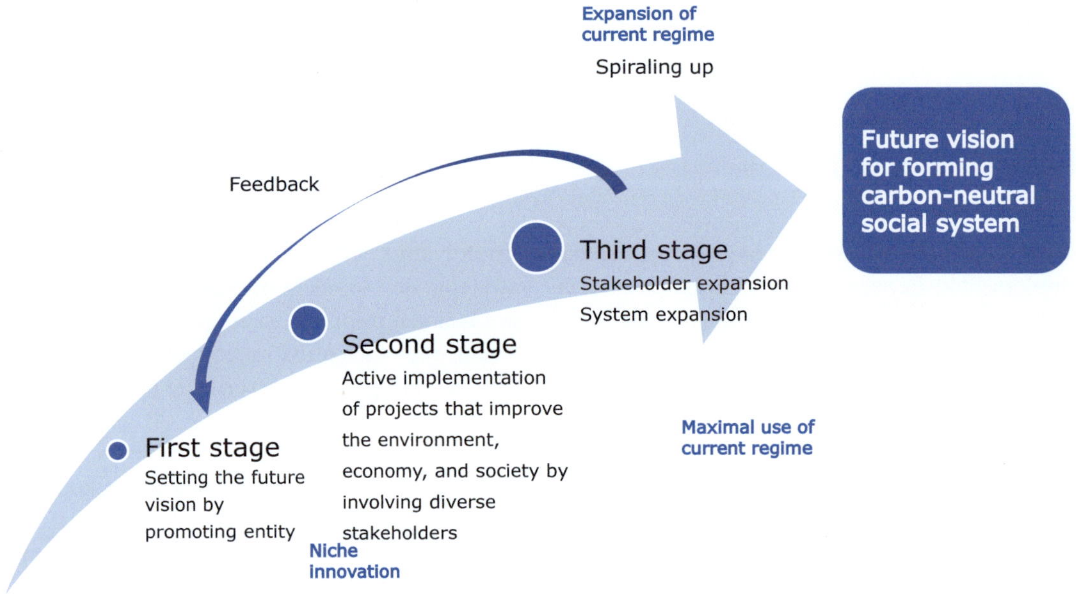

Fig. 5.4 Formation of a carbon-neutral society

it is desirable for an intermediate organization to assist in the development of leaders; GEOC and EPO, which assist in the formation of the circular and ecological sphere, can be understood as such intermediate organizations. Additionally, as previously indicated (Loorbach, 2010; Matsuura, 2017), promoting entities require a broad perspective; for this, it is desirable for members with diverse backgrounds to participate as promoting entities. Additionally, for the second stage, an innovative perspective is necessary to "improve the environment, society, and economy." In doing so, the use of digital technology that utilizes ICT will likely bring about easy-to-understand results. As a result of the third stage providing feedback on the results and refining the concept, the project can be expanded, rather than ending after a single goal has been achieved. However, in order to achieve this goal, it is crucial to establish a system for intermediate project evaluation that can effectively evaluate results and offer guidance on required improvements.

5.3 Social System for Building Rural Areas that Utilize RE

In this section, we will examine the current system for supporting the building of rural areas that utilize RE in particular, based on the flow diagram in Fig. 5.4.

5.3.1 Formulation of Local Government Action Plans Based on the Act on Promotion of Global Warming Countermeasures[2]

According to the Act on Promotion of Global Warming Countermeasures (Warming Countermeasures Act), local governments are required to formulate plans to reduce greenhouse gas emissions in accordance with the national plan for global warming countermeasures. Policies related to regional decarbonization projects were added in a revision in financial year (FY) 2021; thus, a system relating to regional decarbonization projects was added to the local government action plan system. However, according to previous research (Nomura Research Institute, 2023), Nagano and Tokushima prefectures are the only two prefectures that have formulated standards relating to establishing such promotion areas; the 29 organizations currently considering such formulations only account for two-thirds of the total.

The purpose of the regional decarbonization promotion project is to "promote the introduction of RE projects that co-exist with the region by establishing a smooth consensus, giving appropriate consideration to the environment, and leading to advantages for the region." Under this system, (1) municipalities formulate local government action plans and (2) regional decarbonization promotion project plans are certified by planning municipalities. The (1) plan formulation is to be conducted in discussion forums (e.g., councils) in which stakeholders participate and (2) certified regional decarbonization projects do not require related permissions and other procedures. Furthermore, the (2) regional decarbonization promotion projects must be "initiatives that contribute to the sustainable development of the regional economy and society"; examples of such initiatives are shown in Table 5.2. Efforts are also being made to promote initiatives that contribute to the healthy development of agriculture, forestry, and fisheries industries by collaborating with the Rural RE Act.

Summarizing the above in relation to Fig. 5.4, local government action plans are formulated (i.e., future visions are set) by councils, and niche innovation projects that contribute to carbon neutrality are implemented. If a plan is certified as a regional decarbonization promotion project at this time, then the necessary provisions are applied, permitting the implementation of maximum activities within the scope of the current regime. Therefore, the steps up to the second stage in Fig. 5.4 have been institutionalized. However, the process of progressing to the third stage, expanding the number of stakeholders and refining the initial concept based on feedback from the project results, has not been incorporated into the system.

5.3.2 Rural Renewable Energy Act[3]

The Act on Promoting Generation of Electricity from Renewable Energy Sources Harmonized with Sound Development of Agriculture, Forestry, and Fisheries (Rural RE Act), which is also included in Warming Countermeasures Act-based local government action plans, is a law promoting the generation of renewable energy in harmony with the healthy development of agriculture, forestry, and fisheries. This law aids municipalities to effectively use degraded farmland and channel the revenue generated from electricity sales toward fostering sustainable growth of local agriculture, forestry, and fisheries. To this end,

[2] (MOE, 2023b)

[3] Please refer to MOAF (2023a).

Table 5.2 Examples of initiatives that contribute to the sustainable development of the regional economy and society

Advantage to region	Example of initiative
Contribution to regional economy	Initiatives to supply inexpensive RE within the region and promote economic circulation within region
	Initiatives to create local jobs, develop local human resources such as maintenance inspections, share technology, and provide educational programs related to RE projects
	Initiatives to involve business and financial entities, such as local business entities and regional financial entities
	Initiatives to attract industries such as data centers in combination with RE introduction
Addressing regional social issues	Initiatives that can be used to solve issues in other policy fields, such as the utilization of RE as a disaster power source during emergencies, EV sharing, and introduction and utilization of green slow mobility
	Initiatives such as utilizing residual heat generated from RE projects for greenhouse horticulture and using combustion residue as organic fertilizer
	Initiatives such as monitoring services and transportation support for elderly people using profits
	Initiatives against wild animals by utilizing abandoned cultivated land and degraded farmland
	Support for local activities in municipalities

Source Ministry of the Environment (2023b)

municipalities are required to convene councils organized by local stakeholders to formulate and disseminate fundamental strategies related to the implementation of renewable energy infrastructure that will effectively utilize local resources and contribute to local agriculture, forestry, and fisheries. The formulation of a fundamental framework in a council facilitates the process of reaching an agreement in the region. Incorporating basic policies relating to the implementation of projects that promote the transfer of agricultural and forestry land ownership into the basic plan enables the one-time transfer of rights to agricultural and forestry land, which in turn enables effective land use. Municipalities certify equipment improvement plans prepared by equipment maintainers based on the basic plan, and implement the projects. Municipalities also check the progress of equipment maintainers' initiatives and regularly review the content.

In summary, basic plans are formulated by councils, goals are clarified, agricultural and forestry land ownership transfers are coordinated, and projects progress smoothly. Consequently, the steps up to the second stage in Fig. 5.4 are institutionalized. There is also a process in which "municipalities check the progress of equipment maintainers' initiatives and review the content constantly," and the process that corresponds to the third stage of obtaining feedback from the project and refining the initial concept is also included in the system.

5.3.3 Green Food System Strategy Promotion Grant[4]

The purpose of the Green Food System Strategy, formulated by the Ministry of Agriculture, Forestry and Fisheries of Japan in 2021, is to promote initiatives during material procurement, production, processing/distribution, and consumption, as well as innovations to reduce environmental impact, such as carbon neutrality, with the goal of building a sustainable food system. In 2022, the Green Food System Act was enacted, whereby plans by producers who align with the basic plans created by prefectures and municipalities, or businesses that align with national basic policies are granted a Green Certification and have special measures implemented. These include simplifying administrative procedures associated with technological development and prioritization for Green Food System Strategy Promotion Grants. The Grant involves

[4] Please refer to MOAF (2023b, 2023c).

various support measures, with the following items including energy-related support.

- Regional recycling energy system that utilizes local resources in the Comprehensive Measures for Promoting Green Food System Strategy
- Demonstration of cultivation by new technologies that contribute to reduce fossil fuel consumption, and introduction of energy-saving equipment, materials, and self-consumption power generation systems in the Establishment of SDGs-Compliant Greenhouse Horticulture
- Support for agrivoltaics model initiatives and countermeasure surveys to promote the use of unused resources in the Construction of Regional Recycling Energy Systems
- Demonstration of a recycling-based production technology system using rice husks in the Grain Greening Transformation Promotion Project

The main purpose of projects implemented under this strategy is not RE introduction, but rather, to improve productivity and sustainability in the food, agriculture, forestry, and fisheries industries through innovations. However, if RE is introduced, it is likely that this would be incorporated into the regional economy within a broader vision, and it is expected that the probability of project continuation or success would increase. The Key Performance Indicator (KPI) to be achieved through the Green Food System Strategy has been set as a target for 2030 or 2050, and project evaluations are reported to the director of the regional agricultural administration bureaus through the prefectural governor.

Considering Fig. 5.4, all steps prior to the second stage of implementing projects that improve the environment, society, and economy toward a goal are institutionalized. Here, a variety of support measures are prepared that can be provided according to the future vision of each region. The system itself includes a broad perspective, as well as individual benefits to businesses and agriculturalists. The aim is not necessarily to develop a project system for the entire region, with a feedback mechanism for continuous spiral-up improvement.

5.4 Examples of Flows for Building Rural Areas That Utilize RE

In this section, we discuss examples of flows for building rural areas that utilize RE based on Fig. 5.4.

5.4.1 Case Study of Nishiawakura Village, Okayama Prefecture

Since 1999, the government has promoted nationwide municipal mergers to establish the administrative and financial foundations of basic local governments. However, Nishiawakura Village refused to merge with surrounding municipalities in 2004, choosing to become independent. Figure 5.5 shows a timeline of Nishiawakura's independence initiatives. The village has continued to promote the expansion of projects based on strengthening the profitability of forest business, which is the village's main industry (100-year forest concept).

Figure 5.6 shows the method of forest business management established in Nishiawakura Village. Integrated forest management requires establishing a foundation that would separate usage rights of forest land from ownership to enable efficient forest management. Therefore, the village allowed both the local government and trust banks to manage forest land by receiving land entrustments or trusts from forest owners. Hyakumori Co., Ltd. is a management company that receives operating fees from the village office and is involved in forest land management. Orders for forestry operations and material production are made to a forestry association under a comprehensive management plan using a forest analysis information system that utilizes ICT. The materials produced are sold to wood-related businesses in the village, where they are processed

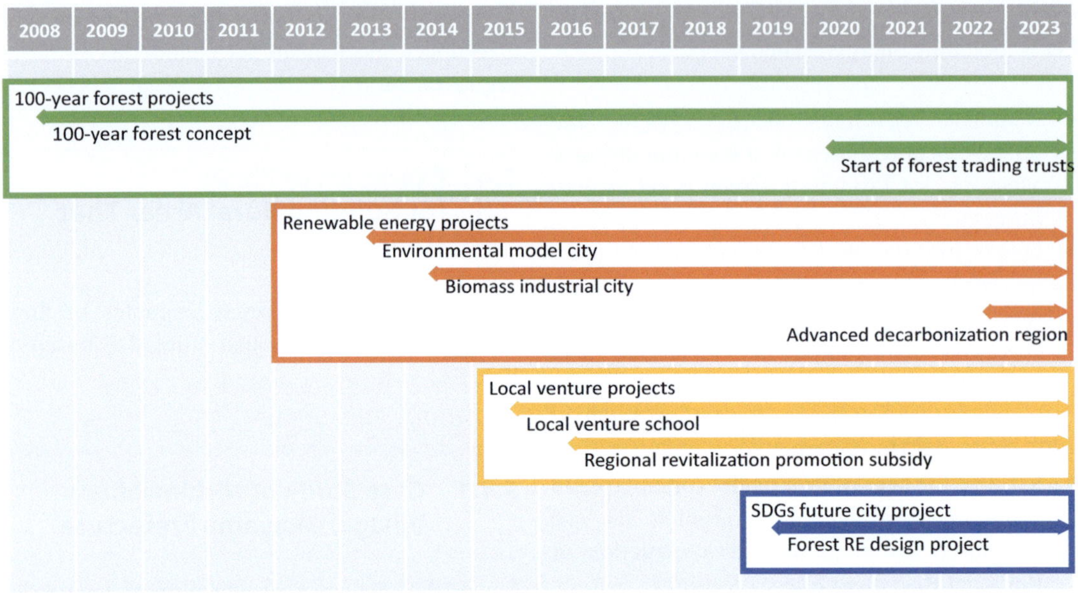

Fig. 5.5 Flow of initiatives in Nishiawakura Village, Okayama Prefecture

and sold as niche products with added value both inside and outside the village. Additionally, thinned wood and offcuts are used as fuel by the regional heating company in the village (RE project development). These profits are then distributed to both forest owners and the village office. This effectively enables the village office to recover a portion of the operational costs. Wood-related businesses diversify beyond wood processing as local ventures and strive to improve profits (local venture projects). The expansion of local venture projects increases the influx of people from outside the village, which, in turn, increases tax revenue and state subsidies allocated to the village. Recently, local venture projects have been expanded to include tourism, welfare for the elderly, and new electric power projects, which aim to improve the welfare of all villagers (SDGs Future City Project). In these series of flows, the village office focuses on planning and management, whereas business operations are outsourced to the private sector.

Regarding RE use, total small hydropower generation, photovoltaics, and woody biomass power generation is 545 kW, which generates 4.2 million kWh of the 7.2 million kWh total electricity consumed by the village (87% of household power). Additionally, the wood biomass boiler has an output of 1,290 kW and supplies heat to the village office, library, nursery school, elementary and junior high schools, and facilities for the elderly. The total energy produced through power generation and thermal energy is 21,451 GJ/year (self-sufficiency rate of approximately 15%).[5] Throughout this period, substantial progress has been achieved in the development of automatic dust removal systems for small hydroelectric power generators, as well as advancements in wood biomass chip supply equipment. However, it is important to note that ample opportunities still exist for further technological enhancements and innovations aimed at optimizing the efficiency of small-scale power generation.

Various business developments have been implanted in Nishiawakura Village with a focus on the 100-year forest plan, which aims to strengthen forestry, the village's core industry. In other words, projects have spiraled up as a result of development from the first to third stage in Fig. 5.4, and by repeating these stages. Interviews revealed that teams that cut across each section of the village office are constantly established

[5] From materials provided by Nishiawakura Village.

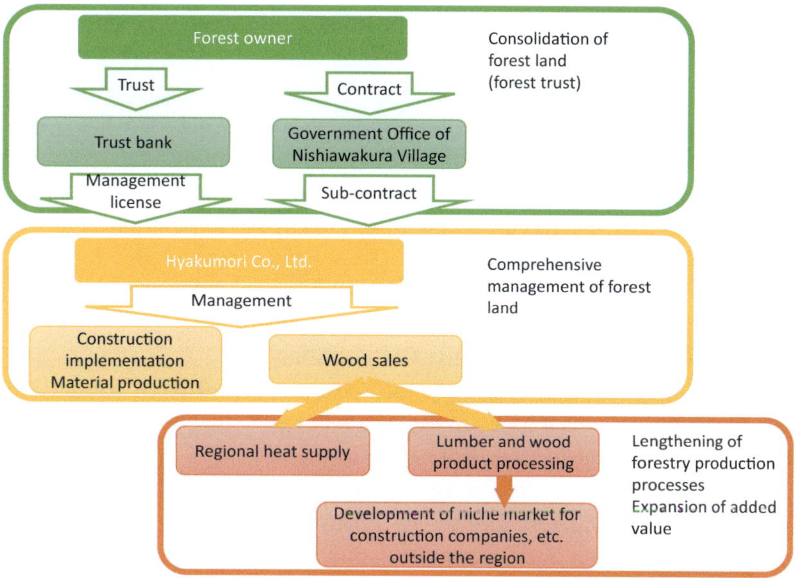

Fig. 5.6 Establishing a comprehensive management method for forestry operations

with the involvement of private businesses. These teams adopt a flexible approach to develop project concepts along with specific business plan for the concepts; this process includes a mechanism for intermediate project evaluation and feedback. Furthermore, various projects are implemented by private businesses based on market principles, ensuring the efficiency of these projects despite being funded through subsidies. In 2022, Nishiawakura Village was certified as an advanced decarbonization region; this can be conceptualized as the mechanism that was already in place at Nishiawakura Village being applied to this system.

5.4.2 Case Study of Sosa City, Chiba Prefecture

Sosa City, Chiba Prefecture, is aiming to achieve decarbonization by implementing agrivoltaics that are compatible with paddy agriculture, the main industry of the region.

Agrivoltaics is a type of photovoltaics, in which pillars are erected on farmland and photovoltaics equipment is installed on top in order to generate electricity while farming continues below. The area ratio of the photovoltaics equipment installation above the farmland is maintained at a level that can secure the amount of sunlight needed for crop growth. This aims to balance power generation and farming projects. Thus, the purpose is to share sunlight between power generation and farming; therefore, it is also called "solar sharing" in Japan.[6] Conversion of farmland is necessary when installing conventional open-field photovoltaics equipment, meaning that ordinary open-field photovoltaics equipment cannot be installed on so-called good farmland. However, in the case of agrivoltaics above the farmland where farming continues, except for pillars to support the photovoltaics equipment, farmland conversion is not necessary. Therefore, this can be implemented even on good farmland (Shibata et al., 2021). The proportion of flat land area in Japan's total land area is the lowest among all major countries, and the installed capacity for photovoltaics per flat land area is already the highest in the world; thus, the amount of suitable land for photovoltaics is decreasing. In this context, farmland has attracted attention as the remaining frontier of photovoltaics, and agrivoltaics has attracted attention as an approach that can address two issues, by enabling photovoltaics to be implemented while farming continues.

[6] Overseas, this is called "agrivoltaics," as a coined word from "agriculture" and "photovoltaics."

However, agrivoltaics is a new initiative, and its position under the Cropland Act was clarified by the Ministry of Agriculture, Forestry, and Fisheries in 2013. Even now, nearly 10 years later, annual installations of agrivoltaics account for only ~2% of the total photovoltaics (non-residential) (FY2020). Thus, it can be positioned as a niche innovation in both the agriculture and renewable energy fields.

On the other hand, there are many citizen-led agrivoltaics projects in Sosa City, Chiba Prefecture. The city is known as one of the most advanced agrivoltaics regions in Japan. Just after the Ministry of Agriculture, Forestry, and Fisheries clarified the position of agrivoltaics (2014), volunteers from citizen groups in Sosa City established a limited liability company and initiated agrivoltaics projects (MOAF, 2023d). The purpose of this initiative was to restore farmland that had been degraded due to aging populations, and to utilize the profits from electricity sales for farmland preservation, regional development, and environmental conservation, with the initiative being praised as "improving the environment, society, and economy" of the region. In 2023, Sosa City hall announced the Sosa City Zero Carbon City Declaration with the aim of a decarbonized society being realized by 2050. Simultaneously, the decarbonization promotion plan centered on agrivoltaics formulated by Sosa City Hall has progressed, and was selected as a Ministry of the Environment "advanced decarbonization region" project (MOE, 2023c). This plan involves the participation the City Hall and local agrivoltaics practitioners, and various stakeholders, including universities, financial institutions, and local chambers of commerce. In relation to Fig. 5.4, this plan is currently moving from the first stage, setting a future vision, to the second stage, involving diverse stakeholders. This plan also aims to expand agrivoltaics, which has conventionally been practiced on dry field agriculture, to paddy agriculture, which is Sosa City's main industry; as well as utilizing agrivoltaics research and human resource development to share knowledge inside and outside the city and increase the number of connections and exchanges outside the city. Therefore, this is aiming for the third stage established in Fig. 5.4, which is to expand the system by expanding the number of stakeholders. Sosa City is conducting a transition experiment to expand the niche innovation of agrivoltaics into paddy agriculture through a subsidy from the Ministry of the Environment, obtaining cooperation from various stakeholders to bring agrivoltaics into the mainstream, and aiming to achieve a transition.

5.4.3 Case Study of Hachimantai City Hachimantai Smart Farm in Iwate Prefecture[7]

Hachimantai Smart Farm is an agricultural facility that uses ICT to control geothermal water, and cultivates basil using hydroponics.

Around 2013, Hachimantai City was faced with the need to replace its hot spring water supply pipes, which was addressed using a depopulated area development bond (depopulation bond).[8] The hot water supply was expected to increase following pipe replacement, and there was an attempt to acquire a new demand entity. For more than 30 years, agriculture in the city has used plastic greenhouses that use hot water supplied by geothermal power plants for heating (i.e., hydrothermal greenhouses); however, these facilities are limited by farmers leaving the workforce due to aging populations as well as the aging of facilities. Under such circumstances, in 2017, a hydroponic cultivation venture company, an information technology (IT) venture company with experience in the regeneration of hydrothermal greenhouses, and Hachimantai City signed a basic agreement for the Smart Farm project, which aimed to rehabilitate abandoned greenhouses and train new farmers. In 2019, Hachimantai Smart

[7] Please refer to VentureTimes (2023).

[8] The depopulated area development bond is a local bond issued by municipalities designated as depopulated areas in 2000, under the Act on Special Measures for Promotion of Independence in Underpopulated Areas. It serves as a financial source for projects based on municipal plans aimed at promoting self-reliance in depopulated areas (MIAC, 2023).

Farm Co., Ltd. was established as a corporation that is eligible for farmland ownership as stipulated in the Cropland Act. This project was introduced in the FY2022 White Paper on Food, Agriculture, and Rural Areas as a case study of initiatives based on the Digital Garden City Nation Initiative. This is an important case study because a regional issue with aging facilities was resolved through the cooperation of a city and venture company; the fact that RE use was meaningfully incorporated into this resolution should be well-regarded. The Second Hachimantai City Comprehensive Strategy for the City, Humans, and Job Creation, which was revised in 2020, set regional revitalization projects that utilize geothermal energy, with the "promotion of smart agriculture using hydrothermal greenhouses" as one of the main measures.

Hachimantai City is home to Japan's first commercial geothermal power plant, and there is an historical understanding of geothermal energy as a region-specific resource. Interest in the effective use of such resources deepened as the hot water pipes were updated. Then, the coincidental establishment of a joint system between a venture company and the local government resulted in the development of a new project. In relation to the flow shown in Fig. 5.4, the process from the first to the second stage occurred naturally in Hachimantai City based on a positive climate toward the utilization of geothermal energy. Thus, a system is being established that follows the current situation to ensure the process to date while promoting further stages. Hachimantai City is also promoting public understanding of geothermal energy, and a spiraling up of these projects is expected in the future.

5.4.4 Status of Building Rural Areas That Utilize RE

It is hoped that each region will consider their individual energy characteristics when transitioning to a low-carbon society, with the aim of a carbon-neutral society. To that end, each region must first clarify its long-term vision in a basic plan and their transition goals (first stage in Fig. 5.4). In doing so, the goals must include the environment, economy, and society. This is because, as shown in the case studies in Sect. 5.4, initiatives for environmental and energy issues are more likely to succeed when they are incorporated into a social context that addresses both the economic and social issues of the region. The promoting entity (e.g., council) plays a central in initiatives and promotes those that include a niche innovation perspective toward a goal (second and third stages in Fig. 5.4).[9] It is believed that a process which includes intermediate evaluation of the project outcomes and the provision of suitable guidance to the promoting entity can be considered effective in ensuring the ongoing expansion of the project. One factor for why Nishiawakura Village has continued consistent initiatives based on the 100-year forest concept is the thorough governance of the village office in relation to the reform of the village as a whole, and its history of identifying future directions based on the results of previous projects.

The current social system described in Sect. 5.3 includes a mechanism for executing the processes up to the first and second stages as shown in Fig. 5.4. However, further innovations for mechanisms that spiral up initial efforts are needed. Specifically, this is a mechanism that involves intermediate evaluations of project progress and provides feedback with relevant advice.[10] Simultaneously, there is a need to foster an expert group to conduct intermediate evaluations, an information platform to ensure the objectivity of intermediate evaluations, and an intermediate organization that is responsible for this platform.

[9] Here, successful niche innovations that are not a main part of the current regime require the support of administrative and local governments (e.g., application of special provisions).

[10] According to Iida et al. (2022), rural development initiatives in Germany have established a system for monitoring and evaluating projects at the regional and federal levels, and are said to be functioning effectively.

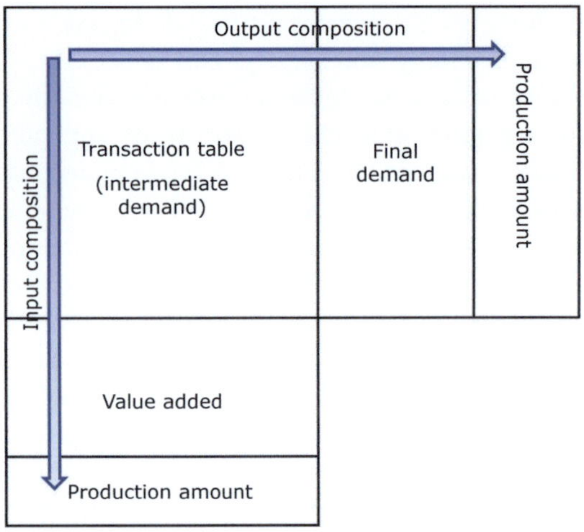

Fig. 5.7 Conceptual diagram of an input–output table

5.5 Carbon-Neutral Industrial Structure

In this section, we introduce the input–output analysis, and use it to explain the industrial structure that will emerge when building rural areas that utilize RE.

5.5.1 Input–Output Analysis

Figure 5.7 is a conceptual diagram of an input–output table. The input–output table is a statistical table developed in the 1930s by Wassily Leontief (1986), who won the Nobel Prize in Economics for his work. This table is a matrix that describes intersectoral transactions in the economy over a 1-year period. The vertical columns show the input amount (in monetary units) for intermediate goods purchased by each sector from other sectors, and the horizontal rows show the output amount from each sector to other sectors. The interdependence between economic sectors can be revealed through the use of an input–output table that simultaneously describes the relationship between the output of one sector and the input of another sector.

Due to limited computing power, Leontief rearranged the rows and columns of input–output tables to minimize the calculations needed to derive numerical solutions. Consequently, he identified two characteristics of transactional relationships between sectors, as follows:

(1) Triangularity: hierarchical relationship from composite goods to basic material goods. Here, a composite good refers to a "good that is composed of many intermediate goods, but does not itself become an intermediate good for other goods," and a basic material good is a "good that is input as intermediate goods to many goods, but does not itself require other inputs to be produced."
(2) Block independency: the economy includes industrial blocks with strong interdependence. The interdependence within blocks is strong, but the relationships between blocks are weak.

Detailed analyses using Japanese censuses of manufacturers Ozaki, (2004) showed that: (1) triangularity was a property observed based on the supply chain from base materials to products, and (2) block interdependency was a property based on differences in the starting raw materials. According to Ozaki (2004), an industrial structure shown in Fig. 5.8 was established in Japan following rapid economic growth in the 1960s. Five blocks exist in the manufacturing industry (machines/metals, food, cement/ceramics, textiles, and petrochemicals) based on four raw materials (metal ores, agricultural/forestry/fisheries products, non-metal ores, and crude oil), and a triangulated structure based on the respective supply chains is shown.

5.5.2 Economic Growth and Changes in Industrial Structure

According to Ozaki (2004), the relatively strong economic growth in Japan until the 1980s can be explained as follows. The machinery industry at the top of the triangle experienced strong demand for domestic investment in the 1960s. Strong demand for overseas exports of energy-saving machines made in Japan in the 1970s and 1980s had a large economic ripple effect on the industrial

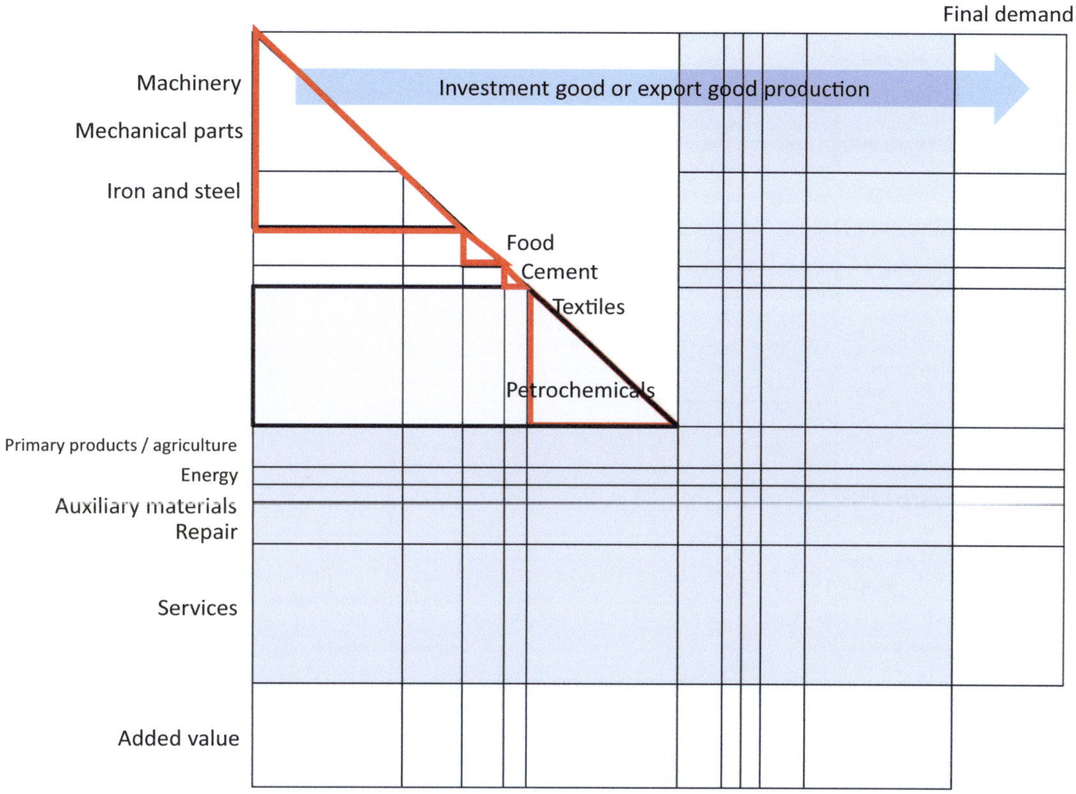

Fig. 5.8 Input–output structure in Japan following a period of high economic growth

groups at the bottom of the triangle. However, the demand traction shown by the arrow in Fig. 5.8 decreased considerably from the 1990s to the 2000s. This was due to an international financial crisis and the rapid economic development of Asian countries surrounding Japan. Due to this loss of traction, economic circulation ended, and the Japanese economy stagnated from the 1990s onwards.

Meanwhile, the development of ICT since the mid-1990s gradually changed the industrial structure of Japan. Figure 5.9 presents a conceptual diagram of the input–output structure of a smart society as a result of ICT development (Nakano & Washizu, 2018). A service industry block based on the information and communications industry emerged at the bottom right of Fig. 5.9, which formed a new supply chain. Thus, this is the supply chain of "industry that provides information infrastructure (primary information service industry) ⇒ services that provide applications using primary information services (secondary information service industry) ⇒ last mile service industry that uses secondary information services." A smart society is one in which this supply chain is firmly established, and outputs of the service industry that utilizes ICT are input into all industries, including the manufacturing industry. This creates new economic circulation and improves the production efficiency of each industry. Additionally, the depiction of the arrow in the upper right corner of Fig. 5.9 differs from that in Fig. 5.8. The machinery sector output (information goods) is input into the information services sector as intermediate input goods for the current period or investment goods for the next period.[11] This indicates the realization of a

[11] Another change is that manufacturers of electronic equipment, which were once classified into manufacturing sectors, are shifting to providing services that utilize electronic equipment, which are now reclassified into service

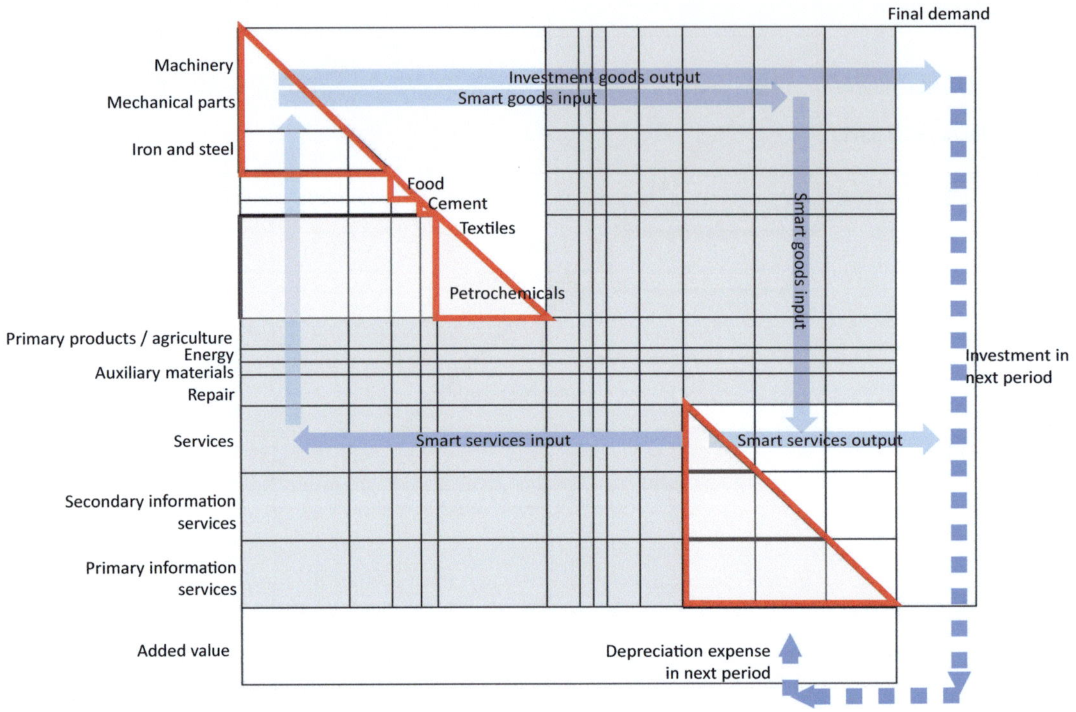

Fig. 5.9 Input–output structure of a smart society

broader economic circulation through the input–output relationship between the manufacturing and service sectors.

There has been increased attention on a shift to renewable energy-based electricity following the Great East Japan Earthquake in 2011 and the associated nuclear disaster. The introduction of photovoltaics was particularly rapid in Japan. ICT-based energy management technology has become essential for effectively utilizing the energy of photovoltaics, which has large output fluctuations. There was also demand for more thorough energy conservation in order to reduce fossil fuel dependence; in this context, thorough energy management helped to achieve such goals (Ishii, 2022). In other words, energy management services, which represents an industry at the top of the service block in Fig. 5.9, are important for building a low-carbon society.[12]

sectors. Thus, the number of entrants to the service industry block at the bottom right of Fig. 5.9 has increased.

[12] Washizu et al. (2024) (Paper accepted by peer-reviewed journal) defined the virtual power plant (VPP) business as an industry that provides these energy management services, where input–output analyses were performed to analyze the future economic and environmental effects of the VPP business.

5.5.3 Industrial Structure of a Carbon-Neutral Society

We have discussed social innovations aimed at the decarbonization of rural areas; the industrial structure that is expected through such innovation is shown in Fig. 5.10. In contrast to Fig. 5.9, a triangular block representing the supply chain of the agriculture, forestry, and fisheries industries is drawn in the middle of the figure, and a link between the agriculture, forestry, and fisheries industries and all industrial sectors is created through the input–output relationship of environmental credits. Social systems for building rural areas that utilize RE were discussed in Sect. 5.3; these recommend that the dissemination of RE contributes to the revitalization of production activities in the agriculture, forestry, and fisheries industries. Additionally, as shown by the example

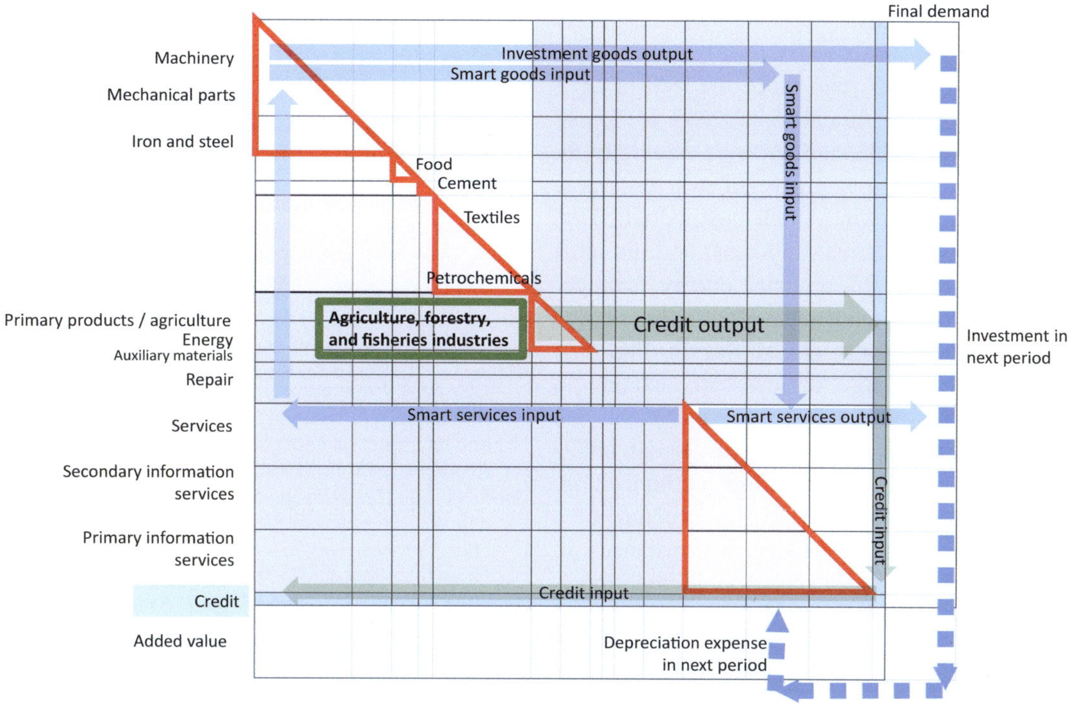

Fig. 5.10 Input–output structure of a carbon–neutral society

of Nishiawakura Village presented in Sect. 5.4.1, the high added value of forestry in the region (i.e., increased processing of forest products and opportunities to generate value added within the region) promotes the use of woody biomass for energy and contributes to decarbonization. The triangular block of the agriculture, forestry, and fisheries sector in Fig. 5.10 shows that extending the supply chain of these industries in rural areas increases the opportunities to generate value added within the region.

If the supply chain of the agriculture, forestry, and fisheries industry block is further expanded and connected to the supply chains of other blocks, the circular structure of the entire economy will be strengthened. Figure 5.10 proposes the generation of environmental credits from the agriculture, forestry, and fisheries industries as a mechanism for creating a new circulation structure. Environmental credits show that the agriculture, forestry, and fisheries sector is an industry that produces "agricultural, forestry, and fisheries products," as well as environmental conservation services, such as low carbonization. The creation of an environmental credits market is consistent with international environmental changes, such as institutionalization of the TCFD and TNFD. Macrolevel changes in the international environment are causing microlevel niche innovations, such as the introduction of RE in rural areas. Future mesolevel regime changes, such as the creation of an environmental credits market,[13] are likely to support transition to the input–output structure of a carbon–neutral society as shown in Fig. 5.10.

[13] Industrial initiatives such as carbon capture, utilization, and storage (CCUS) technologies, hydrogen reduction steelmaking, and circular carbon chemistry are currently expensive, and the products produced through these technologies are not price competitive. In such cases, assuming an environmental credit value in addition to the material value of those products provides an effective investment destination for companies interested in making environment-related investments and will contribute to dissemination of these technologies.

5.6 Conclusion

It has been over 10 years since the feed-in tariff (FIT) system was launched in 2012 to spread the use of renewable energy-based electricity. In recent years, the characteristics of each renewable energy type (solar, wind, small and medium-sized hydropower, geothermal, biomass) and the most effective methods of regional energy management based on those characteristics have become clearer. These results are shown in Fig. 5.1. Given the importance of low carbonization in rural areas for social science, in this chapter, we discussed theories, systems, and current initiatives for the transition to low carbonization. Figure 5.10 shows the industrial structure of a carbon-neutral society under circumstances in which progress has been made in the transition to low carbonization in rural areas. The results also showed the desirability of incorporating the agriculture, forestry, and fisheries industries into the circulation structure of the overall economy by creating an environmental credits market.

Building a carbon-neutral energy system and the industrial structure of a carbon-neutral society to support it are urgent issues. There is also a need to develop methodologies based on a comprehensive understanding of social science to move toward these goals.

Acknowledgements This research was conducted under the Grants-in-Aid for Scientific Research (KAKENHI) (Grant number JP21H03676), Waseda University for Special Research Project (2025C-178, 2025R-037), FY2023 Carbon Recycling Fund Research Grant, Research Institute for Humanity and Nature (IS starting in 2023), and the Environment Research and Technology Development Fund (JPMEERF20241004) of the Environmental Restoration and Conservation Agency provided by Ministry of the Environment of Japan. We would also like to thank Takahiro Ueyama and Keizo Shirahata of the government office of the Nishiawakura Village in Okayama Prefecture for agreeing to interviews described as part of the case study in Sect. 5.4. We greatly appreciate valuable comments from Professor Takashi Watanabe (Nagasaki University).

References

Chen, Y.-C., Shiroyama, H., Sugiyama, M., Aoki, K., Kimura, O., Mori, A., Ohta, K., Matsuura, M., & Matsuo, M. (2022). Sustainability transitions research in Japan: Status and prospects. *Review of Environmental Economics and Policy Studies, 15*(2), 1–11. (in Japanese).

Geels, F. W. (2019). Socio-technical transitions to sustainability: A review of criticisms and elaborations of the multi-level perspective. *Current Opinion in Environmental Sustainability, 39*, 187–201.

Iida, K., Ichida, T., Asai, M., & Suda, F. (2022). The LEADER Programme in Germany: A study of the evaluation/self-assessment by the local action groups and its impact on collective learning. *Journal of Rural Economics, 93*(4), 419–424. (in Japanese).

Ishii, H. (2022). Next-generation power systems suitable for the renewable energy era. In Arimura et al. (Eds.), *Frontiers in Carbon Pricing* (Chapter 8). Tokyo: Nippon Hyoronsha Co., Ltd. (in Japanese).

Leontief, W. (1986). *Input-Output economics.* Oxford University Press.

Loorbach, D. (2010). Transition management for sustainable development: A prescriptive, complexity-based governance framework. *Governance: An International Journal of Policy, Administration, and Institutions, 23*(1), 161–183.

Matsuura, M. (2017). Transition management as an approach to and a method for structural transformation. *Environmental Information Science, 46*(4), 17–22. (in Japanese).

Ministry of Agriculture, Forestry and Fisheries of Japan. (2023a). *Nōsan gyoson saiseikanō enerugīhō ni motozuku kihon keikaku no sakusei-tō no tebiki* (Guidance for creating basic plans based on the rural renewable energy act). Retrieved October 25, 2023, from https://www.maff.go.jp/j/shokusan/renewable/energy/attach/pdf/houritu-15.pdf (in Japanese).

Ministry of Agriculture, Forestry and Fisheries of Japan. (2023b). *Midori no shokuryō shisutemu senryaku* (Green food system strategy). Retrieved October 25, 2023, from https://www.maff.go.jp/j/kanbo/kankyo/seisaku/midori/index.html#midori_budget (in Japanese).

Ministry of Agriculture, Forestry and Fisheries of Japan. (2023c). *Midori no shokuryō shisutemu senryaku reiwa 4 nendo hosei yosan, Reiwa 5 nendo yosan gaisan kettei no gaiyō* (Green Food System Strategy: Summary of the FY2022 Supplementary Budget and FY2023 Budget Decision). Retrieved October 25, 2023, from https://www.maff.go.jp/j/shokusan/renewable/energy/attach/pdf/yosan-40.pdf (in Japanese).

Ministry of Agriculture, Forestry and Fisheries of Japan. (2023d). *Shimin shusshi no einō-gata taiyōkō hatsuden ni yoru kōhai nōchi no saisei* (Examples of renewable energy initiatives in rural areas). Retrieved December 4, 2023, from https://www.maff.go.jp/j/shokusan/renewable/energy/attach/pdf/zirei-88.pdf (in Japanese).

Ministry of Internal Affairs and Communications of Japan. (2023). *Kaso taisaku jigyō-sai* (Depopulated area development bonds). Retrieved October 25, 2023, from https://www.soumu.go.jp/main_content/000666664.pdf (in Japanese).

Ministry of the Environment of Japan. (2023a). *Chiiki junkan kyōsei ken* (Local SDGs circular and ecological sphere). Retrieved October 25, 2023, from http://chiikijunkan.env.go.jp/shiru/ (in Japanese).

Ministry of the Environment of Japan. (2023b). *Chihō kōkyō dantai jikkō keikaku (kuiki shisaku hen) sakutei jisshi manyuaru (chiiki datsutansoka sokushin jigyō hen)* (Local action plan formulation and implementation support website; Local action plan (area policy edition) formulation and implementation manual (local decarbonization promotion project edition)). Retrieved October 25, 2023, from https://www.env.go.jp/policy/local_keikaku/h_manual/s0.html (in Japanese).

Ministry of the Environment of Japan. (2023c). *Datsutanso senkō chiiki sentei kekka (dai 4 kai)* (Decarbonized regional development support website: Overview of results of the selection of advanced decarbonization regions (4th round)). Retrieved December 4, 2023, from https://policies.env.go.jp/policy/roadmap/preceding-region/#selection-4 (in Japanese).

Nakano, S., & Washizu, A. (2018). Induced effects of smart food/agri-systems in Japan: Towards a structural analysis of information technology. *Telecommunications Policy, 42*(10), 824–835.

Nomura Research Institute. (2023). *Reiwa 4 nendo chihō kōkyō dantai ni okeru chikyū ondanka taisaku no suishin ni kansuru hōritsu shikō jōkyō chōsa chōsa kekka hōkokusho* (FY2022 survey on implementation status of act on promotion of global warming countermeasures: survey results report). Retrieved October 25, 2023, from https://www.env.go.jp/policy/local_keikaku/data/dantai/R4/R04_report01.pdf (in Japanese).

Ozaki, I. (2004). *Nihon no Sangyo-kozo* (Industrial Structure in Japan). Keio Gijuku University Press (in Japanese).

Shibata, N., Nishikizawa, S., Nagaoka, A., & Murayama, T. (2021). Awareness of municipal governments to implementation of solar-sharing projects. *Papers on Environmental Information Science, 35*, 167–172. (in Japanese).

Tasaki, T., Kameyama, Y., Masui, T., Takahashi, K., Tsurumi, T., Hara, K., Hotta, Y., & Koide, R. (2023). Evolution of sustainability science in the Era of Anthropocene. *Environmental Science, 36*(2), 53–82. (in Japanese).

VentureTimes. (2023). *Gurīn Ribā Hōrudingusu, MOVIMAS ha Iwate-ken Hachimantai-shi to teikei shi, kōsaku hōki binīru hausu saisei jigyō ni chakushu* (Green River Holdings and MOVIMAS have partnered with Hachimantai City in Iwate Prefecture to begin a project for revitalizing abandoned farmland and greenhouses). Retrieved October 25, 2023, from https://venturetimes.jp/venture-news/food/26288.html (in Japanese).

Washizu, A., Ju, Y., Yoshida, A., Tayama, M., & Amano, Y. (2024). Modeling the distributed energy resource aggregator services in a macroeconomic framework: The application to Japan. *Energy, 312*, 133561. ISSN 0360-5442. https://doi.org/10.1016/j.energy.2024.133561

Weber, M., Hoogma, R., Lane, B., & Schot, J. W. (1999). *Experimenting with sustainable transport innovations: A workbook for strategic niche management*. Universiteit Twente. Retrieved March 1, 2023, from https://research.tue.nl/en/publications/experimenting-with-sustainable-transport-innovations-a-workbook-f

6 Renewables Penetration and Environmental Justice

Satoshi Kurokawa

Abstract

This chapter shows that renewable energy penetration is necessary toward a low-carbon society and explores the justice issues arising from the renewables penetration. It highlights the negative impacts on local environments, which often lead to Not-In-My-Backyard (NIMBY) movements, but also points out that renewable energy is an important local resource to be utilized by local people. Zoning renewables promotion zones would mitigate the conflicts between renewables facilities and residents. This chapter also addresses the challenges of expanding nuclear energy post-Fukushima. The policies and schemes that prompt renewables such as feed-in tariff are analyzed. They benefit rich households but leave poor households struggle in expensive electricity bills. This structure leads to the energy justice problem.

Keywords

Renewable energy · Low-carbon society · Environmental justice · Energy justice

S. Kurokawa (✉)
Waseda University, Tokyo, Japan
e-mail: kurokawa@waseda.jp

6.1 Introduction

This chapter explores the justice issues that will arise in the process of renewable energy electricity (renewables) penetration toward a low-carbon society. Wind farms and mega-solar generation facilities (mega-solar plants) often encounter not-in-my-backyard (NIMBY) movements because they tend to have a negative impact on the neighboring living environment. However, renewable energy sources are valuable local resources that help develop regions and increase their resilience. Therefore, local governments are legislating statutes to regulate and prompt renewables facility siting. Legal schemes that prompt renewables penetration may sometimes lead to unfair treatment of underprivileged people. These schemes, such as feed-in tariffs, benefit rich people but force the underprivileged population to expend more on electricity. This perspective of distributive justice is necessary to prompt renewables toward a low-carbon society.

6.2 Background

6.2.1 Long Road to a Low-Carbon Society

Japan has declared its commitment to achieve carbon neutrality by 2050. In addition, it has

K. Akao and A. Washizu (eds.), *Climate Change Issues and Social Sciences*,
Sustainable Development Goals Series, https://doi.org/10.1007/978-981-96-8951-4_6

a nationally determined contribution (NDC) to reduce greenhouse gas emissions by 46%–50% below 2013 levels by 2030. To achieve these goals, it is important to reduce CO_2 emissions resulting from energy use. According to NIES (2024, Esp5), CO_2 emissions related to energy use accounted for approximately 87.1% of total GHGs emissions, which amounted to 113.5 million tons equivalent CO_2. According to the ANRE (2024), approximately three-quarters of electricity is generated in thermal power plants by burning fossil fuels. Among power sources, renewable energy, including hydropower with large dam sites, accounted for 21.7%, nuclear power accounted for 5.5%, and thermal power, excluding biomass, accounted for 72.8% in 2022. More than 70% of Japan's final energy consumption occurred in forms other than electricity. In 2022, electricity accounted for only 27.4% of the total final energy consumption, whereas the energy consumption from the combustion of fossil fuels and their products, such as gasoline, accounted for approximately 65%. To progress toward a low-carbon society, it is necessary to shift energy consumption from fossil fuel combustion to electricity, similar to shifting from gasoline-fueled cars to electric cars.

This shift will lead to a dramatic increase in electricity demand. To meet the growing demand, the supply of electricity from zero-emission power sources must be increased toward a low-carbon society. In 2022, renewable energy sources, including hydropower at large dam sites, accounted for 11% of the primary energy supply, whereas nuclear power accounted for 3%. Unless the total volume of energy used changes, electricity shortages will occur. To avoid electricity shortages, reducing energy use is critical, in addition to increasing the electricity supply from renewable and nuclear energies.

6.2.2 Unreliable Nuclear Energy and the NIMBY Against Renewables Facilities

It is not easy to expand nuclear energy because it has lost people's trust in terms of safety since the severe accident at Fukushima Daiichi Nuclear Power Plant in 2011. Resuming nuclear power plants that ceased operations after the accident faced difficulties.

Moreover, it is difficult to expand renewables rapidly. Wind farms and mega-solar plants have caused conflicts with residents owing to noise, vibration, and landscape destruction. In other words, they tend to be the targets of the NIMBY campaigns by residents living near the proposed sites. Local governments also enacted local statutes (=ordinances) to regulate the construction of wind farms and mega-solar plants.

The NIMBY movement against wind farms and mega-solar plants seems justifiable at first glance, because they might be a source of nuisance with vibration, noise, reflection, or flicking shadows. Siting these facilities forcefully has the same structure as siting waste management facilities. This relates to the environmental justice issue in which poor people or marginalized communities are harmed unproportionally by unfavorable facilities.

6.3 Expansion of Renewable Energy

6.3.1 Loss of Trust in Nuclear Energy

Before the severe accident at the Fukushima Daiichi Nuclear Power Plant in March 2011, the promotion of nuclear power was the driver of Japan's climate change policy. Prior to the Fukushima disaster, 54 nuclear reactors were in commercial operation, and more than a quarter of the electricity used in Japan were produced in nuclear power plants. The 2010 Strategic Energy Plan called for the construction of at least 14 new reactors to mitigate climate change. However, climate policies that rely on nuclear power have failed. After the Fukushima disaster, more than half of the 54 reactors were decommissioned or decided to be decommissioned. It costs a lot to meet new stricter regulations which apply to all reactors including reactors that have operation licenses. As of

November 2024, only 13 reactors had resumed operations.

The 2021 Strategic Energy Plan set the goal that the share of nuclear power in the power-supply mix should be 20%–22% by 2030. However, achieving this goal is difficult because the generation capacity of the remaining reactors is not sufficient, and constructing new reactors is extremely difficult in Japan.

Even before the Fukushima disaster of 2011, residents living near nuclear power plants filed numerous lawsuits throughout Japan seeking injunctions against the construction and operation of nuclear power plants or the revocation of a license to install nuclear reactors. Except for two lower court decisions (Kanazawa district court March 24, 2006, and Nagoya appellate court Kanazawa branch January 27, 2003), the plaintiff residents lost their cases because of the dominant "nuclear safety myth" that Japan's nuclear power plants were safe and would not cause a serious accident. The myth was unreasonable, but those in charge of nuclear regulation and operation were unable to talk about the risk of serious accidents. Some residents who were sensitive to risks, opposed the installation of nuclear power plants in their neighborhoods to avoid accepting disproportional risks. Therefore, the NIMBY against nuclear power plants can be justified from the perspective of environmental justice. After the Fukushima disaster, some courts ruled to halt reactor operations (e.g., Hiroshima Appellate Court January 17, 2020) or to revoke the installation license (e.g., Osaka District Court December 4, 2020).

Nuclear power plants have nuclear safety agreements with regional governments. Under the provisions of the agreement, the nuclear power plant must acquire consent from the regional government in advance to resume reactor operation. Therefore, unless trust is rebuilt between residents and the nuclear power plant, the local government will be reluctant to provide consent. If residents find environmental injustice in a nuclear power plant location, they will no longer accept it because the safety myth disappeared in 2011. In addition to the risks of nuclear power plant operations, there is another environmental justice issue related to nuclear power. The 2000 High-Level Radioactive Waste Final Disposal Act stipulates that HLW generated during the reprocessing of spent nuclear fuels should be buried in a stratum no less than 300 m underground. However, the site location remains to be determined because inviting the site causes severe political conflict in the area because it is a NIMBY facility.

6.3.2 Expectations for Renewable Energy and Conflicts with Residents

Japan can no longer rely on nuclear power as a tool to build a low-carbon society. Therefore, the promotion of renewable energy is a pressing policy issue. The share of electricity generated from renewable energy sources was 21.7% in 2022:7.6% (76.8 billion kwh) from hydropower, 9.2% (92.6 billion kwh) from solar, 0.9% (9.3 billion kwh) from wind power, and 3.7% (37.2 billion kwh) from biomass. Japan must construct more wind farms and mega-solar plants toward a low-carbon society. However, the construction of wind farms and mega-solar plants often involves the NIMBY movement of residents. Local governments impose stricter regulations on wind farms and mega-solar plants in response to the NIMBY movement of residents.

Wind turbines degrade the living environment in their vicinity through noise and vibrations, especially low-frequency vibrations. Photovoltaic (PV) power generation facilities also generate noise and vibrations owing to inverters that convert DC current to AC current. There is also the problem of light pollution, where sunlight reflects off solar panels and glares. Large-scale ecosystem destruction can occur when forests are developed for wind farms and mega-solar plants. Wind turbines cause bird strikes, which have serious negative impacts on ecosystems. The construction of wind farms and mega-solar plants on slopes increases the risk of landslides and other disasters. Furthermore, wind farms and mega-solar plants cause landscape modifications and destroy landscape aesthetics. According to Fujii and Yamashita (2021), in 2014, 2.9% of municipalities reported that they were currently experiencing problems with the operation of renewable energy power facilities. In the 2020 survey, 11.8% of the respondents answered that they had not had problems in the past but were concerned that they might occur in the future. In the 2014 survey, 11.1% of the respondents answered that they had had problems in the past but did not have problems at that time. A survey in 2020 revealed that the most common problems were landscaping (54%), noise (33.4%), and landslides (32.9%).

Residents concerned about the negative impacts on the local environment often oppose the construction of wind farms and mega-solar plants. Local governments sometimes oppose construction on behalf of their residents. For example, the plan of the Kansai Electric Power Company to construct a wind farm (Kawasaki Wind Farm) in the Zao Mountain Range in Miyagi Prefecture was abandoned in 2022 due to strong opposition from the local population, municipal governments, and prefectural government. The plan to construct a wind farm (Kurikoyama Wind Farm) was canceled in Yonezawa City, Yamagata Prefecture, owing to the opposition of local cities and prefectures.

Residents sometimes file lawsuits seeking injunctions against renewable energy facilities. The Yokohama District Court (April 18, 2012) granted the request of neighbors to remove rooftop solar panels installed on the new building, which was located south of the plaintiffs' location, but the Tokyo Appellate Court (March 31, 2013) reversed the decision. The Oita District Court (November 11, 2016) dismissed a petition filed by residents seeking injunctions against the construction of a mega-solar plant to protect their scenic interests. The Nagoya District Court Toyohashi branch (April 22, 2015) dismissed a claim for an injunction to stop the wind turbines and paying damages. Even offshore wind farms cause disputes with fishermen (Hiroshima Appellate Court, June 26, 2019). The decision of the Kochi District Court (January 23, 2024) dismissed the petition to revoke the denial of a mega-solar installation license on land along the Shimanto River, one of Japan's most famous clear streams. In this case, the mayor refused to issue a license because he believed that the people in the city opposed the construction of a mega-solar plant along the river.

To prevent such conflicts, a growing number of local governments have enacted ordinances regulating the installation of renewable energy generation facilities. The earliest of these was the 2014 ordinance of Yufu City in Oita Prefecture. As of August 29, 2024, 167 ordinances targeted only solar power facilities, while 123 ordinances regulated wind, biomass, geothermal, and other renewable energy generation facilities, as well as solar power generation facilities (RILG, 2024b). Furthermore, ordinances such as those in Miyagi Prefecture impose an annual tax on renewable energy facilities based on the amount of electricity they generate.

6.3.3 Recognizing Renewable Energy as a Local Energy Resource

While renewable energy plants have negative impacts on residents in their vicinity, as described above, they also have positive economic impacts. They exploit local natural energy resources. Iida City in Nagano Prefecture enacted an interesting ordinance in 2013. It declared the right of local communities to local renewable energy resources (Sect. 6.3). It defines the right of the local community to use local renewable energy resources and survive in a harmonious living environment. The statute is based on the notion that local renewable energy is a local resource, and that local people have an environmental right to use it. The notion of local environmental rights for renewable energy sources leads to local production and consumption policies for renewable energy. Even if renewables is sold to outside businesses, the profits from the sale of renewables should belong to the local community. If wind farms and mega-solar plants are built and operated by companies outside the area, the local community will be negatively impacted by the renewable energy facility but will not profit from renewable energy resources. This structure might make it a rational course of action for residents to adopt a negative attitude toward the construction of wind farms and mega-solar plants.

Wood biomass power generation is a popular example of local efforts to utilize local energy resources in mountainous regions with abundant forest resources. Wood waste generated during the processes of thinning or making lumber from wood is used as fuel to generate electricity. Aged logger populations and high forestry costs have discouraged the management of production forests in Japan. However, biomass power generation compensates for part of the cost of managing forests and endorses the development of towns and villages in mountainous areas.

In contrast to nuclear reactors, PVs and wind turbines are suitable distributed energy sources and can be installed in places close to electricity consumption areas. Although a single nuclear reactor can generate as much as 1 gigawatt of electricity, a significant risk of environmental pollution exists in the event of an accident. Therefore, large-scale nuclear power plants are located far from populated areas. Although each renewable supply may be small and unstable, aggregators are expected to aggregate and smooth out the variations. A local power company (community power) is often described as an ideal model to be an aggregator that sells local renewables to households and businesses. For example, a community power company named Ikoma Shimin Power purchases electricity generated by PVs and small hydroelectric facilities in the community and sells it to local businesses and households. Some of these projects are certified by the Ministry of the Environment as model projects for establishing regional decarbonization promotion entities.

6.3.4 Distributed Energy Sources

In addition, the government promotes zero-emission house (ZEH) and zero-emission building (ZEB) policies. In a ZEH, the electricity generated by rooftop PVs is consumed by the household, and surplus electricity is sold to electricity companies. Agrivoltaics, in which solar panels are installed on farmlands to supplement farmers' agricultural income, are also being promoted by the government.

The aggregation of distributed power supplies leads to local power grid formation and mitigates the effects of power outages in the event of a disaster. This increases community resilience. Therefore, PVs systems and wind farms are valuable local facilities and should not be excluded as nuisance facilities. The conflict regarding the siting of wind turbines and PVs cannot be viewed as a simple environmental justice issue. It is a complex issue in which global climate change issues are intertwined with local energy resource management, local resilience, local living environment, natural environment, and landscape. Therefore, procedural justice is important for communities to be receptive to the installation of renewable energy facilities (Motosu et al., 2012).

6.3.5 RPS, FIT, and FIP

In the 1960s, hydropower stations supplied more than half the electricity in Japan. After the 1973 oil crisis, the Japanese government was positive about promoting renewable energy, especially solar power. It established and implemented the Sunshine Project (1974–1992) and New Sunshine Project (1993–2000) successfully. As a result, Japan became the largest PV electricity generator and producer of solar panels in the world early in 2000s. However, the government focused on promoting nuclear energy and decided to use a market-based scheme to encourage renewable energy instead of providing subsidies. The Renewable Portfolio Standard (RPS) Act (Act on Special Measures Concerning New Energy Use by Operators of Electric Utilities) of 2002 was legislated to prompt renewables penetration. The RPS Act requires electricity retailers to make a certain percentage of their retail electricity volume renewables. Electricity retailers were expected to buy renewables from wind farms and mega-solar plants in the market, and the expanded demand would collect investments in wind farms and mega-solar plants. However, the RPS was unsuccessful because the government set an unambitious target for renewables. The legal obligation for renewables was too low to increase their demand. Consequently, it did not encourage investments in wind farms or mega-solar plants. Therefore, only 0.6% of the total domestic electricity supply was generated by PVs (3.5 TWh) and wind turbines (4.0 TWh), even in 2010. To address the failure of RPS, the Energy Supply Structure Sophistication Act of 2009 introduced a net-metering scheme that required electricity utilities to buy surplus renewables generated by rooftop PVs in households and small buildings. This encouraged the installation of PVs on rooftops. After the Fukushima disaster, the FIT Act was enacted in 2011, replacing the unsuccessful RPS scheme with the FIT scheme and absorbing the net-metering scheme for small PVs. Since then, the electricity supply generated by renewable energy sources has steadily increased.

The FIT scheme in Japan made utilities to purchase all eligible electricity generated from renewable energy sources (solar, wind, small-scale hydro, geothermal, and biomass) at a fixed price for a fixed duration. The FIT was particularly successful for PV penetration. The dramatic expansion of PV was possible because the Environmental Impact Assessment (EIA) Act did not require PV facilities to conduct environmental impact assessments. In contrast, wind power electricity generation expanded slowly because EIA was required by law. During those days, the EIA took approximately five years for wind farms. So, the amount of wind power electricity generation only doubled from 4.0 TWh in 2010 to 7.6 TWh in 2019.

Under the FIT scheme, the more renewables generated, the higher is the cost of purchasing renewables at a fixed price. The additional cost was collected as "renewables charge" from those who consumed electricity, such as households and businesses, according to the volume of electricity consumed. The renewables charge was JPY 0.22/kwh in 2012 and JPY 3.49/kwh in 2024. Therefore, the government changed the scheme to curb rising electricity bills by using a market mechanism. The auction for the FIT-eligible capacity of large-scale PVs and onshore winds began in 2017 and 2021, respectively. Price competition in auctions contributed to the lower prices of renewables in these categories. Furthermore, the FIT Act was revised and renamed as the Feed-in Premium (FIP) Act (the Act on Special Measures Concerning Promotion of Utilization of Electricity from Renewable Energy Sources). The FIP scheme started in 2022. It encourages renewables generators to sell electricity in the electricity wholesale market and to accept a "renewable supply promotion subsidy" according to the electricity volume sold in the market. A renewable supply promotion subsidy is usually called a premium. The Organization for Cross-regional Coordination of Transmission Operators (OCCTO) provides renewables generators with premiums collected from electricity users as renewables charge that funds FIT additional costs and FIP premiums.

6.3.6 Balancing Supply and Demand

Electricity supply from wind turbines and PVs is unstable because it depends on weather conditions. Strong winds and strong sunlight lead to significant electricity generation. The wind blows in some places but is weak in other places. In some places, clouds hide the sun. Therefore, the network manager must balance the supply and demand of electricity to meet these fluctuations. The 2020 amendment to the Electricity Business Act provided aggregators with legal status in Japan for the first time. The aggregators aggregate electricity from distributed energy sources such as PVs and wind turbines and smooth out variations in the electricity supply to the networks. They work as adjustment capacities, and as virtual power plants (VPPs). Aggregators play an indispensable role in the penetration of renewable energy. In the future, aggregators will control electricity consumption in households and offices through smart meters, home energy management systems (HEMS), or building energy management systems (BEMS). They used demand response programs to match the electricity supply and demand.

6.4 Emission Trading Scheme

An emission trading scheme (ETS) is a popular tool for controlling CO_2 emissions in many countries. This has contributed to the rapid penetration of renewables, particularly in European countries. In 2010, the cabinet of Japan submitted the Bill of the Global Warming Countermeasures Policy Act, which planned to introduce the ETS in 2013, to the Japanese parliament. However, the bill did not pass Parliament and was scrapped. Despite the failure of the national government, the Tokyo Metropolitan government introduced a cap-and-trade system in 2010 by an ordinance, which targeted the energy use of office buildings. Saitama Prefecture followed Tokyo in 2011.

The 2023 Green Transformation Act stipulates that the ETS will start in 2033 at the national level, and that a carbon emission charge scheme will start in 2028. The ETS and carbon emission charge scheme put a price on CO_2 emission. Therefore, they increase the price of electricity from thermal power plants and make electricity from thermal power plants less competitive. In contrast, renewables will be more competitive in the market. This accelerates the stranding of thermal power plants. This can exacerbate power shortages and increase electricity price. This situation could lead to an energy justice problem in which poor households cannot access the necessary energy at an affordable price.

6.5 Prompting Regional Development of Renewable Energy

The 2013 Act on Promoting Generation of Electricity from Renewable Energy Sources Harmonized with Sound Development of Agriculture, Forestry, and Fisheries prompted the installation of PVs and wind turbines on farmlands and fishing villages. Electricity generation from wood biomass in mountainous villages has also been investigated. Because agricultural productivity is low in Japan, young people are reluctant to take over family farming businesses. This has led to a large amount of abandoned farmland in Japan. However, with combined income from farming and electricity sales, farmers can obtain sufficient income to continue farming. MAFF (2024) reported that 3,217 sites (1,209.3 ha) got permission to use farmland for agrivoltaics by 2022.

Ordinances promoting renewable energy facilities in their areas appeared in 2000. By 2023, 42 local governments including eight prefectures enacted ordinances that used "renewable energy use promotion" in their name. According to RILG (2024a), eight local governments (including five prefectures) enacted ordinances to promote renewables by 2005. Since 2004, the number of ordinances that have "global warming countermeasures" in their name has increased. These ordinances have prompted the penetration of renewables. As of July 26, 2024, 52 ordinances of this kind were enacted. After 2011, legislating the ordinances on renewable energy promotion boomed. However, only three were enacted after 2017. In contrast, ordinances regulating the installation of renewable energy facilities began to appear in 2014, and 301 ordinances were enacted as of July 26, 2024. Thus, at the local government level, renewable energy facilities are no longer simple promotional targets, and their siting needs to be coordinated with the various interests of the local community. When siting wind farms and mega-solar plants, it is important to talk with the residents in the planned areas. Local governments coordinate various interests by promulgating local land use plans.

Ordinances intended to protect the natural environment and preserve landscapes also regulate the construction of renewable energy facilities. For example, the Kochi District Court case (January 23, 2024) involved a mega-solar plant planned to be installed in an area of special importance to preserve the ecosystem and landscape under the ordinance.

The Law on the Promotion of Global Warming Countermeasures requires local governments to promulgate a local greenhouse gas emission reduction plan and establish promotion zones that are subject to local decarbonization promotion projects (Article 21). The installation of renewable energy facilities in promotion zones is entitled to preferential treatment in regulatory procedures. It can enjoy a one-stop application process for permits and approvals under the River Act, the Forest Act, the Agricultural Land Act, and other nature protection laws. It also includes a simplified environmental impact assessment procedure. Separating promotional areas from other areas mitigates conflicts of interests. The ordinance of Miyagi Prefecture, named the Renewable Energy Regional Symbiosis Promotion Tax Ordinance, levies a renewable energy regional symbiosis promotion tax on renewable energy facilities. However, it does not levy tax on renewable energy facilities in the promotion zones. This segregated treatment by the ordinance enhances the effectiveness of the promotion zones.

6.6 Offshore Wind

The 2018 Act on Promoting the Utilization of Sea Areas for the Development of Marine Renewable Energy Power Generation Facilities promotes offshore wind farms. The 2021 Strategic Energy Plan expects wind power to generate 51 TWh of electricity by 2030, compared with 9.4 TWh in 2021. Offshore wind is expected to generate 17 TWh by 2030, accounting for approximately 2% of the power source mix in 2030. The government plans to set up a total of one GW capacity project every year by designating offshore wind promotion zones under this legislation. The government anticipates that the accumulated offshore wind power generation capacity will reach 10 GW by 2030 and 30–45 GW by 2040, whereas offshore wind farms had only 51.6 MW capacity as of the end of 2021.

6.7 Energy Justice Issues

6.7.1 FIT and Energy Justice

Under the FIT scheme, rich households that afford to install rooftop PVs accept incentives and enjoy cheap electricity bills, but poor households that do not afford to install rooftop PVs pay the renewables charge. Thus, money moves from the poor to the rich, under the FIT scheme. It has created an effective incentive and contributed significantly to the expansion of renewables. However, this has caused problems with distributive justice. This injustice structure remains in the FIP scheme because households that do not afford to install rooftop PVs continue to pay the renewables charge more than rich households with rooftop PVs pay.

6.7.2 Zero Emission House

The government is promoting the construction of all-electric homes with high insulation efficiency, equipped with rooftop PVs and home batteries. Such houses do not need to buy electricity from power companies because the electricity generated by rooftop PVs can meet all household electricity demands. Therefore, these houses were referred to as zero-emission houses (ZEHs). All-electric homes without PVs must buy electricity from power companies. It contributes to CO_2 emissions because fossil fuels account for over 70% of power sources averagely in terms of electricity from power companies.

The government provides subsidies for the construction of ZEHs and more subsidies for the construction of life-cycle carbon minus houses. This policy not only prompts distributed power sources but also builds a stabilizer of electricity in the network. Through smart meters and HEMS, ZEHs can absorb excess power in the network by storing hot water in electric water heater tanks and charging BEVs. In addition, ZEHs supply power to the network from home batteries in the case of an electricity shortage. Thus, ZEHs contribute to increasing the capacity of networks to accept renewables from wind farms and mega-solar plants. However, this subsidy policy only benefits the rich, who have their own houses and install rooftop PVs.

6.8 Conclusion

This chapter elaborated on schemes to prompt renewables penetration and the distributive justice issues incurred by them. Wind farms and mega-solar plants can be NIMBY facilities but can contribute to the utilization of local energy sources such as wind and solar power. Regional planning, which includes renewables promotion zones, attempts to solve conflicts in relation to renewable energy facilities. Renewable energy promotion schemes such as the FIT benefit the rich, but leave the poor to pay higher electricity bills. Therefore, energy justice must be considered when formulating renewable energy promotion policies.

References

Agency for National Resources and Energy (ANRE). (2024). *Sougou Enerugi Toukey*. Comprehensive Energy Statics of FY2022. https://www.enecho.meti.go.jp/statistics/total_energy/

Fujii, K., & Yamashita, H. (2021). Chiikiniokeru saiseikanoueneruglriyouno jittaitokadai;Dai3kaizenkokushichouson ankehtono kekkakara (Actual conditions and challenges of renewable energy use in local communities: Results of the third national survey of municipalities). *Hitotsubashi Keizaigaku* (*Hitotsubashi Economics*), *12*(1), 10–13.

Ministry of Agriculture, Forestry and Fisheries (MAFF). (2024). *Einougata taiyoukou hatsudensetsubi secchijoukyou nitsuite Reiwa 4nennmatsugenzai* (Status of installation of PVs on Farmlands (as of the end of FY2022)). https://www.maff.go.jp/j/nousin/noukei/totiriyo/attach/pdf/einogata-52.pdf

Motosu, M., Maruyama, K., Iida, M., & Arakawa, C. (2012). Huuryokuhatsudenno shakaitekijuyousei (Social acceptance of wind power generation). *Kankyoushakaigakukenkyuu* (*Journal of Environmental Sociology*), *18*, 193. https://doi.org/10.24779/jpkankyo.18.0_190

National Institute for Environmental Studies (NIES). (2024). *National Greenhouse Gas Inventory Document of Japan 2024*. https://www.nies.go.jp/gio/en/archive/nir/pi5dm3000010ii0r-att/NID-JPN-2024-v3.0_gioweb.pdf

Research Institute for Local Government (RILG). (2024a). *Saiseikanoueneruginο riyounosokushinnikansuru jourei, Reiwa6nenn7gatu26nichi koushin* (Ordinance on Promoting the Use of Renewable Energy, as of July 26, 2024). http://www.rilg.or.jp/htdocs/img/reiki/122_renewable_energy.htm

Research Institute for Local Government RILG (RILG). (2024b). *Taiyoukou Hatsudensetsubino kiseinikansuru jourei, Reiwa6nenn10gatu25nichi koushin* (Ordinances Regulating PV facilities, as of October 25, 2024). http://www.rilg.or.jp/htdocs/img/reiki/005_solar.htm

7 Nuclear Power in a Carbon-Neutral Society: An Analysis of the Nuclear Damage Compensation System

Akira Yoshida

Abstract

This chapter focuses on the nuclear damage compensation system and discusses the creation of an important yet often forgotten mechanism for facilitating the transition to carbon neutrality in the nuclear sector. The key mechanism of this compensation system enables those exposed to a shared risk to contribute money in the event of an accident. This chapter discusses the importance of this mechanism through three examples: the Brussels Supplementary Convention, the Retrospective Rating Plan in the United States, and the general contribution mechanism created in Japan after the Fukushima accident. This chapter asserts that a mechanism to collect money among nuclear operators who are exposed to the shared risk of a nuclear accident should be created. In Japan, a mechanism to collect money from nuclear operators who are exposed to a common risk (burden sharing) was created after the Fukushima nuclear accident. This resulted in a lack of understanding of the structure of the system and raised doubts. The Japanese case is a useful precedent for managing nuclear power generation in a carbon-neutral society.

Keywords

Nuclear damage compensation system · Brussels Supplementary Convention · Retrospective Rating Plan · General burden

7.1 Introduction

This chapter rests on the assumption that nuclear power will continue to be used in the future. To facilitate the transition to carbon neutrality within the nuclear sector, it is important to establish a mechanism within the current compensation system for nuclear damage that enables those who are exposed to a shared risk to contribute money in advance to prepare for any potential accident. Establishing such a mechanism will make nuclear operators aware of the risks involved in their economic activities by using figures. Moreover, it will also clarify the responsibilities of nuclear operators who are exposed to shared risks, even if they are not involved in an accident.

This chapter discusses the place of nuclear power within a carbon-neutral society while focusing on the nuclear damage compensation system. Nuclear power generation presents one

A. Yoshida (✉)
Waseda University, Tokyo, Japan
e-mail: akira.yoshida@aoni.waseda.jp

K. Akao and A. Washizu (eds.), *Climate Change Issues and Social Sciences*,
Sustainable Development Goals Series, https://doi.org/10.1007/978-981-96-8951-4_7

option for achieving a carbon-neutral society. However, nuclear power can also cause severe accidents, such as the Fukushima Daiichi nuclear accident that occurred on March 11, 2011. If nuclear power is to be used in the future, nuclear operators, who are also at risk when a severe accident occurs, must make the necessary preparations. In Japan, a compensation system for nuclear damage has been established; however, because the preparedness system was created after the 2011 nuclear disaster, it has been accompanied by considerable controversy. However, the United States has a system known as the Retrospective Rating Plan, which notifies nuclear operators exposed to a shared risk about the preparedness measures that must be taken. Moreover, the plan also stipulates how much money is collected from each operator in the event that an accident occurs. This study will clarify what was discussed through the legislative debate process, what implications the debate had, and what should be considered when transitioning to a carbon-neutral society.

7.2 The Paris Convention and the Brussels Supplementary Convention

7.2.1 Overview of the Paris Convention and the Brussels Supplementary Convention

When considering the rationality of a nuclear damage compensation system, it is necessary to discuss why nuclear operators that have not caused nuclear accidents are obliged to bear the costs when accidents occur. Outside of Japan, there are systems that require nuclear operators and the state to bear the costs of an accident even when they have not caused it, such as the Brussels Supplementary Convention and the retrospective rating plan in the United States.

The Brussels Supplementary Convention was a convention held on "January 31, 1963, supplementing the Paris Convention of July 29, 1960, as amended by the Additional Protocol of January 28, 1964, and the Protocol of November 16, 1982." This agreement went into effect in 1974. Article 1 of the Brussels Supplementary Convention states that the Supplemental Treaty of Brussels supplements the Paris Convention on Third Party Liability in the Field of Nuclear Energy (hereafter, "Paris Convention"). The Paris Convention, which was adopted in 1960 and entered into force in 1968, was led by the Nuclear Energy Agency of the OECD. The Paris Convention was designed to deal with any potential nuclear damage that crosses national borders. The Paris Convention has 15 signatories, including Germany and the United Kingdom. Excluding Greece, Portugal, and Turkey, the signatories to the Paris Convention also signed the Supplementary Convention, making for a total of 12 signatories. The Paris Convention and the Brussels Supplementary Convention were subsequently renegotiated as the Revised Paris Convention and the Revised Brussels Supplement. However, these revised agreements are not included in this study because they have not yet entered into force.

Next, this chapter will discuss the ratification of the Paris Convention and the Brussels Supplementary Convention. In particular, this chapter focuses on Article 7 of the Paris Convention and Article 3 of the Brussels Supplementary Convention, both of which are concerned with liability. The financing of compensation involves three steps: The first step is based on Article 7 of the Paris Convention; the second step is based on Article 3b (ii) of the Brussels Supplementary Convention; and the third step is based on Article 3b (iii) of the Brussels Supplementary Convention.

The term special drawing rights (SDR) is used when discussing financing methods. In this study, 1 SDR is calculated as approximately 150 yen. The term "compensation measure" refers to a nuclear reactor operator's provision of compensation payments via insurance and other means prior to the occurrence of an accident.

In the first step, if the total compensation can reach 5 million SDR (750 million yen), the nuclear reactor operator that caused the accident makes compensation payments out of the funds provided by the compensation measures.

If the resulting sum after subtracting 5 million SDR (750 million yen) from the total amount of compensation is positive, the second stage is initiated. The second stage is for cases where the total amount of compensation ranges from 5 million SDR (750 million yen) to 175 million SDR (26.25 billion yen). During the second stage, compensation funds come from the public funds of the contracting party located at the nuclear power facility owned by the nuclear reactor operator that caused the accident. During the second stage, the party located at the nuclear facility that caused the accident is responsible for paying the costs.

If the sum that results after subtracting 175 million SDR (26.25 billion yen) from the total amount of compensation is positive, then the third stage is initiated. The third stage is for cases where the total amount of reparations is between 175 million SDR (26.25 billion yen) and 300 million SDR (45 billion yen). The compensation funds during the third stage come from public funds provided by the contracting parties to the Brussels Supplementary Convention; moreover, the contracting parties to the convention are responsible for bearing the costs.

The third step is, at first glance, counterintuitive. Contracting parties are asked to bear the costs despite having played no role in causing the accident. Therefore, it is necessary to ask why the Brussels Supplementary Convention requires contracting parties unrelated to an accident to participate in cost sharing.

First, Article 3 of the Brussels Supplementary Convention is a mechanism to increase funding sources for compensation. Regarding the Brussels Supplementary Convention, Dougauchi (1993) stated that a fund is created; providing relief to victims (p. 158). In addition, Dussart-Desart (2005), in regard to the financial contributions of the parties in Article 3 of the Brussels Supplementary Convention, emphasized that the effectiveness of the convention would be limited without a large number of parties (p. 28).

In this context, it is reasonable to assume that the primary objective of the Brussels Supplementary Convention was to increase monetary sources for reparations, including the viewpoint of victim protection. The Brussels Supplementary Convention is a mechanism for collecting and pooling funds from various countries. During the design stage of the convention, it was expected that many parties would participate in the convention and that, therefore, the burden placed on each country during the third stage would be small. However, the number of contracting parties ultimately was 12, which is much lower than the number initially anticipated when the system was designed. In Europe, where countries are geographically proximate to one another, there is a risk that a nuclear accident in one country could cause damage to multiple surrounding countries. The Brussels Supplementary Convention was a precursor to subsequent schemes designed to share the costs of a nuclear disaster among at-risk countries, and the purpose of the convention was to protect victims.

7.2.2 The Lessons of the Paris Convention and the Brussels Supplementary Convention

The Brussels Supplementary Convention highlighted key issues that must be addressed when generating nuclear energy in a carbon-neutral society. The Brussels Supplementary Convention is a treaty that recognizes risks related to nuclear power and specifies what actions should be taken when a hazardous decision by one country poses risks to other countries. However, the scope of the agreement is limited to cases where two countries are connected by land.

When using nuclear power, it is necessary for those who share a common risk (in this case, nations) to (1) recognize the risks, (2) describe what to do in the event of an accident, and (3) establish a system that clearly states that those who share a common risk should help each other before an accident occurs. However, nuclear power is often understood as a national issue. In the case of an island nation such as Japan, it is unclear whether the Brussels Supplementary Convention represents the best method of establishing an international treaty. Therefore, this has led Japan to ask the question, "Can we do something on our own?" The United States has developed a mechanism similar to the Brussels Supplementary Convention agreement known as the Retrospective Rating Plan.

7.3 The US Nuclear Damage Compensation System

7.3.1 Overview of the US Nuclear Damage Compensation System

The US nuclear Damage Compensation system was established in 1957 pursuant to the Price-Anderson Act. This act was part of the Atomic Energy Act of 1954, which had been amended to protect the nascent nuclear industry. In 2005, when the Price-Anderson Act was incorporated into the Comprehensive Energy Policy Act, its expiration date was extended to 2025.

At the time of enactment of Price-Anderson Act, the maximum liability limit (total protection) for nuclear operators was set at $560 million, including up to $60 million from private insurance purchased by a nuclear reactor operator and up to $500 million from indemnity contracts with the government.

In 1975, the scheme changed substantially. In particular, the Retrospective Rating Plan was introduced, and as of the 1975 revision, this plan has limited the liability of nuclear operators to the private liability insurance purchased by the nuclear operator (Tier I) and the retrospective rating Plan (Tier II). When the second stage is reached, the federal government implements a compensation plan. At the beginning of 1975, the total amount of compensation available during the first and second phases was $560 million, and the larger amount was set as the nuclear operator's liability limit.

The mechanism of retrospective rating plan is intended to collect monetary contributions from all nuclear operators with nuclear reactors when the damages have not been paid by the private liability insurance company in the first stage. Pursuant to the 1975 amendment, a maximum of $5 million per nuclear reactor can be paid. This $5 million was increased to $63 million in the 1988 amendment and $95.8 million in the 2005 amendment. What was the intention behind the retrospective rating plan?

7.3.2 Discussion of the Retrospective Rating Plan

The intent of the retrospective rating plan can be surmised by analyzing a statement made by Senator Thurmond in the US Senate minutes for December 16, 1975 (Government Publishing Office 121 Cong. Rec.[Bound]—Senate: December 16, 1975):

> The report of the Joint Committee on Atomic Energy directs the Nuclear Regulatory Commission to establish a new retrospective premium schedule sufficient to end Government indemnification by 1985. After that date, it is expected that total public protection will surpass the current $560 million and exceed $1 billion when more nuclear powerplants are built and operating. / Mr. President, the liability limit of the Price-Anderson Act is balanced by a valuable public protection, the "no-fault" provision. The provision enables citizens to recover financially from a nuclear incident by merely showing their damage from such an incident and the dollar amount of damage. / Additionally removal of the liability limit would subject electric utilities and others to unacceptable financial exposure, with serious threats to the public interest. Many architects, engineers, and component suppliers would withdraw from the nuclear industry because their potential profit from nuclear-related business is insufficient to warrant placing all of their assets at risk. This would reduce competition and diminish sources of supply. Passage of this amendment would result in increased cost of investment capital for nuclear projects because of the decreased financial security of investors. This lack of capital would relegate the smaller utilities to fossil fuels, which are in short supply, because they could no longer qualify financially for the Federal licenses needed to build and operate nuclear plants. (pp. 40983)

In the passage above, Senator Thurmond made the following key points: (1) The Joint Committee on Atomic Energy's report states that the Nuclear Regulatory Commission was directed to establish a new retrospective Rating Plan schedule sufficient to eliminate government compensation. (2) The total amount of public protection would exceed the current $560 million, rising to over $1 billion. It is clear that the elimination of liability limits would result in a serious threat to the public interest, including electric utilities, as the legislation is limited to 10 years and, therefore, would need to be discussed every decade. However, the statement above shows that even if held at 10-year intervals, there would be careful discussion regarding limit amount and the question of how to deal with the nuclear industry.

This careful discussion is also essential for the use of nuclear power in a carbon-neutral society. Since the advent of the nuclear power industry, continual discussions have been necessary. In addition, the statement above does not use the word "risk." However, it does highlight how to make operators aware of risks. Simultaneously, the discussion of minutes is also aware of the fact that nuclear operators will benefit from discussions. The minutes of the meeting from December 18, 1975, read as follows (Government Publishing Office 121 Cong. Rec. [Bound]—December 18, 1975):

> Nuclear accident indemnity.-Provides for the phase-out of government indemnity under the Price-Anderson Act, which was enacted in 1957 to protect the public and the emerging nuclear industry by assuring the availability of funds up to $560 m1llion for the payment of claims and by protecting the nuclear industry against unlimited liability in the unlikely event of a catastrophic nuclear accident; requires participation by 11-censees in an insurance retrospective rating plan under which, in case of accident, each licensee would be assessed a deferred premium which would be a prorated share of the damages over the amount of Government indemnity and eliminates Government indemnity at the point where the total of primary insurance and the number of licensees paying the deferred premiums is sufficient to provide $560 million for claims payment without using Government funds; provides that the $560 million limit on liability shall rise thereafter to correspond with the increasing total level of primary insurance and deferred premium funds as more licenses are issued; extends indemnity coverage outside the territorial limits of the United States for certain activities conducted by Energy Research and Development Administration (ERDA) contractors or involving licensed nuclear facilities or licensed activities, none of which involve indemnity for any shipment of nuclear technology abroad under an agreement for cooperation with nations or groups of nations; extends the Act, which is scheduled to expire on August 1, 1977, for a period of 10 years, and contains other provisions.

The minutes clarify the following points: (1) The Price-Anderson Act of 1957 clearly stipulated the phasing out of government compensation. (2) The Price-Anderson Act required nuclear operators to participate in the retrospective rating plan. (3) The policy of protecting the nuclear industry from a catastrophic nuclear accident remained unchanged.

In 1975, there were still plans to establish a compensation system to protect the nuclear industry. As an alternative to abolishing government compensation (state expenditure), a system of retrospective premiums was created. In addition, as pointed out by Senator Thurmond, there was an expectation that the construction and operation of nuclear reactors by new nuclear operators would increase. However, there was no clear rationale for making nuclear operators who had nothing to do with an accident bear the costs of the accident.

The maximum amount of the retrospective rating plan is a factor of the number of reactors and the maximum retrospective premium per reactor. The money collected via this system is then used as a source of compensation. An increase in the construction and operation of new reactors by new nuclear operators results in an increase in the maximum amount of the retrospective rating plan and in the monetary sources of compensation. Moreover, an increase in the construction and operation of reactors by new nuclear operators simultaneously leads to a decrease in the burden per reactor—that is, a decrease in the burden per nuclear operator. This conforms with the idea of the Brussels Supplementary Convention. The document "Reauthorization of the Price-Anderson Act" (Government Publishing Office S. Rept. 108–218) states the following:

> Under current law, in the event that losses from a nuclear incident exceed the required amount of private insurance, the NRC would levy an assessment on its licensees (both public and private) to cover the shortfall in damage coverage. Section 102 would increase the maximum retrospective premium from $84 million to $94 million as well as increase the maximum annual premium from $10 million to $15 million. CBO has determined that raising both the maximum total premium and the annual premium would increase the costs of an existing mandate and would thereby impose both intergovernmental and private-sector mandates under Uv RA. Because the probability of a nuclear accident resulting' in losses exceeding the

> amount of private insurance coverage is low, CBO estimates that the annual costs of complying with the mandates (in expected-value terms) would not be substantial over the next 5 years. (pp. 15)

The amount of insurance funds to be disbursed by the nuclear operator's own private liability insurance (Phase I) is $200 million. In addition, the retrospective rating plan (Phase II) will be used to fund damages over $200 million. The Congressional Budget Office highlighted that if a nuclear accident were to occur, the amount of damage compensation would be less than the $200 million of the first phase.

In this context, the price increase of the second stage in 2003 was intended to account for the eventuality of damages exceeding $200 million. The US Congress has recognized that the estimated amount of damages from a nuclear accident is less than $200 million, and it can be inferred that the retrospective rating plan is designed to prepare for such a contingency. The retrospective rating plan was recognized by the US Congress in 1975, and it was still in place in 2003 as a proxy for state expenditures.

7.3.3 Lessons Learned from the US Nuclear Damage Compensation System

The US system highlights points that should not be overlooked when using nuclear power in a carbon-neutral society. The US system is (1) capable of requiring cost sharing in advance for risky activities not only via treaties but also via domestic laws (hereinafter referred to as the "domestic system"). It is also (2) capable of flexibly adapting to the circumstances of the times and (3) possesses an effective means of reviewing its approach every ten years.

Point 1, which can be achieved through treaties, is synonymous with the meaning that the contents similar to those of the previous chapter (Brussels Supplementary Convention) can be realized in domestic laws (institutions in Japan). The main focus of the Brussels Supplementary Convention was transboundary pollution, and it sought to provide mutual assistance to those at risk of transboundary contamination. The US system can be described as a system that brings together operators who are exposed to a shared risk from a nuclear accident and then imposes a cost-sharing burden on all parties to prepare for a potential accident. That is, the system is a form of mutual aid whereby the operator who caused the accident is supported financially by operators who did not cause the accident.

Point 2 refers to the change in the weight of the burden on nuclear operators. The weight of the burden changed in accordance with an emphasis on "protecting the nuclear industry." It should be noted that there were nuclear accidents, such as the Chernobyl accident, during the period when the weight of the burden changed. It is also apparent that social events were considered.

Point 3 highlights that the legislation was limited in duration, which naturally led to parliamentary deliberations and other discussions in other forums immediately prior to the 10-year time limit expiring. This triggers periodic reviews of the law and allows for changes to be made to the law in response to changing circumstances.

If the law did not have an expiry date, the law would be administered as it was in the historical background when it was first enacted. This means that social changes cannot be considered. As a result, the law will become outdated. Expiry dates for laws are beneficial for preventing "outdated" laws.

7.4 Nuclear Damage Compensation in Japan: The General Contribution System

The Fukushima nuclear accident occurred in 2011. The Japanese Government enacted the Act on the Japan Nuclear Damage Liability Facilitation Fund (currently known as Act on the Japan Nuclear Damage Liability Facilitation Fund) to promptly and appropriately compensate parties for the nuclear damage they suffered. The law established a general burden payment system.

Article 38 of the JNES Act imposes an obligation on nuclear operators to bear the costs for each fiscal year in which they operate nuclear reactors. This cost burden is referred to as the "general burden." All nine electric power companies, including Japan Atomic Power Company and Japan Nuclear Fuel Limited, are required to contribute to the general burden fund. TEPCO, which caused the nuclear accident, is obliged to make special contributions in addition to general contributions. Because the main topic of this chapter is the general burden, I will omit details of the special burden.

Various discussions were held in the Japanese Diet concerning the logic of the general contribution. The government defined the general contribution system as a mutual assistance mechanism. During the deliberation process, the structure of the general contribution system was not discussed. There was no discussion as to why the government introduced the concept of "mutual assistance," which is different from the concept of the previous system (e.g., Act on Asbestos Health Damage Relief), which required third parties unrelated to the accident to bear the costs. As a result, the general burden payment was perceived as an unreasonable system that required third parties who were not contributors to the accident to bear the costs.

Dougauchi (2018) stated that the *Basic Guidelines for Accelerating Fukushima Recovery from Nuclear Disaster*, which was issued on December 20, 2016, was an extension of the general burden payment. Dougauchi continued, stating that "originally, nuclear operators should have created a mutual assistance framework and gathered compensation payment resources in preparation for accidents." The report also pointed out that "it has been determined that nuclear operators should have created a framework of mutual assistance and should have saved up funds to provide compensation in the event of an accident." Dougauchi (2018) also approved the ex-post introduction of the general burden payment system but warned that "it should not be relied upon unnecessarily in the future."

Dougauchi (2018) makes clear criticisms; however, without a more in-depth analysis of the cost-sharing structure, it is unlikely to be understood by many. Dougauchi's (2018) phrase "gathered compensation payment resources in preparation for accidents" provides a clue to understanding the structure of the system. Dougauchi (2018) saw the retrospective creation of the system as problematic.

7.4.1 Lessons from the Japanese System

The decisive difference between the general burden system, on the one hand, and the Brussels Supplementary Convention and Retrospective Rating Plan, on the other, is that the general burden system was created after an accident. When nuclear power is used in a carbon-neutral society, creating a compensation mechanism after an accident may raise doubts about the mechanism itself. Therefore, it is preferable to create the mechanism in advance, before an accident.

It is important that the system not be perceived as an unreasonable system that requires third parties not involved in the accident to bear the costs. It is difficult to analyze the composition of the system amidst the impact of an accident. In addition, it is difficult to reach a consensus on requiring third parties to bear the costs after an accident occurs. This is because the key issue must be how to make the parties involved in the accident take responsibility. A structural analysis (structural explanation) regarding the general burden must consider why third parties that are not contributors to the accident should be asked to bear the cost. This is why it is necessary to hold these discussions before an accident actually occurs. The Japanese case brings to light the problems caused by creating a compensation structure in the wake of an accident. In this way, the Japanese case can serve as a lesson for the future.

7.5 Concluding Remarks

This paper was based on the assumption that nuclear power will be used in the future. Moreover, it uses the three examples of the Brussels Supplementary Convention, the Retrospective Rating Plan in the United States, and the general contribution mechanism in Japan, which was created after the Fukushima accident, to illustrate the importance of establishing a nuclear damage compensation system that enables parties who share a common risk to contribute money to a fund to be used in the event of an accident. This is a crucial component of the transition to a carbon-neutral society.

The case of Japan is of particular importance. The Fukushima Daiichi Nuclear Power Plant disaster was a severe accident. Moreover, a new mechanism of burden sharing was created despite the existence of a compensation system. However, the general burden did not make much progress in understanding the system. This is because the answer to the question, "Why are nuclear operators who have not caused accidents asked to bear the burden of an accident?" It has been shown that the nuclear damage compensation system in place before the Fukushima Daiichi Nuclear Power Plant disaster was insufficient in the context of a carbon-neutral society.

This inadequacy can be compensated for via a mechanism for collecting money among nuclear operators who share a common risk before an accident occurs. It is impossible to reduce the risk to zero. In a carbon-neutral society, it is necessary to "always assume risk when doing something," "consider how to reduce risk before risk occurs if the level of risk is high," and "establish a mechanism to reduce risk and review it periodically." We are at the beginning of the process of fulfilling these requirements.

References

Dougauchi, M. (1993). Liability for nuclear disasters that transcend national borders. *Jurist, 1015*, 157–163.

Dougauchi, M. (2018). *Fukushima-Daiichi-Gennsiryokuhatsudennsho jiko ni yoru sonngai no baisho ni taiou suru tameno houseido no kouchiku to sono unnyou* (Establishment and performance of the Japanese new legal system to cope with damages arising out of the accident of the Fukushima Dai-ichi Nuclear Power Plant). *Waseda Law School Journal,* 3, 45–92.

Dussart Desart, R. (2005). The reform of the Paris Convention on Third Party Liability in the Field of Nuclear Energy and of the Brussels Supplementary Convention. *Nuclear Law Bulletin, 75*, 7–33.

Government Publishing Office 121 Cong. Rec. (Bound)—Senate: December 16, 1975. https://www.govinfo.gov/content/pkg/GPO-CRECB-1975-pt31/pdf/GPO-CRECB-1975-pt31-4-2.pdf, 20/8/2024 referred

Government Publishing Office 121 Cong. Rec. (Bound)—December 18, 1975. https://www.govinfo.gov/content/pkg/GPO-CRECB-1975-pt32/pdf/GPO-CRECB-1975-pt32-2.pdf, 20/8/2024 referred

Government Publishing Office S. Rept. 108–218—Reauthorization of the Price-Anderson Act. https://www.govinfo.gov/content/pkg/CRPT-108srpt218/pdf/CRPT-108srpt218.pdf, 20/8/2024 referred

Policy Conflicts and Environmental Orientation Among Japanese Voters

8

Masahisa Endo and Kohei Tabei

Abstract

This chapter will empirically elucidate the current state of environmental orientation among contemporary Japanese voters to anticipate a transition to a carbon-neutral society. Previous studies have asserted that environmental orientation among Japanese citizens is declining and has become increasingly associated with ideological and political conflicts. To explore which individuals exhibit higher levels of environmental orientation, this study will investigate the relationship between media consumption patterns and the intensity of environmental orientation in Japan, where environmental issues tend to receive limited media coverage. This chapter demonstrates that individuals who have obtained information from Internet-based media tend to exhibit a lower degree of environmental orientation compared to those who rely on traditional media sources, such as newspapers and television. Given the limited media coverage of environmental issues and the high level of environmental orientation among traditional media consumers, the low level of environmental orientation among web media consumers can be attributed to insufficient exposure to environmentally related information. To enhance the environmental orientation of voters and guide them toward a carbon-neutral society, it is imperative that the media coverage of climate change be improved.

Keywords

Environmental orientation · Mass media · Internet media · Social media · Ideology · Climate change in the media

8.1 Introduction

The pursuit of a carbon-neutral society necessitates the implementation of diverse measures. However, when economic incentives prove an insufficient drive for the adoption of such strategies, legislative action becomes imperative. In democratic societies, the public perception of parliamentary decisions will assume critical importance when substantial climate change policy reforms are necessary, irrespective of whether such changes stem from international organizational directives or global trends. The perpetual concern of elected officials regarding public sentiment toward their

M. Endo (✉) · K. Tabei
Waseda University, Tokyo, Japan
e-mail: endo@waseda.jp

K. Akao and A. Washizu (eds.), *Climate Change Issues and Social Sciences*, Sustainable Development Goals Series, https://doi.org/10.1007/978-981-96-8951-4_8

performance, coupled with the potential impediments to implementing policies that lack popular endorsement, underscores the intricate relationship between political governance and societal approval.

This chapter will empirically elucidate the current state of environmental orientation among contemporary Japanese voters to anticipate a transition to a carbon-neutral society. Specifically, assuming Japanese individuals' environmental orientation has become associated with ideological and political conflicts in Japan, the study will investigate the relationship between media consumption patterns and the intensity of environmental orientation among Japanese citizens, as environmental issues receive limited media coverage in the country.

In the subsequent section, this study will review previous research on environmental orientation in Japan. Section 8.3 describes the public opinion survey data and variables utilized in this study. Section 8.4 will report the results of the analysis.

8.2 Environmental Orientation Among Japanese Voters

The relationship between citizens' environmental orientation and political values has been the subject of comparative study since the 1970s. While discussing value transformation in post-industrialized societies, Inglehart (1977, 1990) posited that economic development leads to the emergence of post-materialistic values in advanced democracies, which emphasize quality of life rather than the materialistic priorities of economic and physical security. Environmental protection emerged as a key component of postmaterialist values during the period in which environmentalism gained prominence as a new social movement. These values have subsequently been utilized to elucidate the formation of numerous environmental parties across Europe (see Kitschelt, 1995). Environmental orientation has therefore been integrated into ideological conflicts and become a fundamental aspect of political conflicts, despite manifesting differently across countries and regions.[1] Dalton (2006) observed that the correlation between environmental orientation and ideology is evident in Europe, the United States, Eastern Europe, and Africa, but tends to be less pronounced in other regions.

In Japan, the relationship between environmental orientation and ideology is relatively weak. Unlike many Western countries, Japan's ideological framework is not centered on economic conflict, but rather on attitudes toward security and constitutional issues (Jou & Endo, 2016; Kabashima & Takenaka, 2012). The only environmental issue that demonstrates a strong ideological correlation is the question of reactivating nuclear power plants. Since the Fukushima Daiichi nuclear accident of 2011, the issue of nuclear power plant has become significantly associated with the conservative-liberal dichotomy. However, attitudes toward the subject align more closely with the perspectives for and against nuclear energy within the context of peace and civil movement, rather than the framework of environmental protection. Consequently, this issue has assumed a highly partisan nature (Jou & Endo, 2016).

Meanwhile, broader environmental orientation has not been a significant ideological axis of conflict in Japan until recently. Consequently, there has been a paucity of analyses of the relationship between environmental orientation and ideological conflict in the country. Nevertheless, recent studies indicate that environmental orientation is increasingly becoming associated with partisan debates. Kato (2021), who analyzed data until the late 2010s, found that long-term trends among Japanese voters demonstrate a decline in environmental orientation. However, the study also revealed partisan differences: supporters of the liberal opposition parties exhibited greater concern for environmental protection compared to supporters of the dominant (and conservative)

[1] Several studies have investigated the relationship between environmental orientation and macro-level factors, such as national affluence and the severity of environmental issues (Diekmann & Franzen, 1999; Dunlap & Mertig, 1995; Inglehart, 1995).

Liberal Democratic Party (LDP) while, independents demonstrated a higher level of concern than LDP supporters regarding the prioritization of the global environment over their personal lives.

Endo and Tabei (2024) also confirmed the emergence of environmental orientation as an axis of policy conflict among Japanese voters in 2023. A correlation, albeit a relatively weak one, was observed between environmental orientation and ideological self-placement, with individuals exhibiting a more conservative ideology demonstrating a somewhat greater reluctance to support environmental policies. Furthermore, in an analysis of political party preferences, environmental orientation exhibited a strong positive correlation with support for liberal opposition parties and a negative correlation with support for the LDP (Endo & Tabei, 2024). These observations suggest that while the international community is being called upon to respond to climate change, conflicts are emerging regarding the most appropriate nature of the response.

Kato (2021) has also found that unlike in the United States, where the younger generation (as represented by Generation Z) is considered more environmentally conscious (Pew Research Center, 2017), the younger generation is less environmentally conscious in Japan. While it is challenging to fully elucidate the reasons for the disparities between Japan and the United States, it may be beneficial to examine the differences in their media consumption patterns when considering generational variations within a single country. This approach is warranted because global environmental issues are topics that can only be comprehended through media exposure, and it is essential to consider how individuals access information based on its presentation in Japanese media outlets. According to Saito (2017), who has analyzed climate change coverage in Japan, reporting on the issue has been declining in Japan, as well as in the United Kingdom and the United States, resulting in an overall decrease in opportunities for voters to learn about environmental issues. Consequently, it is plausible that this is why younger Japanese people, who tend to have less exposure to traditional media, exhibit lower levels of environmental awareness.

Conversely, changes in the media landscape may also facilitate the emergence of new channels of information. The wider dissemination of information via the Internet and social media has enabled citizens to access more information on issues they deem significant, rather than being limited to the topics predetermined by traditional media. Furthermore, traditional media in Japan tends to report facts rather than issuing warnings, and it is less likely to report the climate change issue in association with policy actors compared to their counterparts in the United Kingdom and the United States (Saito, 2017). It is then conceivable that those who are discontent with traditional media reporting and proactively obtain information through the Internet may exhibit greater environmental orientation.

This study examines the environmental orientation of Japanese voters and their patterns of media consumption through public opinion poll data collected in 2023 to ascertain whether individuals with more exposure to traditional media or those with more exposure to social media and the Internet exhibited a higher degree of environmental orientation. The analysis will investigate patterns of voters' engagement with newspapers, television programs, and web-based media, as well as their perceptions regarding media consumption.

8.3 Data and Variables

This study will examine the environmental orientation of the Japanese using the SmartNews, Media, Politics, and Public Opinion (SMPP) Survey utilized by Endo and Tabei (2024). The SMPP Survey is a public opinion poll that was conducted by the SmartNews Media Institute in March 2023. This study was designed to analyze the relationship between Japanese voters' media usage and their political opinions. The survey was conducted simultaneously as a web-based and mail survey. The data from the mail survey was used in this study because it was considered more representative. In the mail survey, questionnaires were sent to males and females aged 18 to 79 residing in Japan. The number of respondents was 1901 out of a sample size of 4460, with a response rate of 42.6%.

Utilizing this dataset, Endo and Tabei (2024) investigated the structure of policy conflicts among voters based on their responses to 16 policy-related questions and found that voters' policy conflicts comprise five dimensions: COVID-19 measures, security, government intervention, gender, and environment. Consistent with previous research, economic conflicts were not prominent; instead, the security dimension was strongly defined. Meanwhile, the environmental dimension has gained prominence as a significant factor in political conflict. This emergence is attributed to the pressing challenges posed by climate change and the imperative for implementing mitigation strategies on a global scale.

To focus on environmental orientation as an outcome variable, this study utilized factor scores on the environmental dimension extracted employing the same methodology as Endo and Tabei (2024). This primarily represents strong factor loadings for the responses toward *the abolition of gasoline car sales* and *the sacrifice of living standards for the environment*.[2] Higher values indicate a greater degree of environmental orientation.

For the explanatory variables, this study utilized several questions related to media contact. Initially, dummy variables were created for subscribers to the newspapers *Asahi Shimbun*, *Yomiuri Shimbun*, *Nikkei Shimbun*, and those not subscribed to any newspapers, as derived from multiple-response questions regarding the regularity of reading print and digital editions. *Asahi Shimbun* and *Yomiuri Shimbun* are national newspapers with particularly extensive readerships, recognized for their moderate liberal and moderate conservative orientations, respectively. *Nikkei Shimbun* meanwhile emphasizes business news, maintains a substantial readership among urban business professionals, and is considered politically centrist. In recent years, the proportion of nonsubscribers has been increasing, with the SMPP Survey indicating that 56% of its respondents did not subscribe to any newspaper.

To analyze patterns of television usage, dummy variables were constructed from a multiple-response question concerning the television program that respondents typically viewed. Those who selected *Nippon Hōsō Kyōkai* (NHK, Japan's public broadcaster) news programs (i.e., standard news broadcasts such as the 7pm. news), commercial news programs, daily talk shows (called *waido shō* in Japanese), dramas, and films, as well as those who indicated no specific television program preference, were each assigned a value of 1. Here, daily talk shows are defined as programs that disseminate information about current events and celebrity trends, and which frequently address political topics and social issues. Variables related to entertainment-oriented content, such as drama or movie viewership, were also generated to compare with the viewers of news program. Approximately 18% of the respondents reported not watching television at all, which is a lower percentage than the nonreaders found with the question on newspaper readership.

For media usage on the web, binary variables were constructed from a question that required respondents to provide multiple responses regarding the media they typically consume, with 1 representing those who utilized *Yahoo! News*, *YouTube*, *Facebook*, and *Twitter*, respectively, and those do not utilize any web-based media. *Yahoo! News* has been recognized as having a dominant market share as a news portal site, with 59% of respondents in this survey indicating frequent usage. Among social media platforms, *YouTube* exhibited a high utilization rate of 63%. *Twitter* was found to have a higher usage rate in Japan than *Facebook*, with 31% and 16%, respectively. Conversely, only 8% of respondents reported not using Internet-based media at all.

[2] Respondents were asked to choose from agree, somewhat agree, somewhat disagree, disagree, or don't know to each of the following statements: "Sales of gasoline-powered vehicles should be discontinued as soon as possible in order to reduce greenhouse gas emissions" and "It is necessary to sacrifice living standards in order to protect the environment." The "don't know" response was included in the analysis as an intermediate response.

In addition to the questions regarding media contact, the study incorporated attitudes toward media usage as an explanatory variable. Initially, the survey addressed perceptions of the importance of news. Specifically, it examined four items: news featured on newspapers' front pages, news featured at the top of the NHK's news programs, news appearing at the top of *Yahoo! News*, and news trending on *Twitter*. The response options were coded as follows: "important," "somewhat important," "not very important," "almost unimportant," and "do not consume this medium," with higher values indicating that respondents considered the news medium important.

The survey also included a question about media viewing habits as an explanatory variable. The following items were given: "I do not enjoy watching news programs," "I dislike missing news broadcasts," "I prefer detailed news coverage to headline news, even if it requires more time," "Most nights, I watch television or use the Internet for drama or variety shows," and "I watch television or use the Internet primarily for entertainment purposes." The response options were coded as "agree," "somewhat agree," "neither agree nor disagree," "somewhat disagree," and "disagree," with higher values indicating a greater likelihood of agreement.

Additional socioeconomic variables included age, gender (female being the binary variable), level of educational attainment, and self-reported ideological position. Ideological orientation was assessed using an 11-point scale ranging from 0 (liberal) to 10 (conservative), with binary variables created for the liberal (0–4), centrist (5), and conservative (6–10) positions. Furthermore, a binary variable was included for respondents who expressed "I don't know" regarding their ideological self-placement. The reference category was represented by an intermediate binary variable. The levels of educational attainment were categorized as junior high school, high school, junior college/technical/vocational school, university (bachelor's degree), master's degree, and doctoral degree.

8.4 Results

Given that environmental orientation, the outcome variable, is continuous, regression analysis was employed to investigate which individuals exhibited higher degrees of environmental orientation. Table 8.1 presents the results of the regression analysis, with newspaper usage serving as the explanatory variable.

In terms of newspaper readership, the findings indicated that environmental orientation varied according to newspaper subscription. Subscribers to the *Asahi Shimbun* demonstrated a higher level of environmental orientation, whereas subscribers to the *Yomiuri Shimbun* exhibited a lower level (Table 8.1 [1] and [2]). While these two newspapers are recognized for their contrasting editorial stances, the correlation between subscription and environmental orientation persisted even after controlling for ideological factors. In the case of *Nikkei Shimbun*, which should have paid considerable attention to the effect of climate change on industrial activities, environmental orientation was observed to be neither particularly high nor low (Table 8.1 [3]). However, for individuals without newspaper subscriptions, the regression coefficient was negative, indicating a tendency toward a lower environmental orientation (Table 8.1 [4]).[3]

Table 8.2 presents the results of the analysis on the relationship between television viewership and environmental orientation. Individuals who watched NHK and commercial news programs both demonstrated a tendency toward high environmental orientation (Table 8.2 [1] and [2]). It is likely that those who viewed programs more likely to provide information on environmental issues, such as news broadcasts, exhibited heightened environmental consciousness. Conversely, no correlation was observed between viewing

[3] Among the responses not included in this analysis, subscriptions to local newspapers were positively correlated with environmental orientation, whereas subscriptions to *Sankei Shimbun* and trade newspapers were negatively correlated with environmental orientation. As with *Nikkei Shimbun*, no statistically significant relationship was observed for *Mainichi Shimbun*.

Table 8.1 Regression results for environmental orientation with newspaper readership

	(1)	(2)	(3)	(4)
Asahi Shimbun	0.099*			
	(0.053)			
Yomiuri Shimbun		−0.087*		
		(0.048)		
Nikkei Shimbun			−0.036	
			(0.058)	
Non-subscriber				−0.071*
				(0.040)
Age	0.006***	0.006***	0.006***	0.005***
	(0.001)	(0.001)	(0.001)	(0.001)
Female	0.207***	0.202***	0.204***	0.211***
	(0.036)	(0.036)	(0.036)	(0.036)
Education level	0.004	0.006	0.007	0.004
	(0.020)	(0.020)	(0.020)	(0.020)
Ideology: liberal	0.052	0.061	0.062	0.059
	(0.058)	(0.058)	(0.058)	(0.058)
Ideology: conservative	−0.172***	−0.171***	−0.172***	−0.174***
	(0.053)	(0.053)	(0.053)	(0.053)
Ideology: DK	0.002	0.003	0.001	0.005
	(0.055)	(0.055)	(0.055)	(0.055)
Constant	−0.374***	−0.380***	−0.381***	-0.300***
Observations	1,774	1,774	1,774	1,774
Adjusted R-squared	0.052	0.051	0.050	0.051

Standard errors in parentheses
***$p < 0.01$, **$p < 0.05$, *$p < 0.1$
Source The SMPP Survey, 2023

daily talk shows and environmental orientation (Table 8.2 [3]). This may be attributed to the fact that while these programs frequently address political topics, they do not typically discuss climate change or environmental issues.[4]

Notably, viewers of programs not directly related to environmental issues, such as dramas and movies, also exhibited environmental consciousness (Table 8.2 [4]). This may be attributed to the fact that these individuals may spend considerable time watching television, potentially including extended exposure to news programs. While it is plausible that they watch news because of their environmental consciousness, it is less likely that they watch dramas for the same reason. Therefore, it is reasonable to postulate that television viewing influences their environmental orientation. Corroborating this, the results indicated that individuals who reported not watching television demonstrated low environmental awareness (Table 8.2 [5]).

[4] NHK and commercial news shows were also included among the response options. However, their selection rate was lower than that of NHK and commercial news programs. Consequently, only the results for the news programs are presented in this chapter. Among the response options not included in the analysis, NHK news shows, variety programs, and educational programs demonstrated a positive correlation with environmental orientation. Conversely, no statistically significant relationship was observed for commercial news shows, discussion programs, and sports programs.

Table 8.2 Regression results for environmental orientations with TV viewing

	(1)	(2)	(3)	(4)	(5)
NHK news programs	0.093**				
	(0.038)				
Commercial news programs		0.084**			
		(0.037)			
Daily talk shows			0.048		
			(0.036)		
Dramas and movies				0.137***	
				(0.036)	
Does not watch TV					−0.291***
					(0.063)
Age	0.005***	0.005***	0.006***	0.006***	0.005***
	(0.001)	(0.001)	(0.001)	(0.001)	(0.001)
Female	0.206***	0.206***	0.201***	0.181***	0.186***
	(0.036)	(0.036)	(0.036)	(0.037)	(0.036)
Education level	0.002	0.004	0.008	0.006	0.003
	(0.020)	(0.020)	(0.020)	(0.020)	(0.020)
Ideology: liberal	0.055	0.057	0.063	0.060	0.064
	(0.058)	(0.058)	(0.058)	(0.058)	(0.058)
Ideology: conservative	−0.176***	−0.177***	−0.175***	−0.176***	−0.178***
	(0.053)	(0.053)	(0.053)	(0.053)	(0.053)
Ideology: DK	0.009	0.000	0.000	−0.003	0.007
	(0.055)	(0.055)	(0.055)	(0.055)	(0.055)
Constant	−0.359***	−0.387***	−0.392***	−0.439***	−0.279**
	(0.107)	(0.107)	(0.107)	(0.108)	(0.109)
Observations	1,774	1,774	1,774	1,774	1,774
Adjusted R-squared	0.053	0.052	0.051	0.058	0.061

Standard errors in parentheses
***$p < 0.01$, **$p < 0.05$, *$p < 0.1$
Source The SMPP Survey, 2023

Table 8.3 presents the results of the analysis of Internet media usage as an explanatory variable. In contrast to newspapers and television, individuals who engage with Internet-based media demonstrated a tendency toward a decreased environmental orientation. Even users of *Yahoo! News*, the predominant news portal site in Japan in terms of market share, exhibited a lower environmental orientation compared to non-users (Table 8.3 [1]). Unlike traditional media, which may allow for passive exposure to news articles or programs, web-based media facilitates the active seeking of specific information. Consequently, individuals who utilized only web media may have limited exposure to information about environmental issues unless they specifically sought it. Conversely, those who did not engage with web media were more likely to obtain news information from traditional sources, and these individuals tended to exhibit a stronger environmental orientation (Table 8.3 [5]).[5]

[5] Among the response options not included in the analysis, only *LINE News* exhibited a positive correlation with environmental orientation. Video distribution services such as *Amazon Prime Video*, *Hulu*, and *Netflix*, as well as *5channel*, *Niconico Doga*, and blogs maintained by individuals, demonstrated negative correlations

Table 8.3 Regression results for environmental orientations with Internet media usage

	(1)	(2)	(3)	(4)	(5)
Yahoo! News	−0.073**				
	(0.035)				
Facebook		0.010			
		(0.048)			
Twitter			−0.182***		
			(0.042)		
YouTube				−0.177***	
				(0.040)	
Does not use Internet media					0.129*
					(0.068)
Age	0.006***	0.006***	0.004***	0.004***	0.005***
	(0.001)	(0.001)	(0.001)	(0.001)	(0.001)
Female	0.200***	0.206***	0.215***	0.188***	0.204***
	(0.036)	(0.036)	(0.036)	(0.036)	(0.036)
Education	0.010	0.006	0.012	0.011	0.009
	(0.020)	(0.020)	(0.020)	(0.020)	(0.020)
Ideology: liberal	0.058	0.060	0.073	0.069	0.059
	(0.058)	(0.058)	(0.058)	(0.058)	(0.058)
Ideology: conservative	−0.171***	−0.173***	−0.156***	−0.160***	−0.171***
	(0.053)	(0.053)	(0.053)	(0.053)	(0.053)
Ideology: DK	−0.005	0.002	0.006	0.010	−0.003
	(0.055)	(0.055)	(0.055)	(0.055)	(0.055)
Constant	−0.342***	−0.382***	−0.256**	−0.183	−0.369***
	(0.108)	(0.107)	(0.110)	(0.115)	(0.107)
Observations	1,774	1,774	1,774	1,774	1,774
Adjusted R-squared	0.052	0.050	0.060	0.060	0.052

Standard errors in parentheses
***$p < 0.01$, **$p < 0.05$, *$p < 0.1$
Source The SMPP Survey, 2023

Regarding social media usage, individuals who frequently utilized *Twitter* and *YouTube* demonstrated lower environmental orientations (Table 8.3 [3] and [4]), whereas *Facebook* usage did not exhibit a correlation with environmental orientation (Table 8.3 [2]). It is noteworthy that while both *Facebook* and *Twitter* may be equally susceptible to the "echo chamber" phenomenon by fostering user-following relationships, only *Twitter* usage appears to have negatively influenced environmental orientation.

Table 8.4 examines the relationship between the perceived importance of news items and environmental orientation. The findings demonstrated partial consistency with the results on viewing patterns. Individuals who attributed more significance to news featured on the front pages of newspapers and at the top of NHK news programs tended to exhibit a high degree of environmental

with environmental orientation. Conversely, no statistically significant relationships were observed with news sites provided by newspapers and TV stations, *SmartNews*, *Gunosy*, *NewsPicks*, *Instagram*, *LINE*, *LINE Openchat*, *Snapchat*, *TikTok*, on-demand broadcast programs provided by commercial key stations and NHK.

Table 8.4 Regression results for environmental orientations with perceived importance of news items

	(1)	(2)	(3)	(4)
News featured on front pages	0.092***			
	(0.016)			
News featured in the NHK's news programs		0.107***		
		(0.015)		
News at the top of *Yahoo! News*			0.017	
			(0.016)	
News trending on *Twitter*				−0.031**
				(0.016)
Age	0.005***	0.005***	0.006***	0.006***
	(0.001)	(0.001)	(0.001)	(0.001)
Female	0.202***	0.201***	0.211***	0.213***
	(0.036)	(0.036)	(0.037)	(0.037)
Education level	0.003	−0.004	0.006	0.008
	(0.020)	(0.020)	(0.020)	(0.020)
Ideology: liberal	0.060	0.052	0.075	0.067
	(0.058)	(0.058)	(0.059)	(0.059)
Ideology: conservative	−0.180***	−0.195***	−0.177***	−0.174***
	(0.053)	(0.053)	(0.054)	(0.054)
Ideology: DK	0.025	0.019	0.009	0.005
	(0.055)	(0.055)	(0.056)	(0.056)
Constant	−0.594***	−0.561***	−0.457***	−0.338***
	(0.111)	(0.109)	(0.116)	(0.118)
Observations	1,735	1,730	1,722	1,713
Adjusted R-squared	0.071	0.080	0.054	0.055

Standard errors in parentheses
***$p < 0.01$, **$p < 0.05$, *$p < 0.1$
Source The SMPP Survey, 2023

orientation (Table 8.4 [1] and [2]). These individuals exhibited confidence in the agenda-setting function of the mass media, and a positive correlation may exist between the importance they attributed to the environmental issues reported in the media and their level of environmental orientation.[6]

No correlation was observed between environmental orientation and the respondents' perceived importance of the news featured on the top page of *Yahoo! News*. The top page of *Yahoo! News* is intended to present a comprehensive array of news from various domains: politics, economics, international affairs, entertainment, and sports, which may not align with the priorities of individuals who prefer more substantive news content. Conversely, individuals who appear to prioritize trending topics on *Twitter* demonstrated a tendency toward lower environmental orientation. While political news can achieve trending status on *Twitter*, which serves as a major platform for partisan discourse, it is less likely to promote environmental orientation.

[6] Among the items excluded from the analysis, the significance of information officially disseminated by governmental or local governmental entities exhibited a positive correlation with environmental orientation. Conversely, the importance attributed to news content delivered via applications such as SmartNews, news shared by acquaintances on social media, and news disseminated by high-profile individuals with substantial follower bases on social media demonstrated no statistically significant relationship.

Table 8.5 Regression results for environmental orientations with media viewing habits

	(1)	(2)	(3)	(4)	(5)
I do not enjoy watching news programs	−0.055***				
	(0.016)				
I dislike missing news broadcasts		0.049***			
		(0.016)			
I prefer detailed news coverage			0.016		
			(0.017)		
I watch television or use the Internet for drama or variety shows				0.027**	
				(0.013)	
I watch television or use the Internet primarily for entertainment purposes					−0.044**
					(0.019)
Age	0.005***	0.005***	0.006***	0.006***	0.005***
	(0.001)	(0.001)	(0.001)	(0.001)	(0.001)
Female	0.214***	0.209***	0.210***	0.199***	0.209***
	(0.036)	(0.036)	(0.036)	(0.036)	(0.036)
Education level	0.005	0.005	0.009	0.011	0.004
	(0.020)	(0.020)	(0.020)	(0.020)	(0.020)
Ideology: liberal	0.058	0.065	0.071	0.070	0.066
	(0.059)	(0.058)	(0.059)	(0.059)	(0.059)
Ideology: conservative	−0.177***	−0.173***	−0.170***	−0.170***	−0.163***
	(0.053)	(0.053)	(0.054)	(0.053)	(0.053)
Ideology: DK	0.011	0.02	0.016	0.004	0.008
	(0.055)	(0.055)	(0.055)	(0.055)	(0.055)
Constant	−0.265**	−0.423***	−0.429***	−0.473***	−0.207
	(0.114)	(0.107)	(0.109)	(0.114)	(0.133)
Observations	1,764	1,764	1,763	1,767	1,768
Adjusted R-squared	0.057	0.056	0.052	0.052	0.052

Standard errors in parentheses
***$p < 0.01$, **$p < 0.05$, *$p < 0.1$
Source The SMPP Survey, 2023

Table 8.5 presents the results of the analysis of media viewing habits. The data indicated a correlation between environmental and news orientations in media consumption patterns. Individuals who did not exhibit a preference for news programs tended to demonstrate a lower environmental orientation (Table 8.5 [1]), whereas those who prioritized the news were more likely to display a high environmental orientation (Table 8.5 [2]). However, it is noteworthy that environmental orientation was not elevated among individuals who sought more comprehensive news coverage instead of headline news (Table 8.5 [3]).

Of particular interest is the relationship between the questions regarding entertainment and environmental orientation. Individuals who regularly watched television dramas and variety shows on television and the Internet demonstrated a higher environmental orientation (Table 8.5 [4]), whereas those who primarily used television and the Internet for entertainment purposes exhibited a lower environmental orientation (Table 8.5 [5]). The distinction between these two questions may be that the former potentially engaged with additional media while consuming entertainment programs on television and the Internet, while the latter used media exclusively for entertainment purposes. This observation aligned with the finding that drama viewers tend to be more environmentally oriented (Table 8.2 [4]), possibly due to their increased media consumption, which may result in greater exposure to news content and, consequently, a heightened environmental awareness.

When examining the control variables, age and environmental orientation demonstrated a correlation with all the above analyses, confirming the tendency in Japan for older generations to exhibit higher environmental orientations and younger generations to display lower environmental orientations. As previously noted, this trend aligns with the observations made by Kato (2021). Furthermore, females tended to demonstrate higher environmental orientations compared to males. However, educational attainment did not correlate with environmental orientation. This finding diverges from that of Kato (2021), who reported a positive correlation between education level and environmental consciousness.

Regarding ideological self-placement, conservative citizens exhibited lower environmental orientations compared to ideologically moderate citizens, while liberals showed a minimal difference from moderate citizens. In essence, public opinion demonstrated an overall moderate environmental orientation, with only conservatives appearing to prioritize their personal interests over environmental concerns. Nonetheless, this observation corroborates Endo and Tabei's (2024) assertion that environmental orientation correlates with ideology, albeit weakly.

8.5 Conclusion

Public support will facilitate drastic political changes such as the embrace of carbon-neutral policies. In Japan, where environmental consciousness is reportedly declining (Kato, 2021), it is crucial to identify individuals that exhibit both higher and lower levels of environmental orientation. It has been noted that environmental orientation has emerged as a significant axis of policy divergence among voters and is now associated with partisan conflicts (Endo & Tabei, 2024). This chapter examined the relationship between media consumption patterns and environmental orientation using data from the 2023 SMPP Survey, as individuals are presumed to obtain information from media sources when forming their opinions on global environmental issues.

This chapter demonstrated that individuals who obtained information primarily from the Internet exhibited a lower degree of environmental orientation compared to those who relied on the traditional media sources of newspapers and television programs. Given the limited media coverage of environmental issues and the high level of environmental orientation among traditional media consumers, the low level of environmental orientation among web-based media consumers can be attributed to insufficient exposure to environmental information. The observation that individuals who exclusively consumed entertainment-related media exhibited lower environmental orientations, while those who engage in the consumption of diverse media (including entertainment) demonstrated higher environmental orientations, further corroborates the above inferences. Moreover, it is noteworthy that the results of this analysis remained consistent even when controlling for age. Although it is evident that the patterns of media engagement varied with age, age alone did not account for the findings of this analysis.

The findings of this study suggest that media exposure, rather than selective media choice based on pre-existing environmental orientation, influences environmental consciousness among Japanese citizens. This causal interpretation is supported by several key observations. First, the positive correlation between entertainment media consumption (such as drama viewing) and environmental orientation indicates that incidental exposure to information through general media consumption, rather than selective exposure to environmental content, enhances environmental consciousness. Second, the consistent pattern across various types of traditional media exposure demonstrates a broader media effect rather than selective media choice by environmentally conscious individuals.

To enhance the environmental orientations of voters and guide them toward achieving a carbon-neutral society, two key interventions are necessary. First, given the decline in Japanese media coverage of climate change and the predominantly factual nature of reporting (Saito, 2017), traditional media outlets need to increase their coverage and emphasize a cautionary investigative style of reporting, considering the significance of climate change, to implement assertive policies toward carbon neutrality. Merely calling upon the media to do this is insufficient; it is imperative to consider developing incentives for news organizations to promote such reporting.

Second, since voters—particularly those who engage with online media—prioritize information aligned with their interests, it can be inferred that they are less informed about climate change in contexts in which the volume of climate change-related information is already limited. Given that a significant proportion of the information in online media comprises news content produced by traditional media outlets, it is imperative to increase the quantity of the coverage on climate change itself. Moreover, it is also essential to establish a media system that facilitates exposure to incidental information, thereby increasing the probability of exposure among individuals who exclusively obtain information through online channels.

References

Dalton, R. J. (2006). Social modernization and the end of ideology debate: Patterns of ideological polarization. *Japanese Journal of Political Science, 7*(1), 1–22.

Diekmann, A., & Franzen, A. (1999). The wealth of nations and environmental concern. *Environment and Behavior, 31*(4), 540–549.

Dunlap, R. E., & Mertig, A. G. (1995). Global concern for the environment: Is affluence a prerequisite? *Journal of Social Issues, 51*(4), 121–137.

Endo, M., & Tabei, K. (2024). *Seiji tairitsu ha nihon shakai no tairitsu wo kitei shiteiruka: Ideorogii ni yoru bundan* (Does political conflict define conflict in Japanese society: Division by ideology). In K. Ikeda, Y. Maeda, & T. Yamawaki (Eds.), *Nihon no bundan ha dokoni arunoka: Sumato nyusu media kachikan chosa kara kensho suru* (Where is Japan divided?: An examination from the SmartNews Media, Politics, and Public Opinion Survey) (pp. 111–138). Keiso Shobo.

Inglehart, R. (1977). *The silent revolution: Changing values and political styles among Western publics.* Princeton University Press.

Inglehart, R. (1990). *Culture shift in advanced industrial society.* Princeton University Press.

Inglehart, R. (1995). Public support for environmental protection: Objective problems and subjective values in 43 societies. *PS: Political Science & Politics, 28*(1), 57–72.

Jou, W., & Endo, M. (2016). *Generational gap in Japanese politics: A longitudinal study of political attitudes and behaviour.* Palgrave Macmillan.

Kabashima, I., & Takenaka, Y. (2012). *Ideorogii (Ideology).* Tokyo Daigaku Shuppankai.

Kato, N. (2021). Nihonjin no kankyo ishiki: 'Nihonjin no kokuminsei chosa' niyoru sono hensen to kanren yoin no kosatsu (Environmental Awareness of Japanese Citizens: Quantitative analysis of chronological changes and related factors performed: Using the Japanese National Character Survey). *Tokei Suri (Proceedings of the Institute of Statistical Mathematics), 69*(2), 295–314.

Kitschelt, H. (1995). *The radical right in Western Europe: A comparative analysis.* University of Michigan Press.

Pew Research Center. (2017). *Public divides over environmental regulation and energy policy.* Retrieved January 4, 2025, from https://www.pewresearch.org/social-trends/2017/05/16/public-divides-over-environmental-regulation-and-energy-policy/

Saito, S. (2017). Kikohendo hodoni kansuru nichibeiei hikaku: IPCC dai 4 ji dai 5 ji hyoka hokokusho ni shoten wo atete (Comparison of Climate Change Newspaper Articles in Japan, the United States, Britain: Focusing on IPCC Fourth Assessment Report and Fifth Assessment Report). *Kankyo Jyoho Kagaku Gakujyutu Kenkyu Ronbunshu (Papers on Environmental Information Science), 31*, 25–28.

New Sustainability Disclosure Requirements for Listed Companies in Japan

9

Izumi Kawashima

Abstract

After the amendment of the Cabinet Office Ordinance on the Disclosure of Corporate Affairs in Japan, "sustainability-related policy and initiatives" have been added to the annual securities report as a disclosure item. Under this item, the disclosure of "governance," "strategy," "risk management," and "metrics and targets" is required. These are widely used both domestically and internationally as core elements in climate-related financial disclosures based on TCFD Recommendations. This chapter first reviews the content and characteristics of this amended Ordinance, focusing on climate-related disclosure, and then explains the method of description, treatment of forward-looking information, criteria for materiality, and the materiality adopted by the amended Ordinance. Trends in sustainability disclosure standards have also been discussed. After confirming the status of the amended Ordinance in comparison with the UK Companies Act and its guidance, the chapter concludes with a brief outlook on the future. The statutory requirement to disclose listed companies' sustainability-related information is not simply a demand to disclose the current status of the companies. However, it is positioned to encourage them to generate business profits and avoid losses by creating a sustainable business environment for them. This is expected to contribute to the realization of a carbon-neutral society.

Keywords

Disclosure · Sustainability · TCFD · GHG · Scope 3

9.1 Introduction

This chapter concerns the statutory regulations for the disclosure of sustainability-related information, including climate change, by listed companies in Japan. These regulations are not simply a demand to disclose the current situation of a company regarding these matters but to encourage listed companies to generate business profits and avoid losses in their business activities by creating a sustainable business environment, including climate change. This is expected to contribute to the realization of a carbon-neutral society by generating profits while avoiding losses in business activities.

I. Kawashima (✉)
Waseda University, Tokyo, Japan
e-mail: kizumi@waseda.jp

K. Akao and A. Washizu (eds.), *Climate Change Issues and Social Sciences*,
Sustainable Development Goals Series, https://doi.org/10.1007/978-981-96-8951-4_9

On January 31, 2023, the Cabinet Office Ordinance on the Disclosure of Corporate Affairs (referred to as the "Disclosure Ordinance") was partially amended and came into effect on the same day. As a result of this amendment, additional disclosure requirements were added to the annual securities report, which must be submitted to the Prime Minister under the Financial Instruments and Exchange Act. Specifically, a new subsection titled "Sustainability-related policy and initiatives" (referred to as the "Sustainability statement subsection") has been added to "Section 2: Overview of Business" of "Part 1: Corporate Information." In addition, the Financial Services Agency's website provides, as guidance on the Sustainability statement subsection, "Principles for Disclosure of Descriptive Information (Attachment) - Disclosure of Sustainability Information" (referred to as the "Attachment") (FSA, 2022), and at the end of 2023, the Financial Services Authority published on this website "Good Examples of Disclosure of Descriptive Information 2023 (Disclosure of Sustainability-related policy and initiatives) (December 27, 2023)" (FSA, 2023).

This amended Disclosure Ordinance has been applied to annual securities reports for the fiscal year ending March 31, 2023, and all listed companies in Japan have already submitted and disclosed their annual securities reports for the first time since the application. This amendment also enhances the content of disclosures in the "Status of Corporate Governance, etc." subsection and the "Status of Employees, etc." subsection.

Incidentally, the legalization of climate-related financial disclosure was implemented in the United Kingdom in early 2022. In March 2024, the Securities and Exchange Commission in the United States adopted rules requiring the disclosure of greenhouse gases (SEC, 2024).

This chapter first introduces the contents of the amended Disclosure Ordinance in Japan as a disclosure of climate change-related information by listed companies (Sect. 9.2), shows how to describe sustainability information and the relation between forward-looking information and fraudulent statements (Sect. 9.3). It then explains the criteria for the materiality of sustainability information disclosure (Sect. 9.4) and the trends in the sustainability disclosure standards and these setting bodies (Sect. 9.5). In Sect. 9.6, the author addresses some basic issues with this amendment to the Disclosure Ordinance in comparison to the UK legislative method. The following reviews the state of sustainability disclosure legislation in Japan and the issues it faces, considering some of the actual requirements and conditions of the Disclosure Ordinance.

9.2 The Sustainability Statement Subsection and Its Regulatory Features

9.2.1 Description of the Sustainability Statement Subsection

In the "Sustainability-related policy and initiatives" subsection, which was added to "Section 2: Overview of Business" of "Part 1: Corporate Information" in the annual securities report, listed companies are required by the Disclosure Ordinance to provide the information on each company's approach to sustainability and the status of its initiatives based on four components consistent with international disclosure frameworks: "governance," "strategy," "risk management," and "metrics and targets."

Regarding the specific content to be described, from the viewpoint that listed companies need a framework to recognize sustainability information[1] and judge its importance based on their own business conditions, business environment and its impact on corporate value, "governance," and "risk management" should be disclosed by all listed companies. It is desirable to disclose "strategy" and "metrics and targets." However, each listed company is required to disclose them based on its judgment of their importance (materiality) through the framework of "governance" and

[1] Sustainability information could include, for example, matters related to the environment, society, employees, respect for human rights, anti-corruption, anti-bribery, governance, cybersecurity, and data security (Note 1 in the Attachment).

"risk management." For example, in response to climate change, each listed company is required to decide whether to disclose by judging the materiality from the perspective of investors' investment decisions through the framework of "governance" and "risk management." The same is true for other sustainability items, as explained by the person in charge of the FSA (Agari et al., 2023).

Notably, "importance," or in another term "materiality," here does not refer to the importance of the issue as a social issue. However, it should be judged based on whether it is important for investors' investment decisions. According to the explanation of the person in charge of the FSA, it is desirable to judge based on the impact the matter has on corporate value and performance (Agari et al., 2023). Whether "strategy" and "metrics and targets" should be disclosed depends on the importance of the relevant sustainability information to the listed company's corporate value and performance "from the perspective of investors' investment decisions," and the importance of the relationship with corporate performance is emphasized.

However, regarding "human capital and diversity," also included in the "Sustainability statement subsection," regardless of the judgment of importance, listed companies are required to describe their policies on human resource development, including ensuring diversity of human resource and policies on the internal environment in "strategy." The indicators related to these policies and the targets and achievements using these indicators are required to be described in "metrics and targets."

Form 3 (Annual Securities Report) attached to the amended Disclosure Ordinance includes the following notes: "governance" refers to "the governance processes, controls and procedures for monitoring and managing sustainability-related risks and opportunities," and "risk management" refers to "the processes for identifying, evaluating and managing sustainability-related risks and opportunities" (Note 10-2, described in accordance with Note 30-2 in Form 2). Similarly, "strategy" refers to "the process for addressing sustainability-related risks and opportunities that may affect the management policies and strategies of the consolidated entity over the short, medium and long term," and "metrics and targets" refers to "information used to evaluate, manage, and monitor the performance of the consolidated entity over time with respect to sustainability-related risks and opportunities over time." Thus, business risk is considered as a counterpart to business opportunity.

9.2.2 Regulatory Characteristics of the Sustainability Statement Subsection

The amended Disclosure Ordinance is characterized by the fact that it does not prescribe detailed items to be included in the sustainability statement subsection but rather provides a framework that allows for flexible disclosure according to the current efforts of each listed company. Among the four components, "strategy" and "metrics and targets" are treated as a matter for each company to decide whether to disclose or not based on importance (materiality). In the Attachment, it is stated that "strategy" and "metrics and targets" should be disclosed by each company based on its judgment of importance through the framework of "governance" and "risk management" and, in addition, as an effort toward desirable disclosure, even if a company decides not to include "strategy" and "metrics and targets," it is expected to disclose the reason and the basis for that judgment.

Two months after the promulgation and enforcement of the amended Disclosure Ordinance, the application of the amended Ordinance began. Therefore, it seems that the amended Disclosure Ordinance adopted a framework in which listed companies can flexibly adapt to and gradually enhance their disclosure content consistent with the progress of each company's sustainability efforts. However, even before sustainability-related information disclosure became legally obligatory, many listed companies had already begun to disclose climate-related information in their integrated reports, sustainability reports, or other voluntary disclosure documents based on the recommendations of the Task Force on Climate-related Financial Disclosures (TCFD). Such companies are expected to actively disclose "strategy" and "metrics and targets" in the newly established Sustainability statement subsection.

The Japanese Corporate Governance Code (JPCG Code), which was revised in 2021, states that "The board of directors should consider the following issues in its consideration of sustainability, such as global environmental issues like climate change, respect for human rights, the health and working environment of employees, fair and appropriate treatment of employees, and fair and appropriate transactions with business partners, and risk management in response to natural disasters and other disasters. The board of directors should recognize that addressing the sustainability issues mentioned above is an important management issue that reduces risks and leads to profit-earning opportunities" (Supporting Principle 2-3①), "…in particular, companies listed on the Prime Market of the Tokyo Stock Exchange should collect and analyze necessary data on the impact of climate change-related risks and profit opportunities on their business activities, earnings, etc., and …should consider and enhance the quality and quantity of disclosure based on the TCFD or equivalent framework" (Supporting Principle 3-1③), "The board of directors should formulate a basic policy on the company's sustainability initiatives from the perspective of improving corporate value over the medium to long term" (Supporting Principle 4-2②).

Although the JPCG Code is a soft law that adopts the principle-based approach and the method of "comply or explain," it is assumed that the companies listed on the Prime Market of the Tokyo Stock Exchange, regardless of their business format or management environment, have already formulated a basic policy on their sustainability initiatives, considered the positive and negative impacts of climate change on their business activities and profit opportunities. They have already collected and analyzed relevant data before the amendment of the Disclosure Ordinance this time.[2]

9.3 Sustainability Information and Forward-Looking Information in Relation to Misrepresentation

9.3.1 Method of Description

The "Sustainability-related policy and initiatives" subsection is a flexible framework that allows each listed company to describe its initiative status. According to the FSA's views on the comments, it is possible to define the four components (governance, strategy, risk management, and metrics and targets) together without separate itemization of each (Agari et al., 2023).[3] However, it is also useful for investors to understand which of the four components is clearly described so that they can easily understand the information.

In addition, before the amendment of the Disclosure Ordinance, some listed companies had already disclosed sustainability information in voluntary disclosure reports such as integrated or sustainability reports. Therefore, it is permissible for listed companies to include the sustainability information required in their annual securities reports and then refer to other public documents for detailed information that supplements sustainability information in their annual securities reports.

[2] According to the Financial System Council Disclosure Working Group Report released on June 13, 2022, as for Scope 1 (direct greenhouse gas (GHG) emissions by businesses themselves) and Scope 2 (indirect GHG emissions from the use of electricity, heat, and steam supplied by other companies), listed companies are expected to actively disclose their GHG emissions.

[3] Financial Services Agency, "Summary of Public Comments on the Draft Cabinet Office Ordinance Partially Amending the Cabinet Office Ordinance on Disclosure of Corporate Information, etc. and the FSA's Views on the Comments", Nos. 83 to 87.

The problem in such cases is the occurrence of false descriptions in the referenced public documents. The Financial Services Agency clearly states in its Guidelines for the Disclosure of Corporate Information, etc., that a company is not immediately liable for a misstatement merely because of a misstatement in a reference document, except in cases where a company refers to a document knowing that the referenced document contains an obviously false or misleading representation or where the mere fact that a company relates to a document could itself constitute a material misstatement in its annual securities report (Financial Services Agency, Guideline of Corporate Details Disclosure 5-16-4). It is also assumed that websites can be used as a referenced document (Financial Services Agency, "Summary of Comments and Financial Services Agency's View on Comments" No. 257-No.261).

Regarding voluntary disclosure documents, one concern is that the timing of disclosure of the integrated report or the sustainability report, to which reference is made, is different from that of the annual securities report in conventional practice. For example, for a company whose fiscal year ends in March, the annual securities report is generally submitted after the general shareholders' meeting held in June. However, the publication of the integrated report or the sustainability report often occurs three months to half a year after the submission of the annual securities report. Aligning the publication timing of these documents and enhancing the disclosure of annual securities reports are issues to be addressed imminently.

9.3.2 Forward-Looking Information and Misrepresentation

Sustainability information is about the medium- to long-term sustainability of a company and includes forward-looking information. Not limited to sustainability information, the descriptive information in annual securities reports may contain forward-looking information. However, forward-looking information is not necessarily a definitive matter, and there is no guarantee that the information will actually be realized as described in the forward-looking information. Thus, the relationship between forward-looking information and false statements remains unclear.

On this point, the Financial Services Agency published its view during the 2019 amendment of the Disclosure Office Ordinance in "Summary of Public Comments on the 'Draft Cabinet Office Ordinance Partially Revising the Cabinet Office Ordinance on Disclosure of Corporate Information, etc.' and the Financial Services Agency's Position on the Comments" (January 31, 2019, No. 16), and clarified its position, on the relationship between forward-looking information and misstatements, that "if a specific explanation has been provided within a range generally considered reasonable, the responsibility for false statements will not be questioned simply because circumstances have changed after submission." In the "Financial System Council Disclosure Working Group Report" (June 13, 2022), this view of the Financial Services Agency was quoted, and the next point was emphasized. From the perspective of providing useful information for investors' investment decisions in sustainability disclosure, the disclosure practices should not shrink due to fear of being held liable for false statements if circumstances change after the fact. Therefore, the Financial Services Agency strives to spread this approach to practice. The Working Group Report stated that further clarification of this view in practice and the concept should be considered in the Guidelines for Disclosure of Corporate Information by assuming examples of sustainability disclosure.

However, management may be held liable for misstatement if it is aware that the information is important future information that could affect investors' investment decisions on the date of submission but fails to state it or if unaware that the information is important future information without a reasonable basis and fails to state it, even though it is an important matter that should be noted. In such cases, awareness of the possibility of being held liable for a misstatement is necessary. From these and other legal perspectives, each company must have a system to ensure that forward-looking information is appropriately disclosed.

9.4 Criteria for Materiality and Importance

9.4.1 Materiality of Sustainability Information Disclosure

Prior to the amendment of the Disclosure Ordinance at the end of January 2023, sustainability information, including climate change, was subject to voluntary disclosure in Japan and not legally required. However, as mentioned above, the JPCG Code revised in 2021 requires the companies listed on the Prime Market of the Tokyo Stock Exchange to collect and analyze necessary data on the impact of climate change-related risks and opportunities on their business activities and earnings and to enhance the quality and quantity of disclosure based on the TCFD or an equivalent framework. Therefore, many companies listed on the Prime Market disclose sustainability information in their integrated or sustainability reports, which are voluntary disclosure documents.

In disclosing sustainability information, the evaluation criterion of "materiality" has been emphasized in practice. This word is generally written in katakana and translated into Japanese as "importance" or "important issue." Sustainability information should be disclosed based on the materiality criteria. However, materiality is multifaceted and there is no common understanding.

Often, there are two ways of looking at this criterion: single materiality and double materiality (Oh-ebashi LPC & Partners, 2024). Single materiality refers to the impact of sustainability issues on corporate value. An example of adopting a single materiality is the TCFD recommendations published in 2017 and revised in 2021. TCFD's name reflects that it is a task force on climate change-related "financial" disclosure. Financial information represents the financial value of the company in question. It indicates that the task force is concerned with disclosing material information that affects the corporate value of the company. In addition, according to the TCFD Recommendations, disclosure of climate-related financial information should be based on (1) "governance" (the organization's governance of climate-related risks and opportunities), (2) "strategy" (the actual and potential impact of climate-related risks and opportunities on the organization's business, strategy and financial plans), (3) "risk management" (the organization's way to identify, assess, and properly manage the climate-related risks and opportunities), and (4) "metrics and targets" (the metrics used to assess progress against targets, and the description of the targets used to manage climate-related risks, and to realize climate-related opportunities).

These four elements in the TCFD Recommendations were subsequently adopted by the International Sustainability Standards Board (ISSB), established by the IFRS Foundation and integrated into its International Disclosure Standards. The objective of the ISSB is to help investors, lenders, and creditors, the primary users of general-purpose financial reporting, decide whether to provide resources to an entity (Matsumoto, 2024). In the United Kingdom, the Companies Act was amended in early 2022 to insert new provisions regarding the disclosure of corporate information consistent with the TCFD Recommendations, and this amendment adopts the single materiality approach. The non-binding guidance issued by the Department for Business, Energy, and Industrial Strategy (BEIS), the government agency in charge of this amendment, explains that this legislation is intended to make TCFD-aligned climate-related financial disclosures mandatory (BEIS, 2022).

However, double materiality refers to the importance of the impact of corporate activities on sustainability issues as a criterion for judgment. Some explanations are as follows: Double materiality determines whether sustainability-related information is "material" and requires disclosure, which includes financial materiality (whether sustainability has a significant financial impact on the company) and impact materiality (whether the company significantly impacts people and the environment) (Matsumoto, 2024).

The European Union (EU) has required the disclosure of non-financial information, including sustainability information, through the Non-financial Reporting Directive of 2014. The Corporate Sustainability Reporting Directive (CSRD), which was implemented in January 2023, has extended the scope of the entity applied and enhanced the contents of required disclosure information, according to the European Sustainability Reporting Standards (ESRS). The ESRS sets disclosure standards for each environmental (E), social (S), and governance (G) areas in addition to general disclosure standards. Among these, ESRS E1 has established disclosure standards related to climate change. CSRD's adoption of double-maternity characterizes this approach. When determining whether certain sustainability-related information is material and requires disclosure, financial materiality (that is, whether the matter has a significant financial impact on the company) and impact materiality (the impact of the company on people and the environment) are considered in the evaluation. Under this concept, if a corporate activity had a material impact on the environment, even if it does not affect the company's financial position, the activity is material and must, therefore, be disclosed. In the EU, there is strong advocacy for corporate governance that criticizes the shareholder-first principle that companies should be managed to benefit shareholders and integrates a sustainability perspective (Kansaku, 2022), and the concept of double materiality can be understood in this context (Matsumoto, 2024).

9.4.2 Criteria for Materiality of Sustainability Information in the Amended Disclosure Ordinance

As mentioned in Sect. 9.4.1, in Japan, it is conventionally known, at least among professionals involved in sustainability information disclosure practices and those in charge of such information disclosure at listed companies, that two types of materiality, single and double materiality, exist. What position do you think our Disclosure Ordinance, revised in 2023, will take with respect to how materiality is viewed?

First, from the viewpoint of single materiality, if sustainability information is material to investors' investment decisions in that it affects the financial condition of the company, the disclosure of sustainability information is required. Second, based on double materiality, if a company has a significant impact on the environment or society, even if the sustainability information does not affect the financial condition, the company should be required to disclose such information. In addition to these two positions, the third position requires disclosure if the sustainability information is important for its shareholders' investment decisions in that "I want to invest in a sustainable company" or "I do not want to invest in a company that does not consider sustainability," even if the information is not important for normal investment decisions.

The Financial Services Agency staff explained that, as mentioned in Sect. 9.2, each listed company determines the need for disclosure by assessing materiality from the perspective of investors' investment decisions through the "governance" and "risk management" frameworks and that the decision whether to disclose information is based on whether it is material to investors' investment decisions. Since the decision to disclose information is based on whether it is material to investors' investment decisions, it is considered from the standpoint of single materiality.

It has also been highlighted that adopting a double materiality approach to the Disclosure Ordinance is inconsistent with the purpose of the Financial Instruments and Exchange Act (Matsumoto, 2024), which is a statute whose main objective is to ensure fairness in the market and protect investors. It includes a provision that prescribes liability for damages to investors in the event of misstatements or omissions in annual securities reports (the Financial Instruments and Exchange Act, Article 18). These provisions are based on the premise that an annual securities report discloses information that is material for investors' investment decisions. Investors who are assumed to be material in their investment decisions intend to profit from their investments. Therefore, the Act assumes that investors' investment decisions are based on their knowledge of matters that affect a company's financial condition and require disclosure of such information when it is material to their investment decisions. This is not an investment decision from the perspective of investing in sustainable companies. While there may be various reasons why investors invest in a company, it would be difficult and impractical for a company to ascertain these reasons. Third, what companies should disclose as a matter of statutory disclosure would also vary from investor to investor. In this respect, it is incompatible with the Disclosure Ordinance under the Financial Instruments and Exchange Act.

Until now, many Japanese-listed companies have disclosed sustainability information in voluntary disclosure documents, such as consolidated or sustainability reports. In some cases, sustainability information is disclosed based on double materiality. According to the amended Disclosure Ordinance, the content required to be disclosed in annual securities reports is based on single materiality. This should be emphasized more. It is desirable that sustainable information, which is disclosed by listed companies based on the criterion of whether the company has a material impact on people and the environment, be disclosed continuously in voluntary disclosure documents.

9.5 Trends in Sustainability Disclosure Standards

9.5.1 Disclosure Standard of Sustainability Information

Supporting Principle 3-1③ of the JPCG Code refers to a disclosure framework as "...the quality and quantity of disclosures based on the TCFD or an equivalent framework." The TCFD is an organization established by the Financial Stability Board at the request of the G20 as the standard-setting body for climate-related financial information disclosure; it published its recommendations in 2017 and its revised version in 2021 (TCFD Recommendations). One key feature of the TCFD Recommendations is a strong focus on the risks and opportunities related to the transition to a lower-carbon economy.

Of particular interest in the TCFD Recommendations is the content related to greenhouse gas (GHG) emissions, which are high-profile components of climate-related financial information. Disclosure of GHG emission is included in the "metrics and targets," which covers Scope 1 (all direct emissions by business entities), Scope 2 (indirect emissions from the consumption of electricity, heat and steam purchased by business entities), and Scope 3 (other indirect emissions not covered by Scope 2). It includes emissions upstream (for example, suppliers) and downstream (for example, customers) of the business entity. Scopes 1 and 2 should be disclosed by all business entities regardless of materiality assessment, and Scope 3 should be disclosed according to the standard of materiality. However, all business entities should consider the disclosure of Scope 3.

The UK Companies Act, which implemented the disclosure of climate-related financial information in early 2022 ahead of Japan, includes provisions for the disclosure of sustainability information consistent with the TCFD Recommendations. The UK Companies Act also adopts the four core elements, "governance," "strategy," "risk management," and "KPIs (instead of metrics) and targets," the same as the Disclosure Ordinance, for the disclosure requirements originally recommended by the TCFD.

In addition to the TCFD, various standard-setting bodies, such as the International Integrated Reporting Council, have developed disclosure standards. Amid this situation, in which multiple standards and standard-setting bodies coexist, in November 2021, the ISSB[4] was established by the IFRS Foundation under its

[4] With the establishment of the ISSB, standard-setting bodies were converging as well, with some standard-setting

umbrella. The IFRS Foundation is an established international accounting standard-setting body.[5] In 2022, the ISSB published an exposure draft of the Sustainability Disclosure Standard for public comment. In June 2023, it published IFRS S1, "General Requirements for Disclosure of Sustainability-related Financial Information," and IFRS S2, "Climate-related Disclosures," as the IFRS Sustainability Disclosure Standard (Yasui, 2023b).[6]

IFRS S1 is a comprehensive standard that applies to sustainability information in general, is not limited to specific themes such as climate change, and is used whenever a business discloses sustainability information based on ISSB standards. Within S1, a content called Core Content defines standards to be followed when there is no standard called "thematic requirements" for specific themes, such as biodiversity. This core content is then described in a manner that follows the framework of the four elements of the TCFD Recommendations (governance, strategy, risk management, and metrics and targets). The IFRS S2 establishes a disclosure standard for climate-related information on sustainability-related topics. As for "metrics and targets," IFRS S2 requires to disclose the absolute total amount of GHG emissions, classified into Scope 1, Scope 2, and Scope 3.

The IFRS Sustainability Disclosure Standards are expected to be incorporated into domestic statutory disclosure rules in the near future. The Sustainability Standards Board of Japan (SSBJ) was established in July 2022 under the Financial Accounting Standards Foundation (FASF) to study and develop specific domestic disclosure standards and to contribute to the development of international sustainability disclosure standards. The SSBJ provided comments on the Exposure Draft of the Sustainability Disclosure Standards published in June 2023 by the ISSB, based on requests from Japanese investors and companies (Kawanishi, 2023). In addition, in response to the announcement of the IFRS sustainability disclosure standards in 2023, the SSBJ indicated that it planned to publish the Exposure Draft by the end of fiscal year 2023 (by March 31, 2024, at the latest), and to finalize the domestic disclosure standards by the end of FY 2024 (Yasui, 2023b), and published its exposure draft on March 29, 2024. The SSBJ also anticipates setting Japanese sustainability disclosure standards based on the IFRS Sustainability Disclosure Standards as a starting point, expected to be internationally consistent (Kawanishi, 2023).[7]

9.5.2 Standards of Audit, Assurance, Etc.

The International Auditing and Assurance Standards Board (IAASB), an international assurance standard-setting body, was discussing the development of third-party assurance to disclose sustainability information (Yasui et al., 2024), and approved International Standard on Sustainability Assurance 5000 (ISSA 5000) on September 20, 2024. With this milestone, IAASB will finalize the text of the standard for certification by the Public Interest Oversight Board, and expects to formally publish by the end of the year (IAASB, 2024). In the near future, third-party assurance will be required to ensure the reliability of the disclosed sustainability information including climate change, and other non-financial information.[8]

---

organizations joining and merging with the IFRS Foundation and the TCFD had ceased its activities (Kurosaki, 2024).

[5] Under the umbrella of the IFRS Foundation is the International Accounting Standards Board (IASB), which develops IFRS accounting standards.

[6] Currently, it seems to be understood that the only specific IFRS sustainability disclosure standard to be applied is IFRS S2, and that sustainability-related risks and opportunities other than climate-related risks are subject to the provisions of IFRS S1.

[7] On March 29, 2024, SSBJ published the Exposure Drafts of the Sustainability Disclosure Standards which include Universal Sustainability Disclosure Standard exposure draft, Theme-based Disclosure Standard exposure draft No.1 "General Disclosures," and Theme-based No.2 "Climate-related Disclosures."

[8] Given that this is an area requiring diverse expertise, it is also important to secure a broad range of assurance providers. In March 2024, the first meeting of the

The Securities Instruments and Exchange Act requires that an internal control report, which is a report on the management's assessment of the so-called internal control system, be submitted to the Prime Minister together with an annual securities report. The internal control report must be certified by a certified public accountant or an audit firm. The items stated in the internal control report include the names of the standards used to establish and operate internal controls, which are the "Standards for Assessment and Audit of Internal Control over Financial Reporting and Implementation Standards for Assessment and Audit of Internal Control over Financial Reporting" set forth by the Business Accounting Council. In 2023, the Business Accounting Council revised the "Standards and Practice Standards for Management Assessment and Audit" for the first time in 15 years and the corresponding Cabinet Office Ordinance (Internal Control Ordinance), which is effective for the assessment and audit of internal controls for fiscal years beginning on or after April 2024.

There are various amendments, but one of the most important is that "reliability of financial reporting," which was previously considered one of the objectives of internal control, has been amended to "reliability of reporting," and "reliability of reporting means to ensure the reliability of reporting (including non-financial information) within and outside the organization." The background of this amendment is that the 2013 revision of the Committee of Sponsoring Organizations of the Tredway Commission (COSO) report on the basic framework for internal control expanded "financial reporting" to "reporting (including non-financial reporting and internal reporting)." It is explained that the revision was based on the expansion of the term "financial reporting" to "reporting (including non-financial reporting and internal reporting)" and the recent disclosures about sustainability and other non-financial information (Saitō et al., 2023). However, the Business Accounting Council also states that the medium- to long-term issue to be addressed is the treatment of non-financial information, such as sustainability, in internal control reports.

9.6 Status of Statutory Disclosures on Sustainability

Finally, we address some basic problems with the amended Disclosure Ordinances. The first problem is that the purpose and content of the required disclosures are unclear. Even Note 10-2 in Form 3 (Note 30-2 in Form 2) are unclear about what "materiality," the criterion for whether to include "strategy" and "metrics and targets," means in terms of importance. Considering the Financial Services Agency's view of the public comments on the draft amendment to the Disclosure Ordinance, the intention of the Financial Services Agency is finally clear, that is, its importance from the perspective of investors' investment decisions. The Note 10-2 in Form 3 that "governance" refers to "…the process, control and procedures of governance to monitor and manage" is also not easy to understand. When "strategy" and "metrics and targets" are not disclosed, it would be difficult to know that the disclosure of this fact and the rationale for it is expected without looking at the "Attachment."

The UK Companies Act, amended in 2022, introduced a statutory disclosure regime with a TCFD-aligned non-financial and sustainability information statement. For example, for the disclosure item corresponding "governance," "a description of the company's governance arrangements in relation to assessing and managing climate-related risks and opportunities" (Companies Act 2006, sec. 414CB, subsection 2A(a)), the non-binding guidance of the BEIS states that for subsection 2A(a), "The information should enable a user of the accounts to understand which person or committee has the responsibility for identifying and considering climate-related risks and opportunities, including how frequently those matters are considered," and "The information should also enable a user of the accounts to understand the extent to which information relating the climate-related risks and opportunities is considered by the

---

Financial System Council's Working Group on Sustainability Information Disclosure and Assurance was held, and discussions are ongoing.

Board," are required to be stated, and "If no directors have oversight of climate-related risks and opportunities and /or no person or persons within the company have responsibility for assessing or managing climate-related risks and opportunities, then this should be stated" and so on (BEIS, 2002, pp. 10–11). By explaining the details specifically, we strive to enrich the descriptive content. Also, under the UK Companies Act, items equivalent to "strategy" and "metrics and targets" are allowed to be omitted in whole or in part, when they are not reasonably believed to be necessary for understanding the company's business. However, in such cases, the UK Companies Act requires a clear and reasonable explanation of the reasons for omissions. It is also interesting that explanations of relevant descriptions and other matters are summarized in the non-binding guidance of the BEIS.

Note 2 in the Attachment refers to the description of GHG emissions in the Report of the Disclosure Working Group of the Financial System Council, published on June 13, 2023. That is, considering the fact that GHG emissions are an effective indicator that contributes to constructive dialogue between investors and companies, companies are expected to actively disclose their Scope 1 (direct emission by the company itself), Scope 2 (indirect emission from the use of electricity, heat, and steam supplied by other companies) GHG emission in particular, based on a judgment of materiality, given that the business conditions and management environment of each company, and Scope 3 which refers to indirect emissions (emission from other companies related to the company's activities) other than Scope 1 and Scope 2.

According to a newspaper (the Nihon Keizai Shinbun) report dated August 22, 2024, based on the annual securities reports of 3553 listed companies for FY 2023 (April 2023 to March 2024), 802 companies disclosed Scope 1, and 807 companies disclosed Scope 2 GHG emissions from energy used at their base (both 23% of the total companies). Scope 3 was disclosed by 157 companies (4%). Only 83 companies (2%) received assurance or verification from a third-party certification body to ensure the reliability of their emissions data. The Financial Services Agency is currently working on rules for sustainability information disclosure, including GHG emissions disclosure, and companies listed on the Prime Market of the Tokyo Stock Exchange with a market capitalization of three trillion yen or more will be required to disclose up to Scope 3 by the fiscal year ending March 2027. The number of companies subject to this disclosure requirement is expected to increase stepwise.

9.7 Concluding Remarks

Looking at the annual securities reports after the application of the amended Disclosure Ordinance, there are noticeable examples where "strategy" and "metrics and targets" for climate change are not described, and in these cases the reason for non-description is not mentioned. As a general trend, climate-related disclosure is buried in other disclosure themes such as human capital, diversity, and the gender wage gap.

In addition, it seems to be related to this first problem; it has been pointed out that in many Japanese-listed companies, there are still many cases where the sustainability initiatives department is the traditional CSR department; thus, sustainability-related initiatives are superficially implemented in a manner inconsistent with business strategies (Yasui, 2023a). Therefore, the BEIS guidance approach should be used as a reference. Although it may vary depending on the type of industry and the category of business, how to incorporate sustainability initiatives into business strategies and connect them to profit opportunities and risk reduction—in other words, the creation and improvement of corporate value—is becoming increasingly important in business activities and is also of high interest to investors. Companies that disclose sustainability information and those in charge of disclosure legislation need to be fully aware of this fact.[9]

[9] The Financial Services Agency has also made efforts to further enhance disclosures and raise the level of disclosure content by publishing "Good Practices for Disclosure of Descriptive Information 2023" (disclosure of sustainability-related concepts and initiatives), but it seems

In addition, there are some important issues. In Japan, listed and some other companies must disclose sustainability-related information, including information on climate change, in their annual securities reports. Most unlisted companies, even large ones, are not subject to the Financial Instruments and Exchange Act and are not required to disclose sustainability-related information. However, the UK Companies Act adds large private companies and limited liability partnerships to the list of the companies needed to make mandatory climate-related financial disclosures, along with listed companies and other quoted companies.[10] The Corporate Sustainability Reporting Directive of the EU also expands the scope of its application to large socially influential business entities, in addition to listed companies. In Japan, when the climate-related disclosure requirements for listed companies are expanded to Scope 3, listed companies obliged to disclose their GHG emissions will need to know the GHG emissions of upstream companies (their suppliers) and downstream companies (their retail dealers, distributors, and so on). The need for large unlisted companies to disclose sustainability-related information, including climate-related information, remains an important issue for the future.

It is expected that by addressing these individual issues, the efforts of Japanese-listed companies toward climate change will be reflected in their climate change-related financial disclosures. As climate change-related financial disclosures are enhanced, the entire corporate society will be involved, moving toward the realization of a carbon-neutral society.

that the basic message is lacking or not being communicated enough. On March 29, 2024, the Financial Services Agency released "Matters to keep in mind when Preparing and Submitting Annual Securities Reports (including a collection of sample disclosures that can be used as a reference when addressing issues related to sustainability disclosure) and regarding the Implementation of Annual Securities Report Reviews (2024)."

[10] The climate-related financial disclosure requirements under the UK Companies Act apply to unlisted companies and large LLPs which have more than 500 employees and a turnover of more than 500 million pounds in addition to relevant Public Interest Entities.

References

Agari, S., Kasai, W., & Morioka, K. (2023). Kigyōnaiyōtō no kaizinikansuru naikakuhurei no kaisei (Amendment of the Cabinet Office Ordinance on Disclosure of Corporate Affairs, etc.). *Shōzihōmu (Commercial Law Review), 2320*, 4–15.

Department for Business, Energy & Industrial Strategy (BEIS). (2022). *Mandatory climate-related financial disclosures by publicly quoted companies, large private companies and LLPs/ Non-binding guidance.* Retrieved August 23, 2024, from https://www.gov.uk/government/publications/climate-related-financial-disclosures-for-companies-and-limited-liability-partnerships-llps

Financial Services Agency (FSA). (2022). *Kijutsujōhō no kaizi ni kansuru gensoku (Betten)* (Principles for Disclosure of Descriptive Information (Attachment)). Retrieved June 20, 2024, from https://www.fsa.go.jp/news/r4/sonota/20221107/05.pdf

Financial Services Agency (FSA). (2023). *Kijutsujōhō no kaizi ni kansuru kōjireisyū 2023* (Good Examples of Disclosure of Descriptive Information 2023 (Disclosure of Sustainability-related policy and initiatives)). Retrieved June 20, 2024, from https://www.fsa.go.jp/news/r5/singi/20231227.html

International Auditing and Assurance Standards Board (IAASB). (2024). *IAASB approves ISSA 5000.* Retrieved October 4, 2024, from https://www.iaasb.org/focus-areas/understanding-international-standard-sustainability-assurance-5000

Kansaku, H. (2022). Sasutenabiritii gabanansu o meguru dōkō (Trends in Sustainability Governance). *Shōzihōmu (Commercial Law Review), 2296*, 4–16.

Kawanishi, Y. (2023). IFRS sasutenabiritii kaizikijun no gaiyō to SSBJ no kongo no keikaku (Overview of IFRS Sustainability Disclosure Standards and SSBJ's Future Plans). *Shōzihōmu (Commercial Law Review), 2336*, 16–22.

Kurosaki, S. (2024). Sasutenabiritii-hizaimuhōkoku to hoshō (Sustainability, nonfinancial reporting and assurance). *Kansayaku (Auditor), 759*, 28–37.

Matsumoto, N. (2024). Sasutenabiritii jōhōkaizi o meguru mondai (Issues concerning the sustainability information disclosure). *Jurist, 1598*, 37–42.

Oh-ebashi LPC & Partners. (2024). *Zero kara wakaru ESG sasutenabiritii hōmu Q&A* (Understanding ESG & Sustainability Legal Affairs Q&A). Tokyo: Kin'yū zaiseizijō kenkyūkai.

Saitō, T., Ozaku, K., Matsui, T., Katou, J., & Seo, Y. (2023). Zaimuhōkoku nikakaru naibutōsei no hyōka to kansa nitsuiteno kizjuntō no kaitei nitsuite (Revision of criteria for evaluation and audit of internal control over financial reporting). *Shōzihōmu (Commercial Law Review), 2329*, 14–29.

Securities and Exchange Commission (SEC). (2024). *Immediate Release 2024–31.* Retrieved June 25, 2024, from https://www.sec.gov/newsroom/pres-release/2024-31

Yasui, K. (2023a). Kaiseikaizihureitō o humaeta kigyō niokeru jitumutaiō (Practical response in companies on the amended Disclosure Ordinance). *Shōzihōmu (Commercial Law Review), 2324*, 26–37.

Yasui, K. (2023b). Sihonsijō to mukiau kigyō jōhōkaiji (Corporate disclosure in the face of capital markets). *Shōzihōmu (Commercial Law Review), 2341*, 30–35.

Yasui, K., Katō, Y., & Minatogawa, T. (2024). Sasutenabilitii jōhōkaizi to hosyō o meguru kokusaidōkō (International trends in sustainability disclosure and assurance). *Shōzihōmu (Commercial Law Review), 2360*, 39–44.

10 Sustainability Management in Multinational Companies: Integrating Varied Development Challenges

Shinji Hasegawa

Abstract

This chapter explores the role of multinational companies (MNCs) in advancing the Sustainable Development Goals (SDGs) through localized strategies and innovative sustainability management. The SDGs, a global framework with 17 goals and 169 targets, require addressing the complex interconnections among environmental, social, and economic dimensions. MNCs that leverage their resources, global reach, and technological expertise are key actors in minimizing trade-offs and maximizing synergies among these goals. This chapter emphasizes the importance of circular economy models and region-specific approaches, such as the French concept of "*terroir*," which integrates natural, cultural, and social resources to create sustainable solutions. This highlights how MNCs can adapt to local contexts, mitigate the "liability of outsidership," and foster collaboration with regional stakeholders. By aligning global strategies with local needs, MNCs can contribute to regional revitalization and broader sustainability goals. Despite challenges such as balancing global and local priorities and addressing negative spillovers, this chapter argues that MNCs have a unique opportunity to link local initiatives with global frameworks. By developing scalable and sustainable models, MNCs can play a transformative role in achieving the SDGs and significantly impact sustainable development worldwide.

Keywords

Multinational companies · Circular economy · Terroir · International spillovers · Trade-offs

10.1 Introduction

This chapter examines how multinational companies (MNCs) can advance sustainable development by adopting localized strategies and circular economic models. Leveraging regional resources and managing the complex interactions among the Sustainable Development Goals (SDGs), MNCs can minimize trade-offs, foster synergies, and drive sustainable value creation. By integrating global strategies with region-specific approaches, MNCs can make meaningful progress toward achieving the SDGs, including the transition to a carbon-neutral society.

S. Hasegawa (✉)
Waseda University, Tokyo, Japan
e-mail: hasegawa@waseda.jp

K. Akao and A. Washizu (eds.), *Climate Change Issues and Social Sciences*,
Sustainable Development Goals Series, https://doi.org/10.1007/978-981-96-8951-4_10

The SDGs adopted by the United Nations in 2015 provide a comprehensive global framework for addressing humanity's most pressing and urgent challenges (UN, 2015). Comprising 17 goals and 169 targets, the SDGs encompass all dimensions of sustainability—environmental, economic, and social—while calling for the cooperation of governments, businesses, NGOs, civil society, and individuals under the guiding principle of "leaving no one behind." These goals include eradicating poverty and hunger, promoting education, achieving gender equality, combating climate change, and promoting sustainable consumption and production.

Among the stakeholders, MNCs are expected to play a key role in achieving SDGs (Burritt et al., 2018). With their ability to leverage innovation, capital, market access, and expertise, MNCs are well-positioned to deliver sustainable solutions. They bear the responsibility of addressing not only global challenges but also those unique to specific regions.

However, these challenges are intricately interwoven, and addressing individual SDGs alone cannot resolve them. In particular, trade-offs between goals (e.g., economic growth vs. environmental protection or social equity vs. cost efficiency) have emerged as significant obstacles to achieving SDGs (Kroll et al., 2019). Thus, strategies that comprehend the interconnections among goals and maximize synergies are required. Additionally, the uneven prioritization of issues and progress across countries and regions complicates global efforts to achieve SDGs. MNCs face the challenge of reconciling their dual responsibilities of contributing to SDGs while pursuing economic profits.

Region-specific approaches have become increasingly important for this complex landscape. For example, the French concept of "*terroir*," which emphasizes the use of region-specific natural, cultural, and social resources, demonstrates the potential to create synergies among SDG targets while mitigating trade-offs (Leedon et al., 2021). Such approaches offer effective ways to link regional challenges to global sustainability goals.

This chapter examines how MNCs can contribute to achieving the SDGs. Specifically, it discusses how adopting circular economy models and implementing localized SDGs utilizing regional characteristics can lead to sustainable value creation. Furthermore, this chapter offers new perspectives on aligning sustainability efforts at both the regional and global levels by managing the interactions among the SDGs.

10.2 Sustainable Development and MNCs

Multinational companies (MNCs) operate across borders (Bartlett & Ghoshal, 1989). Borders have always existed, and the history of international business is characterized by overcoming these boundaries to drive growth and evolution (Hasegawa, 2025; Jones, 2004). Today, rising geopolitical risks, regional conflicts, anti-globalization trends, and pandemic outbreaks are creating new borders and fragmenting the global economy. Modern MNCs are increasingly compelled to rethink and rebuild their international businesses across these boundaries.

Sustainable development (SD) presents another type of border that MNCs must navigate. The Brundtland Report defines SD as development that "meets the needs of the present without compromising the ability of future generations to meet their own needs" (WCED, 1987). This concept extends beyond intergenerational equity to intragenerational equity, which requires preserving and improving environmental, economic, and social systems (Milne & Gray, 2013). Operating across diverse nations and facing institutional pressures regarding environmental and social accountability often limit the profit-maximizing behaviors of MNCs (Kostova & Roth, 2002) and affect their financial performance (Eccles et al., 2014). Under SD constraints, MNCs must fundamentally reorganize their location portfolios, governance structures, and supply chains (Brouthers, 2002; Pinkse & Kolk, 2008).

At the same time, MNCs often wield an influence that surpasses that of nation-states, significantly impacting the global economy and societies of their home and host countries. As innovators in products, technologies, and business models, MNCs are expected to play a critical role in addressing global SD challenges (Bocken et al., 2014; Schaltegger & Wagner, 2011; Voegtlin & Scherer, 2017).

However, the pursuit of corporate profits often conflicts with the societal goals of SD. Historically, addressing this dilemma involved viewing economic, social, and environmental dimensions separately. Firms would redistribute economic gains to address environmental and social issues through voluntary or regulatory mechanisms, often as a form of rhetorical corporate social responsibility (CSR) or philanthropy (Jones, 2023).

In the 1990s, it was recognized that some environmental and social values overlapped with economic values. For instance, environmental pollution and health hazards caused by factories were found to harm corporate value itself, leading to the emergence of the "intersection of sustainability," where environmental, social, and economic dimensions converge (see Fig. 10.1). Consequently, the "triple bottom line" framework, which evaluates corporate performance across financial, environmental, and social metrics, gained prominence (Elkington, 1994).

Efforts to preserve and enhance environmental and social values are not merely cost burdens for firms. For example, efforts to reduce pollution often drive innovation, resulting in improved resource efficiency, which "offsets" pollution reduction costs (Porter & van der Linde, 1995). Additionally, environmental strategies can enhance resilience to future risks, such as climate change and resource depletion, and stabilize long-term revenue (Eccles et al., 2014). Creating and expanding "sustainable value" can also enhance shareholder value (Hart & Milstein, 2003).

As the potential for creating sustainable value expanded, the 2010s witnessed a paradigm shift; economic value is now recognized as being entirely embedded within social and environmental values (Giddings et al., 2002; see Fig. 10.1). Environmental and social degradation threatens corporate survival (Hart, 1995), and economic activities must coexist with healthy environmental and social foundations (Bansal & DesJardine, 2014; Hart & Dowell, 2011). Business innovations created by this coexistence can further enhance economic value, paving the way for concepts such as "creating shared value" (CSV) management (Porter & Kramer, 2019) or "true sustainability" management (Dyllick & Muff, 2016), which MNCs must pursue (Hasegawa, 2023).

10.3 Current Status and Challenges of the SDGs

The 17 SDGs and their 169 associated targets highlight a wide range of critical development challenges that encompass environmental, social, and economic dimensions (UN, 2014). Since their adoption by the United Nations in September 2015, the SDGs have been a guiding framework for all stakeholders—nations, regions, corporations, NGOs, and citizens—to implement the 2030 Agenda for Sustainable Development (UN, 2015). From the 1992 United Nations Framework Convention on Climate Change (UNFCCC) in Rio de Janeiro to the 1997 Kyoto Protocol to the 2009 Copenhagen Climate Change Conference (COP15), the unanimous adoption of the SDGs reflects a growing global awareness of the severe risks posed by human activity often referred to as the "Great Acceleration." This includes crossing "Planetary Boundaries," leading to irreversible "Regime Shifts" in Earth's systems and threatening humanity's survival (IPCC, 2023; Rockström & Klum, 2015; Rockström et al., 2009).

10.3.1 Delays in Achieving the 2030 Agenda

Despite this urgency, progress toward the 2030 Agenda has been disappointing. Significant challenges remain in achieving the goals related to climate change and SD. At COP26 in 2021, serious concern was raised that even if all Nationally Determined Contributions were fully implemented, global greenhouse gas emissions in 2030 would still increase by 13.7% compared to 2010

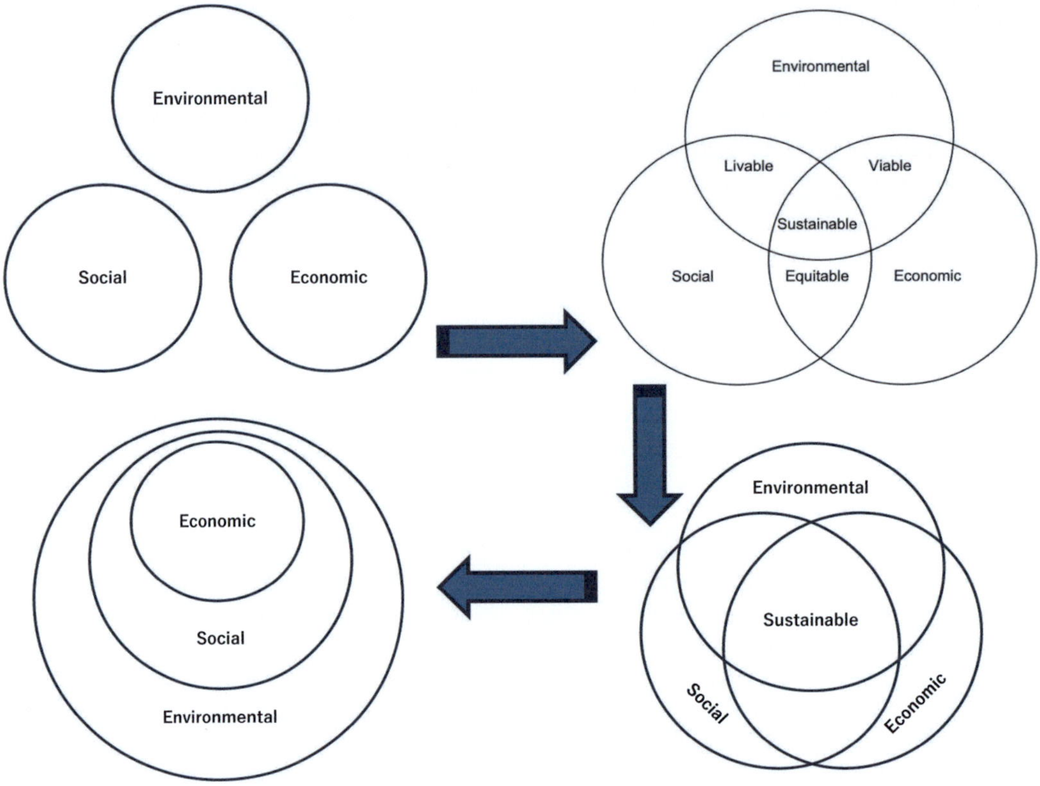

Fig. 10.1 Changing pattern of dimensions of sustainable development. *Source* Author's elaboration based on Tanguay et al. (2009)

levels (UNFCCC, 2021). Moreover, the projected timelines for achieving SDGs continue to be extended (UN, 2024c). In a 2023 expert group evaluation of the progress of the 2030 Agenda, it was again underscored that many goals remain far from being achieved (Nilsson et al., 2024).

In particular, developing countries are facing slow progress in reducing greenhouse gas emissions, protecting ecosystems, and addressing poverty and hunger (UN, 2024b). Voluntary National Reviews shared at the United Nations High-Level Political Forum also reveal significant disparities in SDG implementation between developed and developing nations (GRI, 2023; UN, 2021). Although data inconsistencies make precise comparisons difficult, consistent reports highlight delays in developing nations (Nilsson et al., 2024).

The COVID-19 pandemic has further hindered global progress in achieving the SDGs (Sachs et al., 2024). Recovery packages have often prioritized carbon-intensive industries, while conflicts, wars, and responses to extreme weather events in certain regions have exacerbated setbacks. As of 2024, only 16% of these goals are progressing, whereas the remaining 84% are either stagnant or regressing (Sachs et al., 2024). Particularly, SDG 2 (Zero Hunger), SDG 11 (Sustainable Cities and Communities), SDG 14 (Life Below Water), SDG 15 (Life on Land), and SDG 16 (Peace, Justice, and Strong Institutions) are noted to be "particularly off-track" (Sachs et al., 2024). The UN (2024b) similarly indicates that "the world is severely off track to realize the 2030 Agenda," revealing that achieving the goals by 2030 is nearly impossible (see Fig. 10.2).

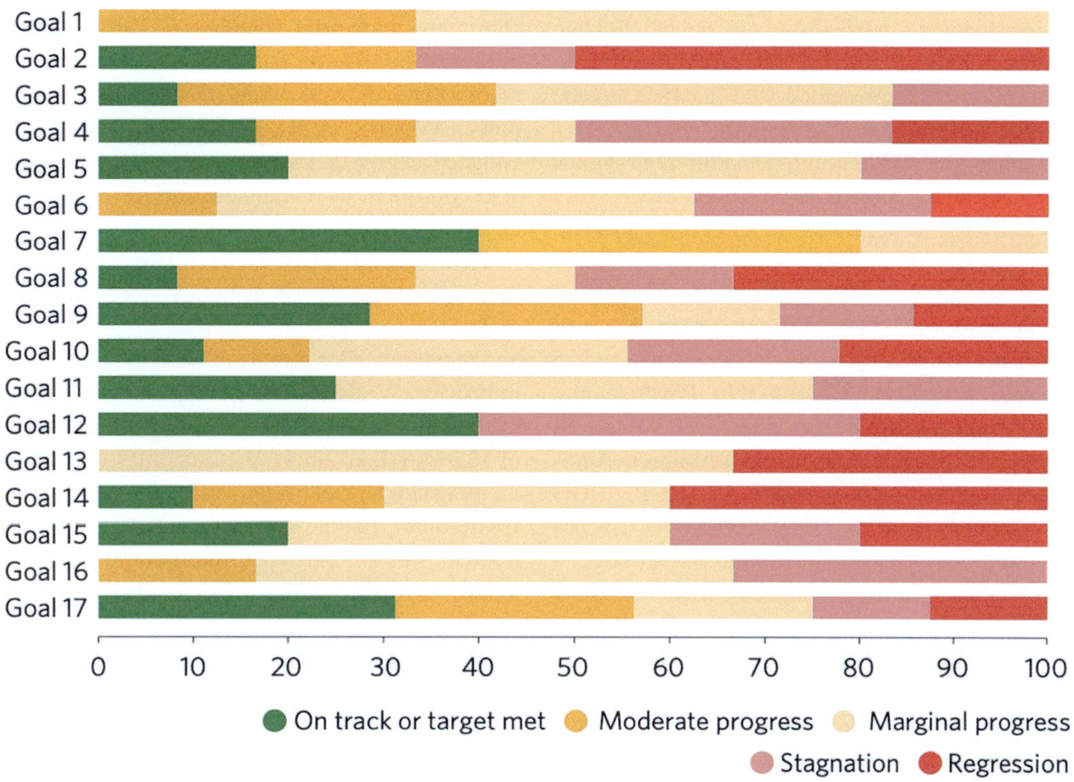

Fig. 10.2 Progress assessment for the 17 Goals based on assessed targets by Goal (%). From The Sustainable Development Goals Report 2024, by UN/DESA, ©2024 United Nations. Reprinted with the permission of the United Nations. This publication is released under a Creative Commons Attribution 4.0 International (CC BY 4.0) license, except for the figure from The Sustainable Development Goals Report 2024, which is used with permission from the United Nations and is not covered under the CC BY license

In response, a group of ten scientists called for intensified efforts to achieve the SDGs and proposed extending the target deadline to 2050, along with a roadmap for these adjustments (Fuso Nerini et al., 2024). Furthermore, there is a growing interest in approaches that focus on the interactions among different goals to break the current impasse. The United Nations' "Decade of Action" initiative emphasizes that focusing solely on individual goals or targets is insufficient; instead, synergy effects should be leveraged by optimizing relationships among goals (Pradhan, 2019; UN, 2024a; UNDP, 2019). Thus, addressing SD requires a holistic approach that minimizes trade-offs and maximizes synergies across goals (Kroll et al., 2019; Nilsson et al., 2018).

10.3.2 Interconnected Goals and Targets

The SDGs and their associated targets are intricately interdependent and form a complex web of linkages. Figure 10.3, derived from network analysis (Le Blanc, 2015), illustrates the connections among the 16 SDGs (excluding SDG 17). These connections are identified based on the presence of references to other goals in the target text. Direct and indirect linkages mediated by specific targets were included in the analysis. For instance, SDG 6 (Clean Water and Sanitation) Target 6.3 references SDG 3 (Good Health and Well-being), and SDG 12 (Responsible Consumption and Production, RCP), creating direct linkages between SDG 6 and these goals. Moreover, Target 6.3 acts as an intermediary, establishing an indirect linkage between SDGs 3 and 12. The strength of these connections is represented in Fig. 10.3 by the thickness of the edges between the goals.

Unlike the Millennium Development Goals (MDGs), which started in 2000, the SDGs were intentionally designed as an integrated system, emphasizing the linkages between goals (Kanie et al., 2017). Figure 10.3 illustrates the complexity of the network. Strong linkages were observed between SDG 4 (Quality Education) and 5 (Gender Equality), SDG 1 (No Poverty) and 10 (Reduced Inequalities), and SDG 10 and 16 (Peace, Justice, and Strong Institutions). Additionally, goals such as SDG 12 (RCP), SDG 10 (inequality), SDG 1 (poverty), SDG 2 (Zero Hunger), and SDG 4 (education) exhibited a high degree centrality, indicating their numerous connections with other goals. Conversely, SDG 14 (Life Below Water), 7 (energy), and 9 (industry and infrastructure) had relatively low centrality.

While these interconnections highlight the interdependencies among the SDGs, they do not indicate whether the effects are positive (synergistic) or negative (conflicting). A closer examination of the specific means required to achieve targets, many of which are embedded in the targets themselves, reveals various options. For example, SDG 7 (Affordable and Clean Energy) Target 7.2 aims to "substantially increase the share of renewable energy." Achieving this requires choosing among energy sources such as solar, wind, hydro, geothermal, and biofuels while also considering Targets 7.3 (improvement in energy efficiency) and 7.1 (universal access to affordable, reliable, and modern energy services) (IEA, 2021; Jacobson & Delucchi, 2011). These decisions are further influenced by technological, economic, and social trends, and require complex policy judgments (Kalghatgi, 2022).

The selected measures often impact other goals and targets (Stafford-Smith et al., 2017). For instance, electric vehicles (EVs) emit no CO_2 during operation compared with conventional internal combustion engine vehicles (ICEVs), aligning with SDG 3 (Good Health) and SDG 13 (Climate Action). Policies in many countries incentivize EV adoption. EVs also strengthen industrial foundations through technological innovation and infrastructure development (SDG 9), and when integrated with solar power systems (V2H: Vehicle-to-Home), they enhance energy self-sufficiency (SDG 7). They also contribute to sustainable and resilient urban development (SDGs 11, 13).

However, the use of EVs remains challenging. Over the lifecycle of batteries, including raw material procurement, manufacturing, disposal, and recycling, EVs rely heavily on CO_2-intensive industries. While estimates suggest that EV lifecycle emissions of CO_2 are less than half of those of ICEVs (IEA, 2024), the outcomes depend on factors such as electricity sources (renewable vs. fossil fuels), vehicle longevity, and usage patterns (Hawkins et al., 2013). Furthermore, mining rare metals for EV batteries (e.g., lithium, nickel, cobalt, and manganese), which is projected to quadruple in demand by 2040 (IEA), poses environmental risks such as deforestation, ecosystem toxicity, human health impacts, labor exploitation, and resource depletion (Slack, 2012). Refining processes for these metals often rely on coal-fired power, further increasing greenhouse gas emissions (Norgate & Haque, 2010). Additionally, the disposal of batteries generates e-waste and chemical waste, which harm both human health and the environment (Gaines, 2014).

While EV adoption is accelerating in emerging economies, the recycling and reuse of battery materials significantly lag behind stricter regulations in regions such as the European Union (EU) and the US. Recycling processes often consume large amounts of energy and chemicals (Neumann et al., 2022). These negative aspects suggest that an unchecked transition to EVs could conflict with SDGs 3, 7, and 8 (Decent Work and Economic Growth), 12, 14 (Life Below Water), and 15 (Life on Land).

Therefore, the SDGs form a highly interdependent system in which trade-offs and synergies coexist among their targets and implementation measures (Alcamo et al., 2020; Nilsson et al., 2016). Achieving the 2030 Agenda requires clarifying these complex interaction mechanisms to minimize trade-offs and maximize synergies (Kroll et al., 2019; Nilsson et al., 2018). The

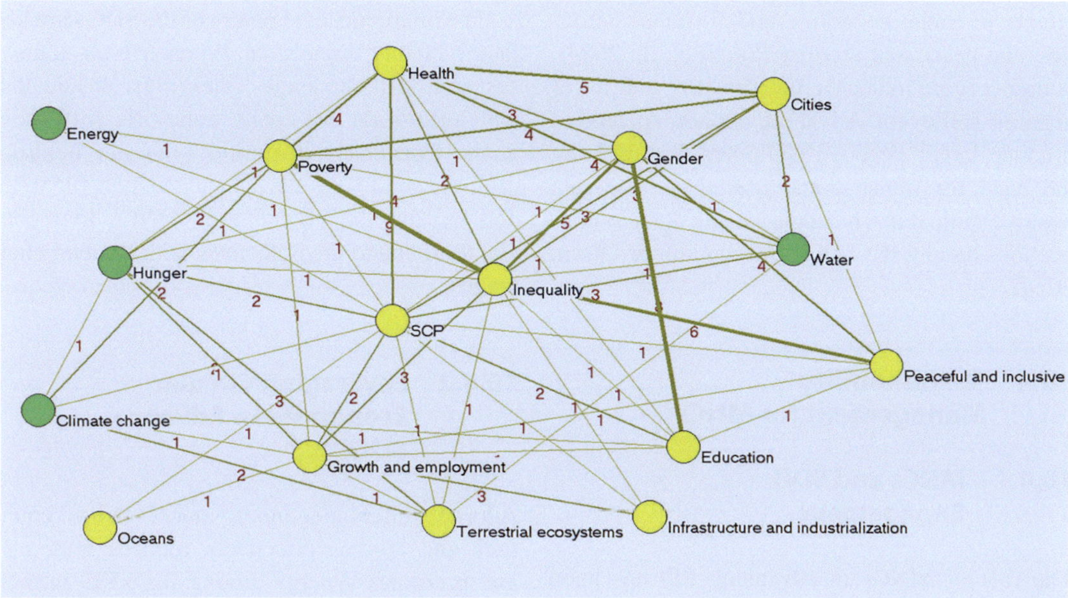

Fig. 10.3 Links between the SDGs through targets: an aggregated picture. *Note* The numbers next to the edges indicate the number of targets linking different goals. *Source* From Towards Integration at Last? The Sustainable Development Goals as a Network of Targets, by David Le Blanc, ©2015 United Nations. Reprinted with the permission of the United Nations. This figure is not available under a Creative Commons license

isolated pursuit of individual SDGs can negatively affect others, making it essential to maintain a balanced approach (Allen et al., 2019; Pradhan et al., 2017).

Efforts to quantitatively assess SDG interactions are numerous but challenging owing to difficulties in capturing and monitoring 232 indicators for SDG targets (Cling & Delecourt, 2022; Renaud et al., 2022). Evaluations also depend on the interpretation of targets within specific contexts and the selection of significant interactions from countless possibilities, leading to variability in the results across goals (Anderson et al., 2022; Coenen et al., 2022; Fuso Nerini et al., 2018; IPCC, 2023; Nilsson et al., 2016; Pradhan et al., 2017; Weitz et al., 2018). These interactions differ not only between countries and regions but also among sectors and stakeholders (Lusseau & Mancini, 2019; Scherer et al., 2018; Warchold et al., 2020).

10.3.3 International Spillovers

International spillovers further complicate the pursuit of SDGs. Activities or policies aimed at achieving SDGs in one country can intentionally or unintentionally impact other countries (Davis & Caldeira, 2010; Gereffi et al., 2005). Spillovers can be positive, such as greenhouse gas reduction or technological innovations that benefit other nations (Engström et al., 2021). However, negative spillovers such as unsustainable consumption, toxic waste exports, and illicit trade from wealthier nations exacerbate deforestation and other environmental and social issues (Engström et al., 2021; Lenzen et al., 2012; Wood et al., 2018).

As shown in Fig. 10.4, the Sustainable Development Solutions Network (SDSN) publishes a spillover index assessing such impacts across three dimensions: environmental and social effects in trade, economic and financial activities, and peace and security (Sachs et al., 2024). A higher score indicates more positive and fewer negative spillovers. Advanced economies tend to produce larger negative spillovers, highlighting the need for transnational initiatives to create positive spillovers while considering the interconnections among the SDGs (Moinuddin & Olsen, 2024).

The interconnected nature of the SDGs is a key factor in this discussion. When MNCs address sustainability challenges, they must ensure that their efforts do not create trade-offs with other goals. Conversely, fostering synergies between different challenges can enhance sustainable value and expand the potential of MNCs to reconcile their economic objectives with societal challenges.

10.4 Sustainability Management for MNCs

10.4.1 MNCs and SDG Engagement

The role of MNCs in advancing SD has been extensively discussed from various perspectives. Kolk et al. (2017) grouped the 2030 Agenda into three categories: (1) poverty and inequality, (2) energy and climate change, and (3) peace; they reviewed prior studies on the role of MNCs. Among these, the most developed area of research pertains to Base/Bottom of the Pyramid (BOP) business models targeting the poor (Prahalad & Hart, 1999). These models explore how MNCs can contribute to SDGs such as SDG 1 (No Poverty), 3 (Good Health and Well-being), 4 (Quality Education), 5 (Gender Equality), 6 (Clean Water and Sanitation), 8 (Decent Work and Economic Growth), and 9 (Industry, Innovation, and Infrastructure), which are often prioritized in developing countries. Efforts in this area aim to balance contributions to SD with the MNCs' pursuit of profitability through innovative business models (Bhatti & Ventresca, 2013; Cañeque & Hart, 2015; Govindarajan & Trimble, 2012; Hasegawa, 2023; London & Hart, 2010).

However, as van Zanten and van Tulder (2018) note, MNCs tend to primarily engage with SDG targets that are actionable within their internal frameworks and focus on avoiding negative impacts on SD. Their contributions have often been narrowed or passive, emphasizing harm avoidance and the mitigation of negative externalities rather than actively advancing SDG.

10.4.2 Leveraging Circular Economies to Advance SDGs

Adopting circular economy models in host countries and regions effectively reduces trade-offs and promotes synergy among the SDG targets. Circular economies align closely with multiple SD challenges by emphasizing resource efficiency and recycling (Pieroni et al., 2019; Schröder & Barrie, 2024).

A circular economy refers to a regenerative system in which resource inputs, waste, emissions, and energy leakage are minimized by slowing, closing, and narrowing the material and energy loops (Geissdoerfer et al., 2017). This is achieved by extending product lifecycles through practices such as sharing, leasing, reusing, repairing, refurbishing, and recycling (European Commission, 2015). Circular economies create synergies between SDG 13 (Climate Action) and 12 (Sustainable Consumption and Production) by improving resource efficiency and reducing waste.

Moreover, circular processes often require specialized skills, leading to job creation (Stahel, 2016). This fosters additional synergies with other SDGs, including SDG 9 (Industry, Innovation and Infrastructure), 8 (Decent Work and Economic Growth), and 4 (Quality Education) (Ghisellini et al., 2016). Enhanced transparency in circular supply chains further supports SDG 8 (fair labor conditions), 11 (pollution reduction), and 14 and 15 (ecosystem preservation) while maintaining economic growth (Genovese et al., 2017).

MNCs are well-positioned to leverage their technological capabilities and global networks to drive the transition from linear economic

Fig. 10.4 International Spillover Index. From The SDGs and the UN Summit of the Future. Sustainable Development Report 2024 (p. 498), by Jeffrey D. Sachs, Guillaume Lafortune, and Grayson Fuller, © 2024 Dublin University Press. Reprinted with the permission of Dublin University Press. *Note* The spillover index, using 16 indicators, assesses spillovers along three dimensions: (1) environmental and social impacts embodied in trade; (2) economy and finance; (3) UN-based multilateralism, peace and security. The score ranges from 0 to 100, where higher scores indicate that the country has more positive and fewer negative spillover impacts. The scores by region are population-weighted averages

models ("make-use-dispose") to circular systems focusing on reducing, reusing, recycling, and repairing. For instance, MNCs can adopt advanced recycling technologies, redesign supply chains to extend product lifecycles, and develop "product as a service" models through sharing or leasing platforms. By doing so, they can address various SD challenges while pursuing sustainable business growth (Lacy & Rutqvist, 2015).

However, a circular economic model is challenging. "Rebound" effects, where improvements in eco-efficiency are offset by increased production and, consequently, consumption can negate potential benefits (Castro et al., 2022). These rebound effects occur through two primary mechanisms (Zink & Geyer, 2017):

1. **Lower-tier products and additional consumption**: Products introduced to the market from a circular economy, such as recycled or refurbished goods, are often of lower quality than newly produced goods (higher-tier goods) produced in a linear economy. Consequently, these lower-tier goods fail to replace higher-tier goods completely and are instead consumed in addition to them, leading to an overall increase in total consumption.
2. **Price effects and consumption expansion**: the "substitution effect" triggered by a price drop in lower-grade goods can increase their consumption. However, the "income effect" (resulting from increased real income due to the price decline of lower-grade goods) may redirect spending toward higher-grade goods, leaving the net effect uncertain. Notably, there is potential for increased consumption of higher-tier goods. If market competition drives down the price of higher-tier goods, the net effect will likely result in expanded consumption of these goods.[1]

Consequently, the original advantage of a circular economy, improving environmental sustainability by enhancing resource efficiency without expanding consumption, is offset.

MNCs can deploy specific strategies to mitigate rebound effects in a circular economy. Advanced recycling technologies can enhance the quality of lower-tier goods, making them viable substitutes for higher-tier products. Additionally, by leveraging their marketing capabilities, MNCs can target consumer segments that are characterized by low price elasticity of demand, avoiding the market-wide price impacts that drive expanded consumption (Zink & Geyer, 2017). By

[1] Depending on the magnitude of the income effect, this could also lead to increased consumption of other goods, further increasing overall consumption (Zink & Geyer, 2017).

harnessing their technological expertise and innovative capacities, MNCs can limit rebound effects, amplifying the sustainable value created through circular economic initiatives and reinforcing their leadership roles in SD.

10.4.3 Adapting to Diverse Sustainability Needs

The importance of sustainability challenges varies significantly depending on the stage of economic development, social and cultural contexts, and normative values in different countries and regions (Pradhan et al., 2017). For example, while energy efficiency and decarbonization are often prioritized in developed countries, basic infrastructure development and poverty alleviation are pressing concerns in developing nations. Social structures and cultural values also influence the prioritization of issues (Adger et al., 2013; Steg & Vlek, 2009). Individualistic societies may emphasize challenges related to personal choices, such as reducing ecological footprints, whereas collectivist societies may focus on community-wide sustainability, such as rural revitalization (Triandis, 1995). Historical and religious values further shape what is considered ethically significant; for instance, in Europe, environmental protection is widely recognized as a moral obligation that influences policy formation (White, 1967).

The interconnections among sustainability challenges also exhibit diversity, reflecting contextual variations (Bali Swain & Ranganathan, 2021; Sachs et al., 2024). In some countries, economic growth may complement environmental conservation, while these objectives may conflict in other countries. European countries, particularly in Northern Europe, exemplify the former (Uğurlu, 2019). Investments in renewable energy, such as wind and solar power (SDG 7), not only help to reduce greenhouse gas emissions (SDG 13) but also stabilize the electricity supply and create jobs through the development of related new industries and markets (SDG 9). Similarly, efforts have been made to reduce waste and improve resource efficiency (SDG 12), lower economic costs (SDG 9), and minimize environmental impacts (SDGs 13, 11, and 6). Initiatives such as the circular economy and sharing economy support sustainable lifestyles (SDGs 12, 13, and 11) while driving consumption and economic growth (SDGs 8 and 9).

In contrast, emerging Asian economies still exhibit trends in which economic development increases the environmental burden. In lower-income and developing countries, sustainability priorities such as addressing poverty (SDG 1) and hunger (SDG 2) often take precedence, frequently resulting in trade-offs with environmental conservation (SDGs 13, 14, and 15) and social equity (SDGs 5, 10, and 16).

Given this contextual diversity, MNCs must adopt flexible adaptation strategies tailored to the circumstances of their host countries and regions. In cases where trade-offs exist among the SDGs, prioritization must consider the urgency of local challenges, the systemic impact of issues, and the availability of resources (Allen et al., 2019; Bandari et al., 2022).

10.4.4 Addressing International Spillovers

As mentioned earlier, SDGs and their targets transcend national boundaries, resulting in international spillovers (Davis & Caldeira, 2010; Gereffi et al., 2005). Similar spillovers also occur between different regions within a country.

For example, MNCs introducing renewable energy technologies or environmentally friendly production methods in developing countries can contribute to SDG 7 (energy) and 9 (industry and innovation). Conversely, global supply chains can generate negative spillovers on SDG 12 (RCP) and 13 (Climate Action), such as increased CO_2 emissions and resource consumption in developing countries (Davis et al., 2011; Peters et al., 2011). Furthermore, while MNCs may generate employment and drive economic growth in host countries, issues such as low wages, harsh working conditions, and child labor can undermine SDG 8 (Decent Work and Economic Growth) and SDG 10 (Reduced Inequalities). Some MNCs also engage in profit-shifting strategies that exploit tax havens, weakening the fiscal base of both the home and host countries and negatively impacting SDG 10 and 17 (partnerships) (Sachs et al., 2024).

To minimize negative spillovers and contribute to global sustainability, MNCs must implement the following actions.

1. **Adhering to and exceeding local regulations**: Establishing sustainable business operations by complying with environmental and labor standards and adopting higher independent benchmarks.
2. **Enhancing supply chain transparency**: Introducing traceability systems and environmentally and socially responsible procurement standards (Porter & Kramer, 2019). For example, implementing mechanisms to ensure that suppliers meet environmental and labor requirements can contribute to responsible consumption and production (SDG 12) (Pedersen & Andersen, 2015).
3. **Collaborating with local stakeholders**: Partnership with local communities, governments, and NGOs to address region-specific social and economic challenges, thereby fostering sustainable operations (Dentoni et al., 2018).

By adopting such comprehensive approaches, MNCs can contribute to improving sustainability in host countries while demonstrating leadership in achieving global sustainability goals.

10.5 Localizing the SDGs: Regional Adaptation and MNC Strategies

10.5.1 Transitioning to Localized Economies: Challenges and Opportunities

The COVID-19 pandemic, which began in Wuhan, China, in 2020, forced a decoupling of global economic activities from linear economies and created pressure to transition toward localized, circular economies (Hasegawa, 2021). Consequently, MNCs were compelled to reconstruct global supply chains that had been stretched to their limits in pursuit of optimal production locations worldwide. As a result, there has been a shift toward embedding redundancy and flexibility into supply chains, transforming them into more resilient systems (Katsaliaki et al., 2022). Actions such as supply chain diversification, shortening, near-shoring, reshoring, friend-shoring, and ally-shoring have emerged. By decentralizing manufacturing and procurement across regions, MNCs increasingly aim to adapt to local environmental regulations and labor markets, improve resource efficiency, and reduce transportation distances and energy consumption (Meyer et al., 2009; Rugman & Verbeke, 2003; Seuring & Müller, 2008).

These changes are also significant for achieving SDGs. Although SDGs are a global set of goals aimed at achieving environmental, social, and economic sustainability, their realization requires action at the regional level (Ordonez-Ponce et al., 2021; Salvia et al., 2019). Each region faces unique challenges shaped by its natural environment and socioeconomic characteristics, necessitating specific adaptive approaches (Bali Swain & Ranganathan, 2021). Additionally, regions possess distinctive resources shaped by their natural conditions, cultures, and historical legacies. Strengthening partnerships with local communities to leverage these resources is essential for addressing regional challenges and creating a sustainable society (Ostrom, 1990). Accumulating successful regional initiatives rooted in local challenges and resources can contribute to achieving the national and global SDGs (Ostrom, 2009; Sachs et al., 2021).

Regional sustainability challenges are diverse. For example, although rural areas in Japan face severe population decline and depopulation at rates exceeding the national average, urban areas struggle with overpopulation, excessive energy consumption, and traffic congestion (ANRE, 2024; MLIT, 2020). Rural areas are experiencing profound challenges owing to population decline and aging, which threaten the survival of communities (MAFF, 2024).

Thus, there is an urgent need to improve agricultural efficiency, increase farmers' incomes, and create new industries to revitalize rural communities. These challenges are not limited to demographic issues, but also impact the preservation of local culture, traditional industries, and Japan's declining food self-sufficiency rate (MAFF, 2024). On the environmental front, reducing dependence on pesticides and chemical fertilizers and reducing greenhouse gas emissions remain significant challenges, requiring solutions that balance productivity with sustainability (MAFF, 2023).

Natural disasters in Japan further highlight the diverse nature of regional challenges. The country faces various disaster risks, including earthquakes, tsunamis, typhoons, heavy rain, floods, and landslides (UNDRR, 2019). These risks vary greatly depending on the regional topography, climate, and social infrastructure (Cabinet Office Japan, 2023). Coastal regions are vulnerable to sea level rise and tsunamis; mountainous regions face severe flooding and landslides; rural areas experience agricultural land erosion and facility damage from typhoons and heavy rains; and urban areas are exposed to building collapses and large-scale fires during earthquakes. Climate change is expected to increase the frequency and severity of these disasters (Mori et al., 2021). Tailored disaster prevention and mitigation measures are necessary, including crop diversification, levee construction, afforestation, seismic retrofitting of buildings, and disaster-resilient urban planning (MOE et al., 2018). Furthermore, each region has a unique natural environment of intrinsic value that requires diverse conservation approaches (MOE, 2021).

These diverse, multifaceted sustainability challenges across environmental, social, and economic dimensions vary by region and require multiple solutions. Efforts must account for the interconnections among these challenges, eliminating trade-offs and creating synergies while advancing region-specific initiatives.

10.5.2 Regional Resources and Challenges in Achieving SDGs

Addressing these varied development challenges also necessitates the utilization of each region's unique resources. Leveraging regional resources creates local employment, revitalizes regional economies and societies, and fosters economic independence. When regions take ownership of their initiatives, actionable solutions emerge, motivation for implementation increases, and community strength increases (Swinburn et al., 2006).

Examples include utilizing abundant water resources for hydropower, promoting forestry through the use of forest resources, and linking traditional crafts to tourism and export industries. Agriculture is particularly important because it affects the interconnections among the SDGs. Hunger and obesity coexist globally, highlighting the urgent need to improve food supply systems and land use (Sachs et al., 2024). Additionally, reducing emissions from agriculture, forestry, and other land uses, which account for approximately one-fourth of the global greenhouse gas emissions, is critical for climate change mitigation (Foley et al., 2005; Smith et al., 2014a, 2014b). Building sustainable agriculture that is tailored to regional realities and effectively utilizing unique resources are keys to SD.

In this context, the "*terroir*" concept, widely recognized in France and Southern Europe, offers valuable insights. *Terroir* refers to the unique characteristics of agricultural products and foods shaped by the natural environment of a specific region as well as the social, cultural, historical, and technical elements constructed over time (Rastoin, 2012; Rastoin & Vissac-Charles, 1999). According to UNESCO (2005), *terroir* is "a delimited geographical space in which a human community builds a collective understanding of production based on a system of interactions between a physical and biological environment and a set of human factors." Within such geographic spaces, "social and technical trajectories give *terroir* products originality, typicality (typicité), and distinctive characteristics, contributing to their reputation" (INAO, 2016a).

For instance, even wines produced from the same grape variety (cépage) can exhibit diverse flavors depending on the natural conditions (soil, climate, and terrain) of specific regions, villages, or vineyards (*climat, lieu dit*), as well as the winemaking techniques historically and culturally developed in these areas. These techniques include harvesting, sorting, destemming, crushing, fermentation, pressing, and aging, all of which contribute to producing highly diverse flavors and qualities (Dion, 2010). Beyond wine, other *terroir* products in France, such as cheese, agricultural goods, and crafts, serve as regional specialties that promote local development and support localized SDG efforts (Delfosse, 2012). The actors involved in constructing *terroirs* include not only producers, agricultural workers, and laborers but also local governments, specialized institutions, educational and training organizations, research and technology centers, consulting services, and financial institutions. Informal networks, such as family ties, friendships, clubs, and self-help groups, also play a significant role (Ditter & Brouard, 2014).

Terroir products are certified under the *Appellation d'Origine Contrôlée* (AOC), a type of geographical indication that ensures their quality and authenticity while granting exclusive rights to the use of their names and protecting them from imitation (INAO, 2016a).[2] AOC is a state certification that was legislated in 1935. It is granted after a rigorous examination by the *Institut National de l'Origine et de la Qualité* (INAO), which is a division within the French Ministry of Agriculture and Food. This examination is based on applications from regional producer organizations.[3] AOC certification specifies detailed standards and procedures across all stages of production, from the origin (cultivation area) and harvesting methods to the processing, preservation, and final production of agricultural goods. These standards are outlined in a comprehensive specification document, or "*cahier des charges*" (INAO, 2016b). The *cahier des charges* is developed by non-profit organizations called *Organisme de Défense et de Gestion* (ODG), composed of producers, technical experts, and others (Cousin & Berche, 2019), and is approved by the government.[4] Recent updates to the *cahier des charges* reflect considerations for environmental sustainability, incorporating requirements for local renewable energy use, waste recycling, carbon footprint reduction, and biodiversity

[2] Other certifications that identify quality and place of origin include IGP (*Indication géographique protégée*: Geographical Indication Protected) and STG (*Spécialité traditionnelle garantie*: Traditional Speciality Guaranteed). STG does not include any requirements regarding geographical origin, and only requires that the ingredients and production/processing methods be traditional.

The main difference between AOC and IGP is that AOC is clearly defined as "quality originating from the *terroir*," whereas IGP is only defined as "quality originating from the relationship with the place of production" (INAO, 2016a). In the case of AOC, the area of production is much more limited than that of IGP, and basically all production and processing must be completed within the same region.

In addition to AOC, IGP, and STG, the INAO also manages the LR (*Label Rouge*), which guarantees excellent quality, and the AB (*Agriculture biologique*) label, which relates to environmentally friendly production methods.

[3] In 1992, the EU established a unified standard for Protected Designation of Origin (PDO), which is equivalent to AOC, and the name in France was changed from AOC to *Appellation d'Origine Protégée* (AOP) accordingly. However, because the label AOC has become widely established in French society, it is still possible for wine to use the AOC label as before, and in many cases it is still used today. With the introduction of unified standards by the EU, French AOC/AOP certified products are now also covered by legal protection in other EU member states, as well as in other countries (including Japan) with which the EU has mutual recognition agreements through bilateral agreements.

For the sake of convenience, this chapter uses the term AOC.

[4] Based on interviews conducted at the INAO-Centre-Est in Mâcon, Burgundy, and with wine producers (Pouilly-Fuissé) and cheese producers (Charolais) in the surrounding area, as well as with ODGs (conducted 30 September–1 October 2024). Also see https://www.inao.gouv.fr/Nos-actualites/visite-delegation-japonaise-bourgogne-octobre2024 (retrieved October 5, 2024).

conservation.[5] Through AOC certification, which embodies the collective dynamics of various actors within the *terroir*, these products achieve high value-added in the market, contributing to job creation, community revitalization, and improved economic and social sustainability. Moreover, promoting resource circulation enhances environmental sustainability.

In many *terroirs*, a core product is complemented by additional regional specialties, services (lodging, restaurants, and tourism), landscapes, and lifestyles. These combinations enhance the value of each component, generating synergistic effects among SDGs (Hirczak et al., 2008; Mollard, 2001; Pecqueur, 2001).

10.5.3 MNCs and Strategies for Regional Sustainability

The concept of "*terroir*" reflects the fusion of natural resources, such as soil and climate, with cultural, historical, and social resources. These diverse regional resources define the characteristics and competitive advantages of *terroir*-based products. Because these resources meet the VRIO (Valuable, Rare, Inimitable, Organization) criteria, they can be considered strategic resources that serve as a source of sustainable competitive advantage (Barney, 1991; Wernerfelt, 1984). Furthermore, the geographic concentration of regional resources enhances "localized learning" through interactions among actors, leading to the accumulation of knowledge and skills and improving the region's competitiveness as a whole. This process is path-dependent (Eskelinen et al., 1998) and is further strengthened through the interplay of geographical proximity and "organized proximity," where shared rules within horizontal and vertical networks in regional clusters regulate competition and cooperation (Ditter et al., 2017). Resources derived from such local social and cultural conditions contribute to regional competitive advantages that support industrial development (Saxenian, 1996).

Such strategic resources are highly attractive to MNCs in their pursuit of sustained competitive advantage. However, these resources are deeply rooted in specific regions and are not easily accessible unless MNCs establish a presence in those regions (Dunning, 1998).

These strategic resources also form a critical foundation for MNCs seeking to commit to circular economies and contribute to achieving the local SDGs. However, as the case of French *terroirs* demonstrates, entering regions where historical, cultural, and social connections among local communities and actors are highly valued poses a challenge for MNCs. Building trust and acquiring "embedded" knowledge within local networks (Granovetter, 1985) can take a significant amount of time, and MNCs may initially be viewed with caution as outsiders (Giuliani & Macchi, 2014; Kolk et al., 2017). Additionally, the global strategies and standardized approaches of MNCs may clash with the localized characteristics of regions, such as *terroirs* (Ghemawat, 2007; Prahalad & Doz, 1987). This lack of familiarity with local networks and trust relationships can create the "liability of outsidership" (LOO) (Johanson & Vahlne, 2009), posing barriers to the implementation of SDG-focused business practices at the local level.

Additionally, the local subsidiaries of MNCs entering a market are also embedded in the multinational corporate network (Ghoshal & Bartlett, 1990; Szulanski, 1996). Managing this dual embeddedness in the internal and external environments presents a significant challenge for MNCs (Andersson et al., 2001; Madhok & Liu, 2006). However, this dynamic co-evolutionary process with both environments is precisely what enables MNCs to create new knowledge and improve their competitive advantage (Nohria & Ghoshal, 1997).

[5] The following website provides a detailed explanation of the procedure for revising the *cahier des charges*. INAO, *Schéma de la procédure de modification AOP/AOC - IGP - IG*. https://www.inao.gouv.fr/Espace-professionnel-et-outils/Les-procedures-d-instruction-pas-a-pas/AOP-AOC-IGP-IG/AOP-AOC-IGP-IG-schema-de-modification/Cahier-des-charges (accessed September 30, 2024).

10.5.4 Scaling Local Initiatives to Global Impact

MNCs must leverage their inherent strengths to overcome LOO, effectively utilize regional strategic resources, and succeed in local sustainability management (Eang et al., 2023).

1. **Enhancing regional capacity through resources and expertise**:
   MNCs can enhance regional capabilities by leveraging access to substantial capital, advanced technologies, human resources, and economies of scale (Sachs, 2012). For example, providing sustainable agricultural technologies or circular economy models can complement and strengthen the natural and constructed resources of *terroirs*, thereby accelerating the achievement of local SDGs. This requires adaptation to local characteristics and collaboration with local producers, governments, and NGOs through joint projects to ensure that MNCs become integral to regional networks, thereby overcoming LOO (Johanson & Vahlne, 2009).
2. **Sharing global knowledge and best practices**:
   MNCs can encourage local producers to adopt new technologies and methods by sharing global insights and successful practices with other countries and regions (Minbaeva et al., 2014). For instance, co-creating projects with regional stakeholders and jointly developing sustainable solutions can strengthen regional resources as strategic assets. By participating in knowledge-sharing platforms, MNC subsidiaries can become embedded in regional and global knowledge networks (Ciabuschi et al., 2014). This integration fosters innovation by linking local knowledge to global advancement and promotes SD.
3. **Providing access to global markets**:
   MNCs can support regional economic revitalization by connecting high-quality, locally produced goods to international markets (Kolk et al., 2017; Sachs, 2012). For example, by developing marketing strategies that emphasize the unique stories and qualities of *terroir* products, MNCs can create ethical brands that resonate globally while fostering connections between consumers and regional producers.

MNCs can contribute to global SDG by accumulating successful local practices. Promoting ecological *terroir* products in global markets and developing sustainable models based on local SDGs and strategic regional resources can have widespread impacts. Transforming local initiatives into globally recognized success stories enables MNCs to generate broad SD outcomes.

10.6 Concluding Remarks

The SDGs serve as a comprehensive framework for addressing global challenges and providing common guidance for countries, communities, businesses, and individuals worldwide. This chapter emphasizes that achieving the SDGs requires meeting individual targets and understanding their complex interconnections to minimize trade-offs and maximize synergies. The role of MNCs in this endeavor is critical.

MNCs possess global reach, technological innovation, financial resources, and international networks required to address sustainability challenges. They are uniquely positioned to manage the intricate interactions among the SDGs and develop adaptive strategies tailored to local needs. By adopting circular economic models, MNCs can efficiently use resources, reduce waste, and contribute to regional economic revitalization and job creation. Furthermore, leveraging the concept of the "*terroir*" allows MNCs to work with regional communities to build sustainable business models that simultaneously enhance economic profits and create social value.

However, MNCs face significant challenges in this process. Tensions may arise between global standards and local characteristics, and building trust in regional communities requires time and effort. Minimizing negative spillovers while contributing to local goals requires flexible strategies and a clear sustainability vision. This chapter highlights the importance of co-creation with regional stakeholders and transparent supply chain management as practical pathways for overcoming these obstacles.

As proposed in this chapter, MNCs can unlock new possibilities for achieving the SDGs by combining their global influence with collaborative local efforts. The SDGs represent challenges that no single country or company can solve alone. However, MNCs can extensively impact SD by connecting local success to global frameworks and scaling sustainable models. Moving toward a sustainable future requires MNCs to demonstrate further leadership and fulfill their critical roles as change agents.

This paper is part of the results of research funded by JSPS KAKENHI (20K01836) and Waseda University Research Grant for Specific Issues (2023C-512, 2024C-534).

References

Adger, N., Barnett, J., Brown, K., Marshall, N., & Hall, P. (2013). Cultural dimensions of climate change impacts and adaptation. *Nature Climate Change, 3*, 112–117. https://doi.org/10.1038/nclimate1666

Agency for Natural Resources and Energy (ANRE). (2024). *Energy White Paper 2024*. Retrieved September 25, 2024, form https://www.enecho.meti.go.jp/en/category/whitepaper/pdf/2024_outline.pdf

Alcamo, J., Thompson, J., Alexander, A., Antoniades, A., Delabre, I., Dolley, J., Marshall, F., Menton, M., Middleton, J., & Scharlemann, J. (2020). Analysing interactions among the sustainable development goals: findings and emerging issues from local and global studies. *Sustainability Science, 15*, 1561–1572. https://doi.org/10.1007/s11625-020-00875-x

Allen, C., Metternicht, G., & Wiedmann, T. (2019). Prioritising SDG targets: Assessing baselines, gaps and interlinkages. *Sustainability Science, 14*, 421–438. https://doi.org/10.1007/s11625-018-0596-8

Anderson, C. C., Denich, M., Warchold, A., Kropp, J. P., & Pradhan, P. (2022). A systems model of SDG target influence on the 2030 Agenda for Sustainable Development. *Sustainability Science, 17*, 1459–1472. https://doi.org/10.1007/s11625-021-01040-8

Andersson, U., Forsgren, M., & Pedersen, T. (2001). Subsidiary performance in multinational corporations: The importance of technology embeddedness. *International Business Review, 10*(1), 3–23. https://doi.org/10.1016/S0969-5931(00)00042-1

Bali Swain, R., & Ranganathan, S. (2021). Modeling interlinkages between sustainable development goals using network analysis. *World Development, 138*, Article 105136. https://doi.org/10.1016/j.worlddev.2020.105136

Bandari, R., Maallemi, E. A., Lester, R. E., Downie, D., & Bryan, B. A. (2022). Prioritising Sustainable Development Goals, characterising interactions, and identifying solutions for local sustainability. *Environmental Science & Policy, 127*, 325–336. https://doi.org/10.1016/j.envsci.2021.09.016

Bansal, P., & DesJardine, M. R. (2014). Business sustainability: It is about time. *Strategic Organization, 12*(1), 70–78. https://doi.org/10.1177/1476127013520265

Barney, J. B. (1991). Firm resources and sustained competitive advantage. *Journal of Management, 17*(1), 99–120. https://doi.org/10.1177/014920639101700108

Bartlett, C. A., & Ghoshal, S. (1989). *Managing across borders: The transnational solution*. Harvard Business School Press.

Bhatti, Y., & Ventresca, M. (2013). *How can 'frugal innovation' be conceptualized?* SSRN Working Paper. https://doi.org/10.2139/ssrn.2203552

Bocken, N. M., Short, S. W., Rana, P., & Evans, S. (2014). A literature and practice review to develop sustainable business model archetypes. *Journal of Cleaner Production, 65*, 42–56. https://doi.org/10.1016/j.jclepro.2013.11.039

Brouthers, K. D. (2002). Institutional, cultural and transaction cost influences on entry mode choice and performance. *Journal of International Business Studies, 33*(2), 203–221. https://doi.org/10.1057/palgrave.jibs.8491013

Burritt, R. L., Christ, K. L., Rammal, H. G., & Schaltegger, S. (2018). Multinational enterprise strategies for addressing sustainability: The need for consolidation. *Journal of Business Ethics, 164*, 389–410. https://doi.org/10.1007/s10551-018-4066-0

Cabinet Office Japan. (2023). *White Paper on Disaster Management in Japan 2023*. Retrieved September 30, 2024, from https://www.bousai.go.jp/en/documentation/white_paper/pdf/2023/R5_hakusho_english.pdf

Cañeque, F. C., & Hart, S. L. (Eds.). (2015). *Base of the Pyramid 3.0: Sustainable development through innovation and entrepreneurship*, Routledge.

Castro, C. G., Trevisan, A. H., Pigosso, D. C. A., & Mascarenhas, J. (2022). The rebound effect of circular economy: Definitions, mechanisms and a research agenda. *Journal of Cleaner Production, 345*, Article 131136. https://doi.org/10.1016/j.jclepro.2022.131136

Ciabuschi, F., Holm, U., & Martín, O. M. (2014). Dual embeddedness, influence and performance of innovating subsidiaries in the multinational corporation. *International Business Review, 23*(5), 897–909. https://doi.org/10.1016/j.ibusrev.2014.02.002

Cling, J.-P., & Delecourt, C. (2022). Interlinkages between the Sustainable Development Goals. *World Development Perspectives, 25*, 100398. https://doi.org/10.1016/j.wdp.2022.100398

Coenen, J., Glass, L.-M., & Sanderink, L. (2022). Two degrees and the SDGs: A network analysis of the interlinkages between transnational climate actions and the Sustainable Development Goals. *Sustainability*

Science, 17, 1489–1510. https://doi.org/10.1007/s11625-021-01007-9

Cousin, D., & Berche, G. (2019). Les Organismes de défense et de gestion (ODG), nouveaux acteurs de la filière viti-vinicole française depuis la réforme de l'INAO. *Pour, 237–238*(1), 155–163. https://doi.org/10.3917/pour.237.0155

Davis, S. J., & Caldeira, K. (2010). Consumption-based accounting of CO2 emissions. *Biological Sciences, 107*(12), 5687–5692. https://doi.org/10.1073/pnas.0906974107

Davis, S. J., Peters, G. P., & Caldeira, K. (2011). The supply chain of CO2 emissions. *Biological Science, 108*(45), 18554–18559. https://doi.org/10.1073/pnas.1107409108

Delfosse, C. (2012). La France et ses terroirs. Un siècle de débats sur les produits et leurs liens à l'espace. *Pour*, N° 215–216(3), 63–74. https://doi.org/10.3917/pour.215.0061

Dentoni, D., Bitzer, V., & Schouten, G. (2018). Harnessing wicked problems in multi-stakeholder partnerships. *Journal of Business Ethics, 150*, 333–356. https://doi.org/10.1007/s10551-018-3858-6

Dion, R. (2010). *Histoire de la vigne et du vin en France. Des origines au XIXè siècle*. CNRS Editions.

Ditter, J. G., & Brouard, J. (2014). The competitiveness of French protected designation of origin wines: A theoretical analysis of the role of proximity. *Journal of Wine Research, 25*(1), 5–18. https://doi.org/10.1080/09571264.2014.871120

Ditter, J.-G., Fensterseifer, J., & Brouard, J. (2017). Proximités et ressources territoriales dans la filière vin: l'exemple du vignoble de Chablis. *Économie Rurale, 362*, 9–28. https://doi.org/10.4000/economierurale.5331

Dunning, J. H. (1998). Location and the multinational enterprise: A neglected factor? *Journal of International Business Studies, 29*(1), 45–66. https://doi.org/10.1057/palgrave.jibs.8490024

Dyllick, T., & Muff, K. (2016). Clarifying the meaning of sustainable business: Introducing a typology from business-as-usual to true business sustainability. *Organization & Environment, 29*(2), 156–174. https://doi.org/10.1177/1086026615575176

Eang, M., Clarke, A., & Ordonez-Ponce, E. (2023). The roles of multinational enterprises in implementing the United Nations Sustainable Development Goals at the local level. *BRQ Business Research Quarterly, 26*(1), 79–97. https://doi.org/10.1177/23409444221140912

Eccles, R. G., Ioannou, I., & Serafeim, G. (2014). The impact of corporate sustainability on organizational processes and performance. *Management Science, 60*(11), 2835–2857. https://doi.org/10.1287/mnsc.2014.1984

Elkington, J. (1994). Towards the sustainable corporation: Win-Win-Win business strategies for sustainable development. *California Management Review, 36*(2), 90–100. https://doi.org/10.2307/41165746

Engström, R. E., Collste, D., Cornell, S. E., Johnson, F. X., Carlsen, H., Jaramillo, F., Finnveden, G., Destouni, G., Howells, M., Weitz, N., Palm, V., & Fuso-Nerini, F. (2021). Succeeding at home and abroad: accounting for the international spillovers of cities' SDG actions. *NPJ Urban Sustainability, 1*, 18. https://doi.org/10.1038/s42949-020-00002-w

Eskelinen, H., Hannibalsson, I., Malmberg, A., Maskell, P., & Vatne, E. (1998). *Competitiveness, localised learning and regional development: Specialization and prosperity in small open economies*. Routledge. https://doi.org/10.4324/9780203030165

European Commission. (2015). *Closing the loop—An EU action plan for the Circular Economy*. Communication From the Commission to the European Parliament, the Council, the European Economic and Social Committee and the Committee of the Regions. Retrieved August 15, 2024, from https://eur-lex.europa.eu/resource.html?uri=cellar:8a8ef5e8-99a0-11e5-b3b7-01aa75ed71a1.0012.02/DOC_1&format=PDF

Foley, J. A., DeFries, R., Asner, G. P., Barford, C., Bonan, G., Carpenter, S. R., Chapin, F. S., Coe, M. T., Daily, G. C., Gibbs, H. K., Helkowski, J. H., Holloway, T., Howard, E. A., Kucharik, C. J., Monfreda, C., Patz, J. A., Colin Prentice, I., Ramankutty, N., & Snyder, P. K. (2005). Global consequences of land use. *Science, 309*(5734), 570–574. https://doi.org/10.1126/science.1111772

Fuso Nerini, F., Mazzucato, M., Rockström, J., van Asselt, H., Hall, J. W., Matos, S., Persson, Å., Sovacool, B., Vinuesa, R., & Sachs, J. (2024). Extending the Sustainable Development Goals to 2050—A road map [comment]. *Nature, 630*, 555–558. https://doi.org/10.1038/d41586-024-01754-6

Fuso Nerini, F., Tomei, J., To, L. S., Bisaga, I., Parlkh, P., Black, M., Borrion, A., Spataru, C., Castan-Broto, V., Anandarajah, G., Milligan, B., & Mulgetta, Y. (2018). Mapping synergies and trade-offs between energy and the Sustainable Development Goals. *Nature Energy, 3*, 10–15. https://doi.org/10.1038/s41560-017-0036-5

Gaines, L. (2014). The future of automotive lithium-ion battery recycling: Charting a sustainable course. *Sustainable Materials and Technologies, 1–2*, 2–7. https://doi.org/10.1016/j.susmat.2014.10.001

Geissdoerfer, M., Savaget, P., Bocken, N. M., & Hultink, E. J. (2017). The circular economy—A new sustainability paradigm? *Journal of Cleaner Production, 143*, 757–768. https://doi.org/10.1016/j.jclepro.2016.12.048

Genovese, A., Acquaye, A. A., Figueroa, A., & Koh, S. C. L. (2017). Sustainable supply chain management and the transition towards a circular economy: Evidence and some applications. *Omega, 66*, 344–357. https://doi.org/10.1016/j.omega.2015.05.015

Gereffi, G., Humphrey, J., & Sturgeon, T. (2005). The governance of global value chains. *Review of International Political Economy, 12*(1), 78–104. https://doi.org/10.1080/09692290500049805

Ghemawat, P. (2007). *Redefining global strategy: Crossing borders in a world where differences still matter*. Harvard Business Review Press.

Ghisellini, P., Cialani, C., & Ulgiati, S. (2016). A review on circular economy: The expected transition to a balanced interplay of environmental and economic systems. *Journal of Cleaner Production, 114*, 11–32. https://doi.org/10.1016/j.jclepro.2015.09.007

Ghoshal, S., & Bartlett, C. A. (1990). The multinational corporation as an interorganizational network. *The Academy of Management Review, 15*(4), 603–625. https://doi.org/10.2307/258684

Giddings, B., Hopwood, B., & O'Brien, G. (2002). Environment, economy and society: Fitting them together into sustainable development. *Sustainable Development, 10*(4), 187–196. https://doi.org/10.1002/sd.199

Giuliani, E., & Macchi, C. (2014). Multinational corporations' economic and human rights impacts on developing countries: A review and research agenda. *Cambridge Journal of Economics, 38*, 479–517. https://doi.org/10.1093/cje/bet060

Global Reporting Initiative (GRI). (2023). *Voluntary National Review (VNR) Analysis 2023*. Retrieved September 30, 2024, from https://www.globalreporting.org/media/vzsl2iyb/gri-vnr-2023-analysis.pdf?utm_source=chatgpt.com

Govindarajan, V., & Trimble, C. (2012). *Reverse innovation: Create far from home*. Harvard Business Review Press.

Granovetter, M. (1985). Economic action and social structure: The problem of embeddedness. *American Journal of Sociology, 91*(3), 481–510. https://doi.org/10.1086/228311

Hart, S. L. (1995). A natural-resource-based view of the firm. *Academy of Management Review, 20*(4), 986–1014. https://doi.org/10.2307/258963

Hart, S. L., & Dowell, G. (2011). A natural-resource-based view of the firm: Fifteen years after. *Journal of Management, 37*(5), 1464–1479. https://doi.org/10.1177/0149206310390219

Hart, S. L., & Milstein, M. B. (2003). Creating sustainable value. *Academy of Management Perspectives, 17*, 56–67. https://doi.org/10.5465/ame.2003.10025194

Hasegawa, S. (Ed.). (2021). *Korona kano Sekai niokeru Keizai to Shakai* (Economic and social perspectives on the world under Covid-19—Through experiences of the lockdown in Paris). Tokyo: Dobunkan Shuppan (in Japanese).

Hasegawa, S. (2023). BoP businesses of multinational corporations and sustainability. In S. Urata, et al. (Eds.), *Sustainable development disciplines for society*. Springer. https://doi.org/10.1007/978-981-19-5145-9_12

Hasegawa, S. (2025). Bundan sareta Sekai ni okeru Kokusai Business no Saikochiku (Redefining international business in a fragmented world for a sustainable future). *Ritsumeikan Business Review, 63*(5), forthcoming.

Hawkins, T. R., Singh, B., Majeau-Bettez, G., & Strømman, A. H. (2013). Comparative environmental life cycle assessment of conventional and electric vehicles. *Journal of Industrial Ecology, 17*(1), 53–64. https://doi.org/10.1111/j.1530-9290.2012.00532.x

Hirczak, M., Moalla, M., Mollard, A., Pecquer, B., Rambonilaza, M., & Vollet, D. (2008). From the basket of goods to a more general model of territorialized complex goods: Concepts, analysis grid and question. *Canadian Journal of Regional Science, 31*(2), 241–260.

IEA. (2024). *Global EV Outlook 2024: Moving towards increased affordability*. Retrieved September 10, 2024, from https://www.iea.org/reports/global-ev-outlook-2024/outlook-for-emissions-reductions/

IInternational Energy Agency (IEA). (2021). *World Energy Outlook 2021*. IEA, Paris. Retrieved September 15, 2024, from https://www.iea.org/reports/world-energy-outlook-2021

INAO. (2016b). *Guide du demandeur d'une Appellation d'origine contrôlée / Appellation d'origine protégée (AOC/AOP) ou d'une Indication géographique protégée (IGP), Secteur viticole*. Retrieved September, 30, 2024, from https://www.inao.gouv.fr/content/download/1840/18727/version/2/file/201611_Guide%20du%20demandeur%20AOPIGP%20viticole.pdf

Institut National de l'Origine et de la Qualité (INAO). (2016a). *L'Appellation d'origine protégée/ contrôlée*. Retrieved Sepember 30, 2024, from https://www.inao.gouv.fr/content/download/893/8162/version/3/file/2016INAO%20-%20fiche%20signe%20-%20AOP%20AOC%20-%20BD.pdf

Intergovernmental Panel on Climate Change (IPCC). (2023). *Climate Change 2023: Synthesis Report. Contribution of Working Groups I, II and III to the Sixth Assessment Report of the Intergovernmental Panel on Climate Change*. IPCC. https://doi.org/10.59327/IPCC/AR6-9789291691647.

Jacobson, M. Z., & Delucchi, M. A. (2011). Providing all global energy with wind, water, and solar power, Part I: Technologies, energy resources, quantities and areas of infrastructure, and materials. *Energy Policy, 39*(3), 1154–1169. https://doi.org/10.1016/j.enpol.2010.11.040

Johanson, J., & Vahlne, J.-E. (2009). The Uppsala internationalization process model revisited: From liability of foreignness to liability of outsidership. *Journal of International Business Studies, 40*, 1411–1431. https://doi.org/10.1057/jibs.2009.24

Jones, G. (2023). *Deeply responsible business: A global history of values-driven leadership*. Harvard University Press.

Jones, G. (2004). Multinationals and global capitalism: From the nineteenth century to the twenty first century. *Oxford University Press*. https://doi.org/10.1093/0199272093.001.0001

Kalghatgi, G. (2022). Is it the end of combustion and engine combustion research? Should it be? *Transportation Engineering, 10*, Article 100142. https://doi.org/10.1016/j.treng.2022.100142

Kanie, N., Bernstein, S., Biermann, F., & Haas, P. M. (2017). Introduction: Global governance through goal setting. In N. Kanie & F. Biermann (Eds.), *Governing through goals* (pp. 1–27). MIT Press.

Katsaliaki, K., Galetsi, P., & Kumar, S. (2022). Supply chain disruptions and resilience: A major review and future research agenda. *Annals of Operations Research, 319*, 965–1002. https://doi.org/10.1007/s10479-020-03912-1

Kolk, A., Kourula, A., & Pisani, N. (2017). Multinational enterprises and the sustainable development goals: What do we know and how to proceed? *Transnational Corporations, 24*(3), 9–32. https://doi.org/10.18356/6f5fab5e-en

Kostova, T., & Roth, K. (2002). Adoption of an organizational practice by subsidiaries of multinational corporations: Institutional and relational effects. *Academy of Management Journal, 45*(1), 215–233. https://doi.org/10.2307/3069293

Kroll, C., Warchold, A., & Pradhan, P. (2019). Sustainable Development Goals (SDGs): Are we successful in turning trade-offs into synergies? *Palgrave Communications, 5*, 140. https://doi.org/10.1057/s41599-019-0335-5

Lacy, P., & Rutqvist, J. (2015). *Waste to wealth: The circular economy advantage*. Palgrave Macmillan. https://doi.org/10.1057/9781137530707

Le Blanc, D. (2015). Towards integration at last? The Sustainable Development Goals as a network of targets. *Sustainable Development, 23*(3), 176–187. https://doi.org/10.1002/sd.1582

Leedon, G., L'Espoir Decosta, P., Buttriss, G., & Lu, V. N. (2021). Consuming the earth? Terroir and rural sustainability. *Journal of Rural Studies, 87*, 415–422. https://doi.org/10.1016/j.jrurstud.2021.09.030

Lenzen, M., Moran, D., Kanemoto, K., Foran, B., Lobefaro, L., & Geschke, A. (2012). International trade drives biodiversity threats in developing nations. *Nature, 486*, 109–112. https://doi.org/10.1038/nature11145

London, T., & Hart, S. L. (2010). *Next generation business strategies for the base of pyramid: New approaches for building mutual value*. FT Press.

Lusseau, D., & Mancini, F. (2019). Income-based variation in Sustainable Development Goal interaction networks. *Nature Sustainability, 2*(3), 242–247. https://doi.org/10.1038/s41893-019-0231-4

Madhok, A., & Liu, C. (2006). A coevolutionary theory of the multinational firm. *Journal of International Management, 12*, 1–21. https://doi.org/10.1016/j.intman.2006.01.001

Meyer, K. E., Estrin, S., Bhaumik, S. K., & Peng, M. W. (2009). Institutions, resources, and entry strategies in emerging economies. *Strategic Management Journal, 30*(1), 61–80. https://doi.org/10.1002/smj.720

Milne, M. J., & Gray, R. (2013). W(h)ither ecology? The triple bottom line, the global reporting initiative, and corporate sustainability reporting. *Journal of Business Ethics, 118*, 13–29. https://doi.org/10.1007/s10551-012-1543-8

Ministry of Agriculture, Forestry and Fisheries of Japan MAFF. (2023). *FY2023 Annual Report on Food, Agriculture and Rural Area in Japan*. Retrieved September 30, 2024, from https://www.maff.go.jp/e/data/publish/Annual_Report/AnnualReportonFoodAgricultureandRuralAreas_FY2023.pdf

MAFF. (2024). *Japan's Strategy for Sustainable Food Systems (Strategy MIDORI), and Climate-Smart Agriculture*. Retrieved September 30, 2024, from https://www.maff.go.jp/e/policies/env/env_policy/04_Strategy_MIDORI_and_Climate-Smart_Agriculture.pdf

Ministry of Land Infrastructure, Transport and Tourism (MLIT). (2020). *Summary of the White Paper on Land, Infrastructure, Transport and Toursim in Japan, 2020*. Retrieved September 30, 2024, from https://www.mlit.go.jp/hakusyo/mlit/r01/hakusho/r02/pdf/English%20Summary.pdf

Ministry of the Environment (MOE), Ministry of Education, Culture, Sports, Science and Technology (MEXT), Ministry of Agriculture, Forestry and Fisheries (MAFF), Ministry of Land, Infrastructure, Transport and Tourism (MLIT), & Japan Meteorological Agency (JMA). (2018). *Synthesis Report on Observations, Projections and Impact Assessments of Climate Change, 2018 "Climate Change in Japan and Its Impacts"*. Retrieved September 25, 2024, from https://www.env.go.jp/content/900449807.pdf

MOE. (2021). *JBO3: Japan Biodiversity Outlook 3*. Working Group for Comprehensive Assessment of Biodiversity and Ecosystem Services, Ministry of the Environment. Retrieved September 30, 2024, from https://www.biodic.go.jp/biodiversity/activity/policy/jbo3/generaloutline/files/JBO3_pamph_en.pdf

Minbaeva, D., Björkman, I., Fey, C., & Park, H. (2014). MNC knowledge transfer, subsidiary absorptive capacity and HRM. *Journal of International Business Studies, 45*, 38–51. https://doi.org/10.1057/jibs.2013.43

Moinuddin, M., & Olsen, S. H. (2024). Examining the unsustainable relationship between SDG performance, ecological footprint and international spillovers. *Scientific Reports, 14*, 11277. https://doi.org/10.1038/s41598-024-61530-4

Mollard, A. (2001). Qualité et développement territorial : une grille d'analyse théorique à partir de la rente. *Économie Rurale, 263*, 16–34. https://doi.org/10.3406/ecoru.2001.5240

Mori, N., Takemi, T., Tachikawa, Y., Tatano, H., Shimura, T. Tanaka, T., Fujimi, T., Osakada, Y., Webb, A., & Nakakita, E. (2021). Recent nationwide climate change impact assessments of natural hazards in Japan and East Asia. *Weather and Climate Extremes, 32*. https://doi.org/10.1016/j.wace.2021.100309

Neumann, J., Petranikova, M., Meeus, M., Gamarra, J.D., Younesi, R., Winter, M., & Nowak, S. (2022). Recycling of Lithium-Ion batteries—Current state of the art, circular economy, and next generation recycling. *Advanced Energy Materials, 12*(17). https://doi.org/10.1002/aenm.202102917

Nilsson, M., Chisholm, E., Griggs, D., Howden-Chapman, P., McCollum, D., Messerli, P., Neumann, B., Stevance, A.-S., Visbeck, M., & Stafford-Smith, M. (2018). Mapping interactions between the sustainable development goals: Lessons learned and ways forward. *Sustainability Science, 13*, 1489–1503. https://doi.org/10.1007/s11625-018-0604-z

Nilsson, M., Griggs, D., & Visbeck, M. (2016). Map the interactions between Sustainable Development Goals. *Nature, 534*, 320–322. https://doi.org/10.1038/534320a

Nilsson, M., Hackmann, H., Sokona, Y., Guilanpour, K., Oni, T., Dzebo, A., Reyers, B., Zusman, E., Olsen, S., & Onoda, S. (2024). *Seeking synergy solutions: Policies that support both climate and SDG action.* Expert Group on Climate and SDG Synergy. Retrieved September 20, 2024, from https://sdgs.un.org/sites/default/files/2024-06/Thematic%20Report%20on%20Climate%20and%20SDGs%20Action-060824.pdf

Nohria, N., & Ghoshal, S. (1997). *The differentiated network: Organizations knowledge flows in multinational corporations.* Jossey-Bass.

Norgate, T., & Haque, N. (2010). Energy and greenhouse gas impacts of mining and mineral processing operations. *Journal of Cleaner Production, 18*(3), 266–274. https://doi.org/10.1016/j.jclepro.2009.09.020

Ordonez-Ponce, E., Clarke, A., & MacDonald, A. (2021). Business contributions to the Sustainable Development Goals through community sustainability partnerships. *Sustainability Accounting, Management and Policy Journal, 12*(6), 1239–1267. https://doi.org/10.1108/SAMPJ-03-2020-0068

Ostrom, E. (1990). *Governing the commons: The evolution of institutions for collective action.* https://doi.org/10.1017/CBO9780511807763

Ostrom, E. (2009). A general framework for analyzing sustainability of social-ecological systems. *Science, 325*(5939), 419–422. https://doi.org/10.1126/science.1172133

Pecqueur, B. (2001). Qualité et développement territorial: l'hypothèse du panier de biens et de services territorialisés. *Économie Rurale, 261*, 37–49. https://doi.org/10.3406/ecoru.2001.5217

Pedersen, E. R. G., & Andersen, K. R. (2015). Sustainability innovators and anchor draggers: A global expert study on sustainable fashion. *Journal of Fashion Marketing and Management, 19*(3), 315–327. https://doi.org/10.1108/JFMM-08-2014-0059

Peters, G. P., Minx, J. C., Weber, C. L., & Edenhofer, O. (2011). Growth in emission transfers via international trade from 1990 to 2008. *Biological Sciences, 108*(21), 8903–8908. https://doi.org/10.1073/pnas.1006388108

Pieroni, M. P. P., McAloone, T. C., & Pigosso, D. C. A. (2019). Business model innovation for circular economy and sustainability: A review of approaches. *Journal of Cleaner Production, 215*. https://doi.org/10.1016/j.jclepro.2019.01.036

Pinkse, J., & Kolk, A. (2008). Multinational enterprises and climate change: Exploring institutional failures and embeddedness. *Journal of International Business Studies, 39*(8), 1357–1378. https://doi.org/10.1057/jibs.2011.56

Porter, M. E., & Kramer, M. R. (2019). Creating shared value. In G. G. Lenssen & N. C. Smith (Eds.), *Managing sustainable business.* Springer, Dordrecht. https://doi.org/10.1007/978-94-024-1144-7_16

Porter, M. E., & van der Linde, C. (1995). Toward a new conception of the environment competitiveness relationship. *Journal of Economic Perspectives, 9*(4), 97–118. https://doi.org/10.1257/jep.9.4.97

Pradhan, P. (2019). Antagonists to meeting the 2030 Agenda. *Nature Sustainability, 2*, 171–172. https://doi.org/10.1002/2017EF000632

Pradhan, P., Costa, L., Rybski, D., Lucht, W., & Kropp, J. P. (2017). A Systematic study of Sustainable Development Goal (SDG) interactions. *Earth's Future, 5*(11), 1169–1179. https://doi.org/10.1002/2017EF000632

Prahalad, C. K., & Doz, Y. L. (1987). *The multinational mission: Balancing local demands and global vision.* Free Press.

Prahalad, C. K., & Hart, S. L. (1999). *Strategies for the bottom of the pyramid: Creating sustainable development.* Ann Arbor and Chapel Hill.

Rastoin, J. L. (2012). The concept of terroir as the basis of corporate strategy in agribusiness: The European social, economic and institutional model. In L. Augustin-Jean, H. Ilbert, & N. Saavedra-Rivano (Eds.), *Geographical Indications and International Agricultural Trade.* Palgrave Macmillan. https://doi.org/10.1057/9781137031907_7

Rastoin, J. L., & Vissac-Charles, V. (1999). Le groupe stratégique des entreprises de terroir. *Revue Internationale PME, 12*(1–2), 171–192. https://doi.org/10.7202/1008655ar

Renaud, F. G., Zhou, X., Bosher, L., Barrett, B., & Huang, S. (2022). Synergies and trade-offs between sustainable development goals and targets: Innovative approaches and new perspectives. *Sustainability Science, 17*, 1317–1322. https://doi.org/10.1007/s11625-022-01209-9

Rockström, J., & Klum, M. (2015). *Big world small planet: Abundance within planetary boundaries.* Yale University Press.

Rockström, J., Steffen, W., Noone, K., Perssn, A., Chapin III, F. S., Lambin, E. F., Lenton, T. M., Scheffer, M., Folke, C., Schellnhuber, H. J., Nykvist, B., de Wit, C. A., Hughes, T., van der Leeuw, S., Rodhe, H., Sörlin, S., Snyder, P. K., Costanza, R., Svedin, U., … Foley, J. A. (2009). A safe operating space for humanity. *Nature, 461*(7263), 472–475. https://doi.org/10.1038/461472a

Rugman, A. M., & Verbeke, A. (2003). Regional multinationals and triad strategy. In A. M. Rugman (Ed.), *Leadership in international business education and research (Research in global strategic management, Vol. 8)* (pp. 253–268). Emerald Group Publishing Limited. https://doi.org/10.1016/S1064-4857(03)08016-1

Sachs, J. D. (2012). From millennium development goals to Sustainable Development Goals. *The Lancet, 379*(9832), 2206–2211. https://doi.org/10.1016/S0140-6736(12)60685-0

Sachs, J. D., Lafortune, G., & Fuller, G. (2024). *The SDGs and the UN Summit of the Future. Sustainable Development Report 2024.* SDSN and Dublin University Press. https://doi.org/10.25546/108572

Sachs, J. D., Schmidt-Traub, G., Kroll, C., Lafortune, G., & Fuller, G. (2021). *Sustainable Development Report 2020: The sustainable development goals and Covid-19 includes the SDG index and dashboards*. Cambridge University Press. https://doi.org/10.1017/9781108992411

Salvia, A. L., Leal Filho, W., Brandli, L. L., & Griebeler, J. S. (2019). Assessing research trends related to Sustainable Development Goals: Local and global issues. *Journal of Cleaner Production, 208*, 841–849. https://doi.org/10.1016/j.jclepro.2018.09.242

Saxenian, A. (1996). *Regional advantage: Culture and competition in Silicon Valley and Route 128*. Harvard University Press. https://doi.org/10.1177/027046769601600314

Schaltegger, S., & Wagner, M. (2011). Sustainable entrepreneurship and sustainability innovation: Categories and interactions. *Business Strategy and the Environment, 20*(4), 222–237. https://doi.org/10.1002/bse.682

Scherer, L., Behrens, P., de Koning, A., Heijungs, R., Sprecher, B., & Tukker, A. (2018). Trade-offs between social and environmental Sustainable Development Goals. *Environmental Science & Policy, 90*, 65–72. https://doi.org/10.1016/j.envsci.2018.10.002

Schröder, P., & Barrie, J. (2024). *How the circular economy can revive the Sustainable Development Goals: Priorities for immediate global action, and a policy blueprint for the transition to 2050*. Chatham House Research Paper, Royal Institute of International Affairs, https://doi.org/10.55317/9781784136222.

Seuring, S., & Müller, M. (2008). From a literature review to a conceptual framework for sustainable supply chain management. *Journal of Cleaner Production, 16*(15), 1699–1710. https://doi.org/10.1016/j.jclepro.2008.04.020

Slack, K. (2012). Mission impossible?: Adopting a CSR-based business model for extractive industries in developing countries. *Resources Policy, 37*(2), 179–184. https://doi.org/10.1016/j.resourpol.2011.02.003

Smith, P., Bustamante, M., Ahammad, H., & Clark, H. (2014). Agriculture, Forestry and Other Land Use (AFOLU). In *Climate Change 2014: Mitigation of Climate Change. IPCC Working Group III Contribution to AR5*. Contribution of Working Group III to the Fifth Assessment Report of the Intergovernmental Panel on Climate Change (IPCC). Retrieved September 30, 2024, from https://www.ipcc.ch/site/assets/uploads/2018/02/ipcc_wg3_ar5_full.pdf

Smith, P., Bustamante, M., Ahammad, H., Clark, H., Dong, H., Elsiddig, E. A., Haberl, H., Harper, R., House, J., Jafari, M., Masera, O., Mbow, C., Ravindranath, N. H., Rice, C. W., Robledo Abad, C., Romanovskaya, A., Sperling, F., & Tubiello, F. (2014). Agriculture, Forestry and Other Land Use (AFOLU). In *Climate Change 2014: Mitigation of Climate Change. Contribution of Working Group III to the Fifth Assessment Report of the Intergovernmental Panel on Climate Change*. Cambridge University Press.

Stafford-Smith, M., Griggs, D., Gaffney, O., Ullah, F., Reyers, B., Kanie, N., Stigson, B., Shrivastava, P., Leach, M., & O'Connell, D. (2017). Integration: the key to implementing the Sustainable Development Goals. *Sustainability Science, 12*, 911–919. https://doi.org/10.1007/s11625-016-0383-3

Stahel, W. R. (2016). The circular economy. *Nature, 24*(531), 435–438. https://doi.org/10.1038/531435a

Steg, L., & Vlek, C. (2009). Encouraging pro-environmental behaviour: An integrative review and research agenda. *Journal of Environmental Psychology, 29*(3), 309–317. https://doi.org/10.1016/j.jenvp.2008.10.004

Swinburn, G., Goga, S., & Murphy, F. (2006). *Local economic development: A primer developing and implementing local economic development strategies and action plans*. World Bank. Retrieved October 10, 2024, from https://documents1.worldbank.org/curated/en/763491468313739403/pdf/337690REVISED0ENGLISH0led1primer.pdf

Szulanski, G. (1996). Exploring internal stickiness: Impediments to the transfer of best practice within the firm. *Strategic Management Journal, 17*(53), 27–43. https://doi.org/10.1002/smj.4250171105

Tanguay, G. A., Rajaonson, J., Lefebvre, J.-F., & Lanoie, P. (2009). *Measuring the sustainability of cities: A survey-based analysis of the use of local indicators*. CIRANO—Scientific Publications, 2009s-02. https://doi.org/10.2139/ssrn.1336649

Triandis, H. C. (1995). *Individualism & collectivism*. Westview Press. https://doi.org/10.4324/9780429499845

Uğurlu, E. (2019). Renewable energy strategies for sustainable development in the European Union. In D. Kurochkin, E. Shabliy, & E. Shittu (Eds.), *Renewable Energy*. Palgrave Macmillan, Cham. https://doi.org/10.1007/978-3-030-14207-0_3

United Nations (UN). (2014). *Report of the Open Working Group of the General Assembly on Sustainable Development Goals*, A/68/970. Retrieved September 30, 2024, from https://digitallibrary.un.org/search?f1=author&as=1&sf=title&so=a&rm=&m1=e&p1=UN.+General+Assembly.+Open+Working+Group+on+Sustainable+Development+Goals+%282014+%3A+New+York%29&ln=en

UN. (2015). *Transforming our world: the 2030 agenda for sustainable development*. United Nations. Retrieved August 30, 2024, from https://sustainabledevelopment.un.org/content/documents/21252030%20Agenda%20for%20Sustainable%20Development%20web.pdf

UN. (2021). *2021 Voluntary National Reviews Synthesis Report*. UN Department of Economic and Social Affairs (UNDESA). Retrieved September 30, 2024, from https://desapublications.un.org/publications/2021-voluntary-national-reviews-synthesis-report?utm_source=chatgpt.com

UN. (2024a). *Synergy Solutions for Climate and SDG Action: Bridging the Ambition Gap for the Future We Want*. UN Department of Economic and Social Affairs. Retrieved September 30, 2024, from https://sdgs.un.org/sites/default/files/2024-08/UN%20Synergy%20Solutions%20for%20Climate%20and%20SDG%20Action-4_0.pdf

UN. (2024b). *The Sustainable Development Goals Report 2024*. United Nations Publications. Retrieved September 30, 2024, from https://unstats.un.org/sdgs/report/2024/The-Sustainable-Development-Goals-Report-2024.pdf

UN. (2024c). *Pact for the Future, Global Digital Compact and Declaration on Future Generations*. Retrieved September 30, 2024, from https://www.un.org/sites/un2.un.org/files/sotf-pact_for_the_future_adopted.pdf

UN Development Programme (UNDP). (2019). *Decade of Action*. Retrieved September 30, 2024, from https://www.undp.org/stories/decade-action?utm_source=chatgpt.com

UN Educational, Scientific and Cultural Organization (UNESCO). (2005). *Rencontres internationales planète terroirs*. Retrieved 2September 30, 2024, from https://unesdoc.unesco.org/ark:/48223/pf0000154388.

UN Framework Convention on Climate Change (UNFCCC). (2021). *COP26: Update to the NDC Synthesis Report*. Retrieved September 30, 2024, from https://unfccc.int/news/cop26-update-to-the-ndc-synthesis-report

UN Office for Disaster Risk Reduction (UNDRR). (2019). *Global Assessment Report on Disaster Risk Reduction 2019*. Retrieved October 5, 2024, from https://www.undrr.org/media/73965

Van Zanten, J. A., & Van Tulder, R. (2018). Multinational enterprises and the Sustainable Development Goals: An institutional approach to corporate engagement. *Journal of International Business Policy, 1*, 208–233. https://doi.org/10.1057/s42214-018-0008-x

Voegtlin, C., & Scherer, A. G. (2017). Responsible innovation and the innovation of responsibility: Governing sustainable development in a globalized world. *Journal of Business Ethics, 143*, 227–243. https://doi.org/10.1007/s10551-015-2769-z

Warchold, A., Pradhan, P., & Kropp, J. P. (2020). Variations in sustainable development goal interactions: Population, regional, and income disaggregation. *Sustainable Development, 29*(2), 285–299. https://doi.org/10.1002/sd.2145

Weitz, N., Carlsen, H., Nilsson, M., & Skanberg, K. (2018). Towards systemic and contextual priority setting for implementing the 2030 Agenda. *Sustainability Science, 13*, 531–548. https://doi.org/10.1007/s11625-017-0470-0

Wernerfelt, B. A. (1984). Resource-based view of the firm. *Strategic Management Journal, 5*(2), 171–180. https://doi.org/10.1002/smj.4250050207

White, L. (1967). The historical roots of our ecologic crisis. *Science, 155*(3767), 1203–1207. https://doi.org/10.1126/science.155.3767.1203

Wood, R., Stadler, K., Simas, M., Bulavskaya, T., Giljum, S., Lutter, S., & Tukker, A. (2018). Growth in environmental footprints and environmental impacts embodied in trade: Resource efficiency indicators from EXIOBASE3. *Journal of Industrial Ecology, 22*, 553–564. https://doi.org/10.1111/jiec.12735

World Commission on Environment and Development (WCED). (1987). *Report of the World Commission on Environment and Development: Our Common Future*. Retrieved September 10, 2024, from https://sustainabledevelopment.un.org/content/documents/5987our-common-future.pdf

Zink, T., & Geyer, R. (2017). Circular economy rebound. *Journal of Industrial Ecology, 21*(3), 593–602. https://doi.org/10.1111/jiec.12545

11 Corporate Response to Sustainability Challenges: Research Review and Future Directions

Toru Yoshikawa and Yi Cheong Eugene Tung

Abstract

This chapter presents the major findings of prior studies on the effects of corporate governance, top executives, and top corporate decision-makers on firm sustainability performance. While there is a view that firms' primary responsibility is to serve shareholder interests, there is a growing awareness that firms should also make contributions in tackling societal problems, including environmental sustainability and carbon neutral society. This is especially true given that some firm activities negatively affect the environment and that firms have the resources to make contributions to address such challenges. Based on our review of previous studies, specific corporate governance practices and top executive attributes are associated with sustainability performance. Research also shows that specific ownership structures such as institutional ownership relate to firm sustainability performance. However, prior studies reveal that such effects are rather complex; some factors, such as executive compensation and board CSR committees, work together. This suggests the importance of examining the effects of corporate governance practices as a "bundle" and thus, firms must design their corporate governance system by choosing multiple practices that reinforce each other when they aim to improve their sustainability performance and contributions to achieve a carbon neutral society.

Keywords

Environmental sustainability · Corporate governance · Board of directors · Top management team · Ownership structure

11.1 Introduction

There is increasing awareness among policymakers as well as the investor and business communities that firms must bear the responsibility for tackling societal problems, including environmental sustainability to achieve a carbon neutral society (George et al., 2016). Even though there remains a view that firms are primarily responsible for delivering financial returns to shareholders (Lazonick & O'Sullivan, 2000), corporate executives now face rising pressures from various stakeholders to allocate corporate

T. Yoshikawa (✉)
Waseda University, Tokyo, Japan
e-mail: toru.yoshikawa@waseda.jp

Y. C. E. Tung
Chinese University of Hong Kong, Sha Tin, Hong Kong, China
e-mail: eugenetung@cuhk.edu.hk

K. Akao and A. Washizu (eds.), *Climate Change Issues and Social Sciences*, Sustainable Development Goals Series, https://doi.org/10.1007/978-981-96-8951-4_11

resources to address societal problems while they pursue financial performance (Business Roundtable, 2019; Fleury et al., 2019).

This chapter examines prior studies that investigated the effects of corporate governance and top corporate executives on firm sustainability performance defined as corporate activities to reduce carbon emissions thereby contributing to achieve a carbon neutral society. Such corporate activities include, e.g., reduction of carbon emissions in their manufacturing process and supply chains and development of green products and technology. Many firms have also started to track, measure, and disclose their carbon footprint. Corporate governance remains the most pertinent field for exploring such corporate efforts and trends in terms of firms' engagement in sustainability-related initiatives, as decision-making bodies manage their relationships with their environments, including pressures to address environmental sustainability issues. By showing what we know about the impact of corporate governance on environmental sustainability and hence, firms' contributions to achieve a carbon neutral society, this chapter aims to offer insights to researchers and practitioners on the role of board of directors and top executives.

Studies on firms' efforts to address environmental sustainability issues have increased over the last decade. While a firm's social responsibility remains a practical, if not ideological, debate (George et al., 2016; Jones & Wicks, 1999), its role in the greater challenge to achieve a carbon neutral society is likely unclear at best, especially for firms whose business does not directly impact the environment. "Where the relationship between means and ends is opaque, (a) internal organizational structures become increasingly complex, (b) organizations persist in a state of perpetual reform, and (c) resources are often diverted away from core goals" (Bromley & Powell, 2012, p. 484; cf. Porter & Kramer, 2006). The pressures on organizations to align their policies and practices have become ever greater. These institutions seem to have "invisibly" figured out the way to channel those pressures to some common if not agreed courses of actions or trajectories—firms now must appear sustainable with appropriate policies, structures, and strategies to survive in the capitalist institutions (de Bakker et al., 2020). Strategic and mimetic (e.g. Dobbin et al., 2011; Kalev et al., 2006; Powell & DiMaggio, 1983) responses have been developed, especially when the means-ends relationship is far from certain (Cyert & James, 1963; Powell & DiMaggio, 1983), to tackle short-to-mid-term problems and survival needs (Centeno & Cohen, 2012; Davis & Kim, 2015). Thus, the role of top decision-makers is extremely important in navigating firms, thereby contributing to tackle environmental sustainability issues and achieve a carbon neutral society.

11.2 The Impact of Corporate Governance on Sustainability

Given recent reviews and meta-analyses (Fifka, 2013; Lu & Taylor, 2016), we note that counting significant results in previous studies does not provide an understanding of the overall relationship between specific corporate governance and sustainability-related dependent variables. This is true with or without taking sample or effect sizes into account (Velte & Stawinoga, 2020). We also note that when institutional systems represent qualitatively different configurations (e.g., Aguilera & Grøgaard, 2019; Yoshikawa et al., 2014), even meta-analytic methodologies have their limitations. Therefore, we do not aspire to add value to the conversation by conducting another comprehensive review. Instead, we seek to examine a few theoretically relevant questions around each element of corporate governance as firms respond to sustainability challenges.

We surveyed literature on firms' responses to sustainability challenges. Specifically, we reviewed corporate governance's impact on environmental performance. While the literature has accumulated a large body of knowledge on the relationship between corporate governance and corporate social responsibility (e.g., Jain & Jamali, 2016; Oh et al., 2018; Yoshikawa et al.,

2021), researchers have only recently begun to pay separate attention to environmental performance. While CSR performance has traditionally been the variable of interest for empirical studies in the field, CSR reporting and disclosure have also recently drawn attention, as they have become more institutionalized and required (Dienes et al., 2016). Thus, we draw from the more established literature but highlight those (newer) studies with the above two specific focuses, where possible.

11.2.1 Corporate Governance and Sustainability Performance

Agency theory posits that boards of directors monitor managers and give incentives to align interests in face of the threat of agency problems arising from the separation of ownership and control (Fama & Jensen, 1983). Where the agents (top managers) know more than the owners (shareholders), opportunistic managers act in self-interest even at the expense of shareholders (Jensen & Meckling, 1976). Stakeholders, not shareholders, manage whether the former enhances the firm's competitive advantages for (long-term) financial performance (Eccles et al., 2014; Surroca et al., 2010).

The Instrumental Stakeholder theory concurs with this stance, arguing that the adoption of prosocial practices could be value-enhancing for the firm (Jones, 1995; Jones et al., 2018). Jensen (2001) accepts stakeholder management's utility when it contributes to maximizing firm value. By contrast, Normative and other versions of stakeholder theories conceptualize the firm as having relationships with a broad set of stakeholders beyond shareholders (Freeman, 1984; Reed, 2002). A corporate governance system must ensure that the firm is managed to benefit all stakeholders (de Graaf & Stoelhorst, 2013; Parmar et al., 2010).

In another dimension, Resource dependence theory seeks to capture the more universally true imperative for firms to manage their relationships with various stakeholders, where a stakeholder controls the resources needed for the firm's survival with limited alternatives (Casciaro & Piskorski, 2005), is in the position of demanding effectiveness from the firm (Mitchell et al., 1997), and may legitimately use power to make the firm comply (Emerson, 1962). On this basis, the board of directors and top managers act as conduits to channel the needed resources (Pfeffer & Salancik, 1978)—the resource provision role (e.g., Boyd, 1990)—and leverage social capital through their social networks (Hillman & Dalziel, 2003). Accordingly, a firm performs sustainably and effectively (Hillman et al., 2001; Luoma & Goodstein, 1999).

In contrast, Institutionalism suggests that socially constructed institutions feature different patterns of dependencies and relationships (Hall & Soskice, 2001; Haxhi & Aguilera, 2017). Thus, different institutions accept shareholder primacy and implicit social responsibility to different extents (Matten & Moon, 2008). Concerning the relationship between corporate governance and sustainability performance, the institutional approach agrees with resource dependence theory, since institutional logic prescribes what is legitimate and how coalitions or actors depend on each other (Friedland & Alford, 1991; Suchman, 1995).

Although the above theoretical approaches are popular for explaining the influence of corporate governance on sustainability performance, this list is not exhaustive. Turning to the respective elements of corporate governance, we relate previous findings in the literature to the above. We screened major management and finance journals using such key words as "corporate governance," "sustainability," and CSR. Table 11.1 shows the selected articles.

11.2.2 Board of Directors and Sustainability Performance

11.2.2.1 Board Composition and Capital

The positive relationship between female directors on the board and a firm's sustainability performance is one of the most reiterated findings. Through a meta-analysis, Endrikat et al.

Table 11.1 Selected articles on corporate governance and sustainability

Authors	Journal (Year)	Title	Findings
Board Composition			
Boulouta	JBE (2013)	Hidden connections: The link between board gender diversity and corporate social performance	Board gender diversity has greater impact in areas that induce higher levels of "empathic caring" among female directors
Byron and Post	CGIR (2016)	Women on boards of directors and corporate social performance: A meta-analysis	The female board representation–social performance relationship is more positive in favorable national contexts that welcome women directors' resources
Galbreath	B&S (2018)	Is board gender diversity linked to financial performance? The mediating mechanism of CSR	Greater representation of women on boards has an indirect, positive impact on financial performance, mediated by higher CSR levels
Harjoto et al.	JBE (2015)	Board diversity and corporate social responsibility	Board diversity is associated with more strengths in CSR and fewer concerns
Hummel et al.	JBE (2019)	The role of sustainability performance and accounting assurors in sustainability assurance engagements	In-depth assurance processes could have appealed to poor sustainability performers to enhance their internal sustainability-related processes and systems. Assurance providers that do not belong to the accounting profession are in turn associated with broader assurance statements
Post et al.	JBE (2015)	From board composition to corporate environmental performance through sustainability-themed alliances	The higher the representation of women or independent directors on board, the more likely the firm is to form sustainability-themed alliances
Rao and Tilt	JBE (2016)	Board composition and corporate social responsibility: The role of diversity, gender, strategy, and decision-making	This review reveals the gaps in conceptualizing CSR and CSR reporting as strategic decision-making processes
Walls and Hoffman	JOB (2013)	Exceptional boards: Environmental experience and positive deviance from institutional norms	Organizations located on the periphery of the network or whose boards possess a high level of environmental experience are more likely to deviate in positive ways
Board Structure			
Arora and Dharwadkar	CGIR (2011)	Corporate governance and corporate social responsibility (CSR): The moderating roles of attainment discrepancy and organization slack	Strong support for behavioral theory of the firm as the association between corporate governance and CSR dimensions depends on the decision-making latitude that originates from relative firm performance
Ayuso et al.	B&S (2014)	Maximizing stakeholders' interests: An empirical analysis of the stakeholder approach to corporate governance	A positive relationship between pro-stakeholder board practices and firm profitability in countries with Anglo-Saxon legal tradition
Ben-Amar et al.	JBE (2017)	Board gender diversity and corporate response to sustainability initiatives: Evidence from the carbon disclosure project	A critical mass of women directors is required for significant positive effects on voluntary climate related disclosures

(continued)

Table 11.1 (continued)

Authors	Journal (Year)	Title	Findings
Burke et al.	JBE (2019)	The heterogeneity of board-level sustainability committees and corporate social performance	Sustainability committees are more effective at impacting relevant strengths when they are focused on a specific stakeholder group, but they do not mitigate relevant concerns
De Villiers et al.	JOM (2011)	The effect of board characteristics on firm environmental performance	Found support for both resource provisional and agency-theoretical roles of the board for higher environmental performance
Endrikat et al.	B&S (2021)	Board characteristics and corporate social responsibility: A meta-analytic investigation	The authors confirm a positive impact of board size, board independence, female board representation, and presence of a CSR committee on CSR
Hussain et al.	JBE (2018)	Corporate governance and sustainability performance: Analysis of triple bottom line performance	Found support for both agency and stakeholder theory. However, no support for any significant relationship between the economic bottom of sustainability performance and CG characteristics
Jain and Jamali	CGIR (2016)	Looking inside the black box: The effect of corporate governance on corporate social responsibility	Corporate governance factors interact across multiple levels of analyses, leading to CSR
Jizi et al.	JBE (2014)	Corporate governance and corporate social responsibility disclosure: Evidence from the US banking sector	Board independence and size are positively related to CSR disclosure; CEO duality is also positively related to CSR disclosure
Lins et al.	JOF (2017)	Social capital, trust, and firm performance: The value of corporate social responsibility during the financial crisis	CSR intensity served as a cushion for stock returns during the financial crisis
Luo et al.	AMJ (2017)	Whose call to answer: Institutional complexity and firms' CSR reporting	Found a specific decoupling response manifested as CSR reporting in face of tension between central and local governments
Oh et al.	JOM (2018)	Complementary or substitutive effects? Corporate governance mechanisms and corporate social responsibility	Blockholder ownership and proportion of outside directors as substituting monitoring mechanisms, while TMT ownership and TMT incentive intensity act as substituting incentive alignment mechanisms
Post et al.	B&S (2011)	Green governance: Boards of directors' composition and environmental corporate social responsibility	A higher proportion of outside board directors, boards with three or more female directors, and boards with an average age of 56 years and a great number of Western European have superior environmental performance
Rodrigue et al.	JBE (2013)	Is environmental governance substantive or symbolic? An empirical investigation	Statistical analyses and interviews show environmental governance mechanisms are largely geared toward protecting the firm from reputational/regulatory harm
Walls and Berrone	JBE (2017)	The power of one to make a difference: How informal and formal CEO power affect environmental sustainability	CEO power based on environmental expertise and formal influence over executives and directors, regardless of shareholder activism, spurs firms toward greener strategies

(continued)

Table 11.1 (continued)

Authors	Journal (Year)	Title	Findings
Walls et al.	SMJ (2012)	Corporate governance and environmental performance: Is there really a link?	A multi-level interactional investigation on board structure to environmental strengths and concerns using fact-based analyses that examined an array of CG IVs and the environmental performance DVs
CEO			
Bénabou and Tirole	Economica (2010)	Individual and corporate social responsibility	An interpretation of the agency problem that leads to higher levels of CSR
Höllerer	JMS (2013)	From taken-for-granted to explicit commitment: The rise of CSR in a corporatist country	Dependence on the international capital market; visibility of corporation in the media; global CSR media coverage; firm age and CEO relationships with established elites
Marquis and Qian	OrgSci (2014)	Corporate social responsibility reporting in China: Symbol or substance?	Political dependence is positively linked to CSR reporting; the higher administrative embeddedness, the more substantive the CSR reporting
Wernicke et al.	AMD (2022)	How much influence do CEOs have on company actions and outcomes? The example of corporate social responsibility	CEOs explain some 30% of variance in CSR. The remainder varies across stakeholder categories and types of CSR/misconducts
CEO Characteristics			
Lewis et al.	SMJ (2014)	Difference in degrees: CEO characteristics and firm environmental disclosure	While CEOs with MBA degrees and newly appointed CEOs are more likely to voluntarily disclose environmental information, those with legal education resist the pressures to disclose
Tang et al.	SMJ (2018)	The differential effects of CEO narcissism and hubris on corporate social responsibility	When board-interlocked peer firms invest less intensively in CSR than a CEO's own firm, it strengthens the positive relationship between CEO narcissism and CSR, and the negative relationship between CEO hubris and CSR
Tang et al.	SMJ (2015)	How CEO hubris affects corporate social (ir) responsibility	CEO hubris is negatively associated with CSR performance and positively with CSiR, but the effect is weakened by higher resource dependence upon stakeholders
TMT			
Dixon-Fowler et al.	JBE (2017)	The role of board environmental committees in corporate environmental performance	A positive association between board environmental committees and CEP, especially when moderated by the presence of a senior-level environmental manager but not separately or for stakeholder representation
Jamali et al.	CGIR (2008)	Corporate governance and corporate social responsibility synergies and interrelationships	The majority of managers among those interviewed conceive of CG as a necessary pillar for sustainable CSR
Kanashiro and Rivera	JBE (2019)	Do chief sustainability officers make companies greener? The moderating role of regulatory pressures	The authors found that CSO has a positive influence on a firm's environmental performance if faced with strict environmental regulations

(continued)

Table 11.1 (continued)

Authors	Journal (Year)	Title	Findings
Lau et al.	JBE (2016)	Corporate social responsibility in China: A corporate governance approach	CSR activities are associated with foreign experience among board members but less so for TMT. CSR performance is positively related to larger board size and state ownership
Peters et al.	JBE (2019)	The influence of corporate sustainability officers on performance	CSO expertise and the firm's existing sustainability performance affect the association between the CSO and post-appointment sustainability performance
Strand	JBE (2013)	The chief officer of corporate social responsibility: A study of its presence in top management teams	Scandinavian firms are significantly more likely than US firms to have such CSR TMT positions, a measure of explicit CSR
Strand	JBE (2014)	Strategic leadership of corporate sustainability	A corporate sustainability TMT position can be an effort to realize external opportunities that may have otherwise gone unrealized, without concerted attention and coordination afforded by a strategic level position
Wiengarten et al.	JBE (2017)	How does sustainability leadership affect firm performance? The choices associated with appointing a chief officer of corporate social responsibility	Under certain conditions, appointing a chief executive of CSR results in financial performance benefits, especially if the appointee is female and has a CSR functional background
Wong et al.	AMJ (2011)	The effects of top management team integrative complexity and decentralized decision-making on corporate social performance	Decentralization has a significant positive effect on corporate social performance, regardless of the integrative complexity
Yin	B&S (2017)	Institutional drivers for corporate social responsibility in an emerging economy: A mixed-method study of Chinese business executives	Implicit ethical corporate culture plays a key role in predicting different CSR aspects, while external institutional mechanisms mainly predict market-oriented CSR initiatives
Ownership Structure			
Aguilera et al.	CGIR (2006)	Corporate governance and social responsibility: A comparative analysis of the UK and the US	Differences in motivations among institutional owners likely would lead to differences in CSR performances
Barnea and Rubin	JBE (2010)	Corporate social responsibility as a conflict between shareholders	Insiders seem to induce firms to over-invest in CSR when they bear little of the cost of doing so
Berrone et al.	ASQ (2010)	Socioemotional wealth and corporate responses to institutional pressures: Do family-controlled firms pollute less?	A socioemotional wealth protection hypothesis: family CEO or CEO duality has no effect. But for non-family firms, CEO stock ownership shows negative environmental impact
Borghesi et al.	JCF (2014)	Corporate socially responsible investments: CEO altruism, reputation, and shareholder interests	Institutional ownership is negatively related to CSR. CEOs receive private benefits from CSR investments. CSR intensity and excess stock returns are negatively correlated

(continued)

Table 11.1 (continued)

Authors	Journal (Year)	Title	Findings
Buchanan et al.	JCF (2018)	Corporate social responsibility, firm value, and influential institutional ownership	During heightened agency conflicts, the costs of ESG/CSR over-investment caused higher-scoring ESG/CSR firms to experience greater declines in firm values
Chen et al.	JFE (2020)	Institutional shareholders and corporate social responsibility	An exogenous increase in institutional holding caused by Russell Index reconstitutions improves portfolio firms' CSR performance, and institutional shareholders can generate real social impact
Dam and Scholtens	CGIR (2012)	Does ownership type matter for corporate social responsibility?	Ownership by employees, individuals, and firms is associated with relatively poor corporate social policies; whereas holdings by banks, institutional investors, and the state appear to be neutral
Dyck et al.	JFE (2019)	Do institutional investors drive corporate social responsibility? International evidence	A causal relationship exists between institutional ownership and E&S performance. Investors increase firms' E&S performance when they come from countries with a strong community belief in E&S issues, but not otherwise
Flammer	JFE (2021)	Corporate green bonds	The author shows that investors respond positively to the issuance announcement, especially for first-time issuers and bonds certified by third parties
Hartzmark and Sussman	JOF (2019)	Do investors value sustainability? A natural experiment examining ranking and fund flows	Sustainability is viewed as positively predicting future performance, but they do not find evidence that high-sustainability funds outperform low-sustainability funds
Khan et al.	JBE (2013)	Corporate governance and corporate social responsibility disclosures: Evidence from an emerging economy	CSR disclosure is positively related to managerial ownership, public or foreign ownership, board independence and audit committees
Lamb and Butler	B&S (2018)	The influence of family firms and institutional owners on corporate social responsibility performance	Family owners and family CEOs promote CSR strengths and reduce CSR concerns. Transient institutional owners have an opposite effect while dedicated institutional owners are positively associated with concerns
Lee and Lounsbury	B&S (2011)	Domesticating radical rant and rage: An exploration of the consequences of environmental shareholder resolutions on corporate environmental performance	A positive effect of environmental shareholder resolutions existed on the targeted firm's pollution management practice. Firms concede to the demands from shareholder activists accordingly to resource dependence theory
Liang and Renneboog	JOF (2017)	On the foundations of corporate social responsibility	Firms from common law countries have lower CSR than companies from civil law countries; Scandinavian civil law firms had the highest ratings
Mackenzie et al.	CGIR (2013)	Do responsible investment indices improve corporate social responsibility? FTSE4Good's impact on environmental management	Compliance depends on how firms are engaged in the FTSE index and the effect of being expelled; time of continuous engagement is a significant factor

(continued)

Table 11.1 (continued)

Authors	Journal (Year)	Title	Findings
Oh et al.	JBE (2011)	The effect of ownership structure on corporate social responsibility: Empirical evidence from Korea	A positive relationship between CSR ratings and ownership by institutions and foreign investors but a negative relationship with shareholding by top managers
Perez-Batres et al.	JBE (2010)	CSR, sustainability and the meaning of global reporting for Latin American corporations	Formal normative pressure mediated by listing on the NYSE
Rees and Rodionova	CGIR (2015)	The influence of family ownership on corporate social responsibility: An international analysis of publicly listed companies	When controlled for governance, closely held equity is no longer associated with environmental and social rankings, but family ownership retains a significant negative association
Tang and Zhang	JCF (2020)	Do shareholders benefit from green bonds?	Firm's issuance of green bonds is beneficial to its existing shareholders, seen from raised stock prices and institutional ownership
Executive Compensation			
Al-Shaer and Zaman	JBE (2019)	CEO compensation and sustainability reporting assurance: Evidence from the UK	Both board-level sustainability committees and sustainability reporting assurance have a positive association with the inclusion of sustainability terms in compensation contracts
Berrone and Gomez-Mejia	AMJ (2009)	Environmental performance and executive compensation: An integrated agency-institutional perspective	Environmental governance mechanisms strengthen how good environmental performance increases CEO pay in polluting industries; pollution prevention strategies have greater effect than end-of-pipe pollution control
Eccles et al.	MS (2014)	The impact of corporate sustainability on organizational processes and performance	High sustainability companies are characterized by expansive direct board involvements and sustainability-linked top executive compensation
Klettner et al.	JBE (2014)	The governance of corporate sustainability: Empirical insights into the development, leadership, and implementation of responsible business strategy	The involvement of the board and TMT in sustainability strategy development and relevant monitoring and incentivizing rewards demonstrate an enlightened shareholder value approach
Kock et al.	JMS (2012)	Corporate governance and the environment: what type of governance creates greener companies?	The divergence between stakeholders' and managers' environmental interests is affected by the presence and nature of both internal and external governance mechanisms
Maas	JBE (2018)	Do corporate social performance targets in executive compensation contribute to corporate social performance?	The use of CSP targets is an effective way to improve CSP, especially to lower CSP weaknesses

Note AMD = Academy of Management Discoveries; AMJ = Academy of Management Journal; ASQ = Administrative Science Quarterly; B&S = Business & Society; CGIR = Corporate Governance: An International Review; JBE = Journal of Business Ethics; JCF = Journal of Corporate Finance; JFE = Journal of Financial Economics; JIBS = Journal of International Business Studies; JMS = Journal of Management Studies; JOB = Journal of Organizational Behavior; JOF = Journal of Finance; JOM = Journal of Management; MS = Management Science; OrgSci = Organization Science; SMJ = Strategic Management Journal

(2021) confirm the positive impact of female board representation on CSR with social responsibility. Giannarakis et al. (2014) find that women's presence on boards determines strategic configurations and disclosure scores. Boulouta (2013) show that boards with greater gender diversity take better care of sustainability-related areas of concern because higher levels of "empathic caring" appeals strongly to female directors. Higher levels of CSR mediate the relationship between greater representation of women on boards and financial performance (Galbreath, 2018). Moreover, the effects of female directors do not appear to be bound by the culture that embeds the board. Areneke et al. (2023) show that female directors influence ethical CG disclosure practices in a highly patriarchal context. Byron and Post (2016) report that females may be better able to bring resources to a motivated board or in cultural contexts where intra-board power distribution may be more balanced owing to higher gender parity. Generally, diversity beyond gender leads to better sustainability performance, especially when diverse resources are provided to the board (de Villiers et al., 2011; Harjoto et al., 2015).

If the above diversity-related determinants largely support the resource dependence premise of board functioning, the impact of the CEO-board relationship could be relatively complicated. Overall, however, a board with better monitoring capability would also likely be one that better monitors firms' decision quality and actions aligning with stakeholders' interests (Dienes et al., 2016). Thus, we cannot assume that CEOs are negatively related to CSR or environmental performance. Higher environmental performance can be found in firms with higher board independence and a lower concentration of directors appointed after the CEO on the board (i.e., stronger monitoring capability of the board), as well as in firms with larger boards where the CEO has active representation and there are more legal experts (i.e., better available resources to tap on when the CEO aspires to lead in positive sustainability performance) (de Villiers et al., 2011). Similarly, Jizi et al. (2014) report a positive relationship between CSR disclosures and board independence (Khan et al., 2013; Oh et al., 2018), size, and CEO duality. A CEO with informal environmental expertise could lead to good environmental performance; however, such a positive impact is amplified when the CEO enjoys formal power over the board (Walls & Berrone, 2017). The authors also find that CEO power acts as a catalyst, transforming shareholder activism into corporate greening. This is likely explained by the mechanisms mentioned in the section on CEO characteristics and aspirations.

11.2.2.2 Board Committees

While the association between board-level sustainability committees and the inclusion of sustainability-related terms in the CEO compensation contract likely demonstrates the board's monitoring role (Al-Shaer & Zaman, 2019; Khan et al., 2013), the role of resource provision also manifests itself in the board committees (de Villiers et al., 2011). The more board standing committees exist, the more likely the firm will conduct voluntary environmental (Ben-Amar et al., 2017) and sustainability performance disclosure (Endrikat et al., 2021). The committee also moderates the relationship between top management team diversity and the firms' adoption of environmental standards (Saeed et al., 2022). Burke et al. (2019) and Dixon-Fowler et al. (2017) find that the link between a sustainability committee and environmental performance can be strengthened when the committee focuses on a specific stakeholder group (Ocasio, 2011), and is managed by a senior-level environmental manager. The presence of a sustainability committee is likely a result of both external and internal pressures, as well as a certain age distribution within the board (Post et al., 2011).

11.2.3 Top Management Team (TMT) and Sustainability Performance

11.2.3.1 CEO Attributes

Idiosyncratic CEO characteristics, values, and beliefs are likely to affect a CEO's inclination toward CSR (Mattingly, 2017). For example, while CEO narcissism correlates positively with

CSR (Petrenko et al., 2016), CEO hubris correlates negatively with CSR (Tang et al., 2015, 2018). charisma (Wowak et al., 2016) and greed (Sajko et al., 2021a) also could together play a role in driving or preventing the CEO from working on improving the firm's sustainability performance because the CEO is the face of the firm (Love et al., 2017), and may receive personal benefit from promoting CSR (Borghesi et al., 2014).

While male and female CEOs likely differ in terms of their inclinations toward sustainability (Glass et al., 2016), the CEOs are also substantially influenced by their "social-class background (Brammer et al., 2007; Kish-Gephart & Campbell, 2015), birth order (Campbell et al., 2019), military training (Koch-Bayram & Wernicke, 2018), the condition of the economy at the time they enter the job market (Bianchi & Mohliver, 2016), or the values prevalent when they grew up (Sajko et al., 2021b)" (Wernicke et al., 2022, p. 48). Educational backgrounds also play a reasonable role. CEOs with MBA degrees and newly appointed CEOs are more likely to voluntarily disclose environmental information, whereas those with a legal education are more sophisticated in understanding legal constraints (Lewis et al., 2014). The above impacts on a CEO's values and beliefs are long lasting (Marquis & Tilcsik, 2013). This is especially true when coupled with the path that the CEO has experienced (Marquis & Qian, 2014), and when CEOs may wish to conserve their legacy by acting sustainably (Kang, 2016).

11.2.3.2 TMT Team Composition

The involvement of TMT (Hambrick, 2007) and the board are crucial in developing sustainability strategies that demonstrate a more enlightened shareholder value approach to cater to stakeholders' needs (Klettner et al., 2014). When integrative complexity is high and the management structure is decentralized, TMTs tend to lead commendable corporate social performance because of their heightened ability to gather information and attend to stakeholders' needs (Carpenter et al., 2004; Finkelstein & Hambrick, 1990). However, high CSR levels are possible in firms with low integrative complexity if decentralization is high (Wong et al., 2011). In particular, the greater the power of female executives, the more likely the firm is to adopt environmental standards in emerging countries (Saeed et al., 2022), a relationship strengthened by the presence of CSR committees.

While the presence of a chief sustainability officer itself signals the firm's commitment to sustainability, Scandinavian firms are significantly more likely than US firms to have CSR TMT positions (Strand, 2013). This could lead to concerted attention and coordination efforts to capture favorable external opportunities. Nevertheless, firms may opt to rely on the bureaucratic implementation of sustainability efforts without such TMT positions (Strand, 2014). The upper echelons perspective and attention-based view suggest that a chief sustainability officer helps keep an eye on negative sustainability issues and channel attention and efforts toward reducing corporate social irresponsibility (Fu et al., 2020). While that sounds like appointing a chief CSR executive may help with CSR performance regardless of the officer's expertise in sustainability (Peters et al., 2019), the resultant financial benefits are greatest only when the appointee is female and has a functional CSR background (Wiengarten et al., 2017).

11.2.4 Executive Compensation and Sustainability Performance

11.2.4.1 Executive Compensation as Rewards and Incentives

Cohesion in the market for corporate governance, and thus social comparison, has resulted in an income surge among CEOs (Kim et al., 2015). However, a great sum of money carries both substance and symbolism (Zajac & Westphal, 1994). In terms of promoting sustainability performance, there is no robust understanding of the substantive or symbolic effects of top executive compensation. Maas (2018, p. 573) find that to boost corporate social performance, "(a) the level of CSP has no effect on the use of CSP targets, (b) the use of CSP targets in general does not automatically lead to better CSP results, and (c) the use of quantitative, hard CSP targets is an effective way to improve CSP results, especially to lower CSP weaknesses." Both the board and the TMT are involved in establishing monitoring and incentivizing rewards, demonstrating a stakeholder orientation (Klettner et al., 2014).

11.2.4.2 Joint Effects with Other Mechanisms

The presence of board-level sustainability committees and sustainability reporting assurance are positively associated with the inclusion of sustainability terms in compensation contracts for top executives (Al-Shaer & Zaman, 2019). Highly sustainable firms often feature direct board involvement in sustainability issues, and top executive compensation is frequently linked to sustainability objectives (Eccles et al., 2014). CEO equity compensation and ownership significantly reduce waste production (Kock et al., 2012). Ultimately, CEO pay could be considered substantive. This is because good environmental performance is rewarded in polluting industries, especially when moderated by environmental governance mechanisms. It is also symbolic, because pollution prevention strategies count (more than end-of-pipe pollution control) and because firms with an explicit environmental pay policy and an environmental committee do not actually reward environmental strategies more than those without such structures (Berrone & Gomez-Mejia, 2009).

11.2.5 Ownership Structure and Sustainability Performance

11.2.5.1 Property Rights and Use Rights

The "separation of ownership and control" (Berle & Means, 1932) suggested that owners forgo a portion of their "use right" when they deem corporate management too costly (Kang & Sørensen, 1999); but one must not mistake that the owners stop governing their properties once they have their managers (Davis, 2005; Zeitlin, 1974). Barnea and Rubin (2010) explicate CSR as a conflict between shareholders. When different owners are conceptualized as having divergent goals and interests (e.g., Li & Qian, 2013), insiders whose reputation is associated with a firm's CSR ratings would probably be the most receptive toward expenditure on sustainability issues, which is arguably an area of non-value-maximizing activities from Fama and Jensen's (1983) perspective. With higher levels of ownership, managers may better pursue their personal agendas through recognized CSR efforts (Bénabou et al., 2010). CSR disclosure is positively related to managerial ownership (Khan et al., 2013). However, seemingly contradictory evidence reveals that managerial ownership is related to poorer CSR policies and ratings (Dam & Scholtens, 2012; Oh et al., 2011). Future studies could contextualize the potential influence of managerial ownership. For instance, Hong et al. (2012) argue that managers are more likely to pursue CSR/ESG ratings when the firm has access to financing and when investors' appetite for risk is higher under macroeconomic conditions.

11.2.5.2 Family Ownership

Family owners also have a reason to promote sustainable performance. The socioemotional wealth protection hypothesis suggests that family-controlled-listed firms have better environmental performance than non-family-controlled firms (Berrone et al., 2010). Lamb and Butler (2018) also find support for families and family CEOs. By contrast, Rees and Rodionova (2015) find that family ownership is negatively associated with ESG performance. The association is strong and consistent across liberal market economies (LME), whereas it is less consistent in coordinated market economies (CME).

11.2.5.3 Institutional Ownership

Holdings by banks, institutional investors, and the state seem to be more neutral in this respect (Dam & Scholtens, 2012), or their effects are generally more institutionally prescribed.

Borghesi et al. (2014) find a negative relationship between institutional ownership and CSR, while Lamb and Butler (2018) report a positive relationship therein. Oh et al. (2011) find a positive relationship between foreign institutional investors' CSR ratings and ownership. Although "doing good by doing well" likely has a legal origin such that firms in civil law countries used to be more responsive to CSR shocks than counterparts from common law countries (Liang & Renneboog, 2017; Matten & Moon, 2008), it seems that there was a turning point around 2010. After this, "foreign institutional owners" seem to become mostly positively related to CSR/ESG in published journal articles (c.f. Bromley & Meyer, 2015). Chen et al. (2020) explain that in CSR areas that are financially material, institutional owners enhance CSR ratings via CSR-related proposals; lower CSR ratings only become visible when shareholders are distracted by exogenous shocks. CSR disclosures are positively related to state and foreign institutional ownership (Khan et al., 2013). International normative pressure seems to be exerted on board members with foreign experience of heightened engagement in CSR activities against concentrated ownership (Lau et al., 2016). A more recent study reported that even domestic institutional ownership increases after a firm issue green bonds. This signals its commitment to sustainability and triggers a positive response in stock prices (Tang & Zhang, 2020).

11.3 Summary of the Impact of Corporate Governance and TMT on Sustainability Performance

Corporate governance has become increasingly complex in the face of sustainability challenges. Boards and corporate executives have been undergoing a series of reforms and have devoted considerable resources and efforts to align with the global sustainability trend. Prior studies have shown that certain corporate governance practices are positively associated with sustainability performance. However, these studies also revealed that such effects are rather complex, as some factors work together. This finding highlights the importance of examining the effects of corporate governance practices as a "bundle." Thus, when firms change and design their corporate governance systems, they must consider multiple practices that may reinforce each other to improve sustainability performance.

Haynes and Hillman (2010) argue from a resource dependence theoretical perspective that with CEO power as the moderator, board human and social capital breadth leads to more strategic changes, while depth leads to fewer. If anything, the board "moderates" during the course of institutional changes. The mental templates governing sustainability diffuse across networks (Palmer et al., 1986). From an outsider's perspective, boards may appear to be designed to fail (Boivie et al., 2016), yet boards are simultaneously grounds on which elites learn to be mentored (McDonald & Westphal, 2011), and where social order is reproduced (Kish-Gephart & Campbell, 2015) or challenged (Palmer & Barber, 2001). In this sense, anything that potentially changes the rule of the game for elites warrants our close attention to demystifying the contemporary decoupling world.

11.4 Discussion on Recent Trends

11.4.1 Political Backlash Against ESG Investments

While the number of institutional investors who use ESG criteria to select their investee firms has been growing globally, political pushbacks to investments that screen investment targets using firms' sustainability performance are also becoming visible. Although institutional investors can play a role in promoting sustainability by facilitating and rewarding firms' sustainability performance, a view also exists that ESG-driven investments constrain firms' activities, thereby damaging firm financial performance (*Financial Times*, June 12, 2023). In the US, for example, politicians from the Republican Party in several states introduced anti-ESG bills, including those that demanded that state pension funds prioritize financial returns over other investment goals. While President Biden vetoed the Republican-led bill to prevent public pension funds from using ESG criteria to make investment decisions in 2023, the political backlash appeared to have influenced some major investment funds to reconsider their ESG investment portfolios. As previously discussed, institutional investors play an important role in promoting firms' sustainability initiatives. One potential implication of such political challenges regarding ESG investments is that capital market and investor pressure on firms to engage in sustainability initiatives may be weakened.

Despite the politicized environment in the US, the EU continues to support ESG activities thanks to political and consumer support for environmentally friendly products. For example, the EU adopted the Corporate Sustainability Reporting Directive in January 2023. This directive requires firms that conduct business in EU countries, or list their securities, to disclose their ESG activities. More recently in April 2024, the EU Parliament has approved the law that will require large firms with more than 1,000 employees to conduct due diligence from 2028 on human rights and environmental abuses in their global supply chains (*Reuters*, April 24, 2024b). In terms of capital market pressure, it has been reported that the capital in sustainable funds in Europe is seven times more than that in the US, and over 70% of pension funds in Europe viewed climate change as a priority in 2023 (*Reuters*, April 12, 2024a). Hence, firms that operate in Europe face far greater pressures to engage in ESG activities especially environmental sustainability initiatives than those in the US. These trends suggest that the perceptions of corporate executives and board directors of the need to address environmental sustainability problems may vary by location. Because this is a critical issue in examining the impact of corporate governance on firms' sustainability performance, further research on the effects of the institutional environment is needed.

11.4.2 Criticism on ESG Rating Agencies

Another issue that may affect the role of institutional investors in promoting sustainability initiatives among investee firms is the role of ESG ratings agencies. According to a survey by the ERM Sustainability Institute (2023), 53% of investor respondents use ESG ratings as their most common source of ESG data. However, it has been pointed out that the ESG data provided by ESG rating agencies have some critical problems, including basic issues such as data errors, lack of transparency in rating methods, comparability of ratings by different ESG agencies, and conflicts of interests. As different rating agencies have different scopes for ESG components and use different measures and weightings for each rating component, the final ratings for the same firm can vary significantly across ESG rating agencies (Berg et al., 2022). It has also been suggested that ESG rating agencies' business models may compromise data reliability (Li et al., 2022; Tang et al., 2022). Li et al. (2022) show that firms with business ties to rating agencies through other services (credit rating services) receive higher ESG ratings. Similarly, Tang et al. (2022) find that firms owned by the same institutions as their rating agencies receive higher ESG ratings. These findings imply that ESG rating agencies may not always be neutral. Yet many institutional investors rely on ESG data provided by those agencies when choosing their investment targets, leading to the potential misallocation of investment funds that promote sustainability. Hence, the challenges posed by ESG rating agencies can negatively affect the role of capital markets and investors in firms' sustainability initiatives, potentially encouraging greenwashing.

11.5 Conclusion

This chapter discusses the main corporate governance drivers—the board of directors, top management team, executive compensation, and ownership structure—of firm sustainability performance. Prior studies reveal that corporate governance factors can improve sustainability performance. As top executives and directors play a critical role in firm resource allocation, their decisions matter when examining the impact of firms' initiatives in addressing environmental sustainability issues thereby contributing to achieve a carbon neutral society. This chapter provides evidence that research on top decision-makers and how firms are governed is an important avenue when discussing the role of business organizations in addressing sustainability issues.

However, there are some limitations to our review. First, the review of prior studies that incorporated institutional factors is limited. Like other social organizations, firms are embedded in a local or national institutional context; hence, top decision-makers are naturally influenced by the context in which they make their decisions. As discussed, the political environment of firms' sustainability initiatives in the US is quite different from that in the EU. Thus, top executives and board members in European firms are likely to be under greater pressure to allocate more resources to ESG initiatives including corporate activities to tackle environmental sustainability issues than top decision-makers in US firms. Given these differences, it is likely that corporate governance factors, such as ownership structure, may function differently. Hence, more comparative studies are needed to examine cross-national differences in the effects of corporate governance on sustainability initiatives.

Second, our review includes prior studies, mostly from management, business ethics, and finance journals. However, firm sustainability performance has been examined in other business disciplines such as accounting as well as in disciplines like law and public policy. Hence, this chapter does not present a comprehensive review of all the relevant disciplines. Depending on the researcher's objectives, it is important to determine what we already know about the role of business organizations in tackling sustainability challenges.

References

Aguilera, R. V., & Grøgaard, B. (2019). The dubious role of institutions in international business: A road forward. *Journal of International Business Studies, 50*, 20–35.

Al-Shaer, H., & Zaman, M. (2019). CEO compensation and sustainability reporting assurance: Evidence from the UK. *Journal of Business Ethics, 158*, 233–252.

Areneke, G., Adegbite, E., Tunyi, A., & Hussain, T. (2023). Female directorship and ethical corporate governance disclosure practices in highly patriarchal contexts. *Journal of Business Research, 164*, Article 114028.

Barnea, A., & Rubin, A. (2010). Corporate social responsibility as a conflict between shareholders. *Journal of Business Ethics, 97*, 71–86.

Bénabou, R., & Tirole, J. (2010). Individual and corporate social responsibility. *Economica, 77*(305), 1–19.

Ben-Amar, W., Chang, M., & McIlkenny, P. (2017). Board gender diversity and corporate response to sustainability initiatives: Evidence from the carbon disclosure project. *Journal of Business Ethics, 142*(2), 369–383.

Berg, F., Kolbel, J. F., & Rigobon, R. (2022). Aggregate confusion: The divergence of ESG ratings. *Review of Finance, 20*(6), 1315–1344.

Berle, A. A., & Means, G. C. (1932). *The modern corporation and private property*. Macmillan.

Berrone, P., Cruz, C., Gomez-Mejia, L. R., & Larraza-Kintana, M. (2010). Socioemotional wealth and corporate responses to institutional pressures: Do family-controlled firms pollute less? *Administrative Science Quarterly, 55*(1), 82–113.

Berrone, P., & Gomez-Mejia, L. R. (2009). Environmental performance and executive compensation: An integrated agency-institutional perspective. *Academy of Management Journal, 52*(1), 103–126.

Bianchi, E. C., & Mohliver, A. (2016). Do good times breed cheats? Prosperous times have immediate and lasting implications for CEO misconduct. *Organization Science, 27*(6), 1488–1503.

Boivie, S., Bednar, M. K., Aguilera, R. V., & Andrus, J. L. (2016). Are boards designed to fail? The implausibility of effective board monitoring. *Academy of Management Annals, 10*(1), 319–407.

Borghesi, R., Houston, J. F., & Naranjo, A. (2014). Corporate socially responsible investments: CEO altruism, reputation, and shareholder interests. *Journal of Corporate Finance, 26*, 164–181.

Boulouta, I. (2013). Hidden connections: The link between board gender diversity and corporate social performance. *Journal of Business Ethics, 113*(2), 185–197.

Boyd, B. (1990). Corporate linkages and organizational environment: A test of the resource dependence model. *Strategic Management Journal, 11*(6), 419–430.

Brammer, S., Millington, A., & Pavelin, S. (2007). Gender and ethnic diversity among UK corporate boards. *Corporate Governance: An International Review, 15*(2), 393–403.

Bromley, P., & Meyer, J. W. (2015). *Hyper organization: Global organizational expansion.* Oxford University Press.

Bromley, P., & Powell, W. W. (2012). From smoke and mirrors to walking the talk: Decoupling in the contemporary world. *Academy of Management Annals, 6*(1), 483–530.

Burke, J.J., Hoitash, R., & Hoitash, U. (2019). The heterogeneity of board-level sustainability committees and corporate social performance. *Journal of Business Ethics, 154*(4), 1–26.

Business Roundtable. (2019, August 19). *Business roundtable redefines the purpose of a corporation to promote an economy that serves all Americans.* Available at: www.businessroundtable.org/business-roundtable-redefines-the-purpose-of-a-corporation-to-promote-an-economy-that-servesall-americans

Byron, K., & Post, C. (2016). Women on boards of directors and corporate social performance: A meta-analysis. *Corporate Governance: An International Review, 24*(4), 428–442.

Campbell, R. J., Jeong, S. H., & Graffin, S. D. (2019). Born to take risk? The effect of CEO birth order on strategic risk taking. *Academy of Management Journal, 62*(4), 1278–1306.

Carpenter, M. A., Geletkanycz, M. A., & Sanders, W. G. (2004). Upper echelons research revisited: Antecedents, elements, and consequences of top management team composition. *Journal of Management, 30*(6), 749–778.

Casciaro, T., & Piskorski, M. J. (2005). Power imbalance, mutual dependence, and constraint absorption: A closer look at resource dependence theory. *Administrative Science Quarterly, 50*(2), 167–199.

Centeno, M.A., & Cohen, J.N. (2012). The arc of neoliberalism. *Annual Reviews of Sociology, 38*, 317–340.

Chen, T., Dong, H., & Lin, C. (2020). Institutional shareholders and corporate social responsibility. *Journal of Financial Economics, 135*(2), 483–504.

Cyert, R., & James, M. (1963). *A behavioral theory of the firm.* Prentice-Hall.

Dam, L., & Scholtens, B. (2012). Does ownership type matter for corporate social responsibility? *Corporate Governance: An International Review, 20*(3), 233–252.

Davis, G. F. (2005). New directions in corporate governance. *Annual Review of Sociology., 31*, 143–162.

Davis, G. F., & Kim, S. (2015). Financialization of the economy. *Annual Review of Sociology, 41*, 203–221.

De Bakker, F. G., Matten, D., Spence, L. J., & Wickert, C. (2020). The elephant in the room: The nascent research agenda on corporations, social responsibility, and capitalism. *Business & Society, 59*(7), 1295–1302.

De Graaf, F. J., & Stoelhorst, J. W. (2013). The role of governance in corporate social responsibility: Lessons from Dutch finance. *Business & Society, 52*(2), 282–317.

De Villiers, C., Naiker, V., & Van Staden, C. J. (2011). The effect of board characteristics on firm environmental performance. *Journal of Management, 37*(6), 1636–1663.

Dienes, D., Sassen, R., & Fischer, J. (2016). What are the drivers of sustainability reporting? A systematic review. *Sustainability Accounting, Management and Policy Journal, 7*(2), 154–189.

DiMaggio, P. J., & Powell, W. W. (1983). The iron cage revisited: Institutional isomorphism and collective rationality in organizational fields. *American Sociological Review, 48*(2), 147–160.

Dixon-Fowler, H. R., Ellstrand, A. E., & Johnson, J. L. (2017). The role of board environmental committees in corporate environmental performance. *Journal of Business Ethics, 140*, 423–438.

Dobbin, F., Kim, S., & Kalev, A. (2011). You can't always get what you need: Organizational determinants of diversity programs. *American Sociological Review, 76*(3), 386–411.

Eccles, R. G., Ioannou, I., & Serafeim, G. (2014). The impact of corporate sustainability on organizational processes and performance. *Management Science, 60*(11), 2835–2857.

Emerson, R. M. (1962). Power-dependence relations. *American Sociological Review, 27*(1), 31–40.

Endrikat, J., De Villiers, C., Guenther, T. W., & Guenther, E. M. (2021). Board characteristics and corporate social responsibility: A meta-analytic investigation. *Business & Society, 60*(8), 2099–2135.

ERM SustanAbility Institute. (2023). *Rate the raters 2023: ESG ratings at a crossroads.* Available at https://www.sustainability.com/thinking/rate-the-raters-2023/

Fama, E. F., & Jensen, M. C. (1983). Separation of ownership and control. *The Journal of Law and Economics, 26*(2), 301–325.

Fifka, M. S. (2013). Corporate responsibility reporting and its determinants in comparative perspective—A review of the empirical literature and a meta-analysis. *Business Strategy and the Environment, 22*(1), 1–35.

Financial Times. (2023, June 12). ESG backlash in the US: What implications for corporations and investors. Available at https://www.ft.com/content/3f064321-138c-4c65-bbb9-6abcc92adead

Finkelstein, S., & Hambrick, D. C. (1990). Top-management-team tenure and organizational outcomes: The moderating role of managerial discretion. *Administrative Science Quarterly, 35*(3), 484–503.

Fleury, B., Delaunay, B., Robé, J. P., & Dunn, G. (2019). *French legislation on corporate purpose.* Harvard Law School forum on Corporate Governance. Available at: https://corpgov.law.harvard.edu/2019/06/08/french-legislation-on-corporate-purpose/

Freeman, R. E. (1984). *Strategic management: A stakeholder approach.* Pitman Publishing Inc.

Friedland, R., & Alford, R. R. (1991). Bringing society back in: Symbols, practices, and institutional conditions. In W. W. Powell & P. J. DiMaggio (Eds.), *New institutionalism in organizational analysis* (pp. 232–266). Chicago: University of Chicago Press.

Fu, R., Tang, Y., & Chen, G. (2020). Chief sustainability officers and corporate social (Ir) responsibility. *Strategic Management Journal, 41*(4), 656–680.

Galbreath, J. (2018). Is board gender diversity linked to financial performance? The mediating mechanism of CSR. *Business & Society, 57*(5), 863–889.

George, G., Howard-Grenville, J., Joshi, A., & Tihanyi, L. (2016). Understanding and tackling societal grand challenges through management research. *Academy of Management Journal, 59*(6), 1880–1895.

Giannarakis, G., Konteos, G., & Sariannidis, N. (2014). Financial, governance and environmental determinants of corporate social responsible disclosure. *Management Decision, 52*(10), 1928–1951.

Glass, C., Cook, A., & Ingersoll, A. R. (2016). Do women leaders promote sustainability? Analyzing the effect of corporate governance composition on environmental performance. *Business Strategy and the Environment, 25*(7), 495–511.

Hall, P. A., & Soskice, D. (Eds.). (2001). *Varieties of capitalism: The institutional foundations of comparative advantage*. Oxford University Press.

Hambrick, D. C. (2007). Upper echelons theory: An update. *Academy of Management Review, 32*(2), 334–343.

Harjoto, M., Laksmana, I., & Lee, R. (2015). Board diversity and corporate social responsibility. *Journal of Business Ethics, 132*, 641–660.

Haxhi, I., & Aguilera, R. V. (2017). An institutional configurational approach to cross-national diversity in corporate governance. *Journal of Management Studies, 54*(3), 261–303.

Haynes, K. T., & Hillman, A. (2010). The effect of board capital and CEO power on strategic change. *Strategic Management Journal, 31*(11), 1145–1163.

Hillman, A. J., & Dalziel, T. (2003). Boards of directors and firm performance: Integrating agency and resource dependence perspectives. *Academy of Management Review, 28*(3), 383–396.

Hillman, A. J., Keim, G. D., & Luce, R. A. (2001). Board composition and stakeholder performance: Do stakeholder directors make a difference? *Business & Society, 40*(3), 295–314.

Hong, H., Kubik, J. D., & Scheinkman, J. A. (2012). *Financial constraints on corporate goodness* (No. w18476). National Bureau of Economic Research.

Jain, T., & Jamali, D. (2016). Looking inside the black box: The effect of corporate governance on corporate social responsibility. *Corporate Governance: An International Review, 24*(3), 253–273.

Jensen, M. (2001). Value maximisation, stakeholder theory, and the corporate objective function. *European Financial Management, 7*(3), 297–317.

Jensen, M. C., & Meckling, W. H. (1976). Theory of the firm: Managerial behavior, agency costs and ownership structure. *Journal of Financial Economics, 3*(4), 305–360.

Jizi, M. I., Salama, A., Dixon, R., & Stratling, R. (2014). Corporate governance and corporate social responsibility disclosure: Evidence from the US banking sector. *Journal of Business Ethics, 125*, 601–615.

Jones, T. M. (1995). Instrumental stakeholder theory: A synthesis of ethics and economics. *Academy of Management Review, 20*(2), 404–437.

Jones, T. M., Harrison, J. S., & Felps, W. (2018). How applying instrumental stakeholder theory can provide sustainable competitive advantage. *Academy of Management Review, 43*(3), 371–391.

Jones, T. M., & Wicks, A. C. (1999). Convergent stakeholder theory. *Academy of Management Review, 24*(4), 206–221.

Kalev, A., Dobbin, F., & Kelly, E. (2006). Best practices or best guesses? Assessing the efficacy of corporate affirmative action and diversity policies. *American Sociological Review, 71*(4), 589–617.

Kang, D. L., & Sørensen, A. B. (1999). Ownership organization and firm performance. *Annual Review of Sociology, 25*(1), 121–144.

Kang, J. (2016). Labor market evaluation versus legacy conservation: What factors determine retiring CEOs' decisions about long-term investment? *Strategic Management Journal, 37*(2), 389–405.

Khan, A., Muttakin, M. B., & Siddiqui, J. (2013). Corporate governance and corporate social responsibility disclosures: Evidence from an emerging economy. *Journal of Business Ethics, 114*, 207–223.

Kim, J. W., Kogut, B., & Yang, J. S. (2015). Executive compensation, fat cats, and best athletes. *American Sociological Review, 80*(2), 299–328.

Kish-Gephart, J. J., & Campbell, J. T. (2015). You don't forget your roots: The influence of CEO social class background on strategic risk taking. *Academy of Management Journal, 58*(6), 1614–1636.

Klettner, A., Clarke, T., & Boersma, M. (2014). The governance of corporate sustainability: Empirical insights into the development, leadership and implementation of responsible business strategy. *Journal of Business Ethics, 122*, 145–165.

Koch-Bayram, I. F., & Wernicke, G. (2018). Drilled to obey? Ex-military CEOs and financial misconduct. *Strategic Management Journal, 39*(11), 2943–2964.

Kock, C. J., Santaló, J., & Diestre, L. (2012). Corporate governance and the environment: What type of governance creates greener companies? *Journal of Management Studies, 49*(3), 492–514.

Lamb, N. H., & Butler, F. C. (2018). The influence of family firms and institutional owners on corporate social responsibility performance. *Business & Society, 57*(7), 1374–1406.

Lau, C., Lu, Y., & Liang, Q. (2016). Corporate social responsibility in China: A corporate governance approach. *Journal of Business Ethics, 136*, 73–87.

Lazonick, W., & O'Sullivan, M. (2000). Maximizing shareholder value: A new ideology for corporate governance. *Economy and Society, 29*(1), 13–35.

Lewis, B. W., Walls, J. L., & Dowell, G. W. (2014). Difference in degrees: CEO characteristics and firm environmental disclosure. *Strategic Management Journal, 35*(5), 712–722.

Liang, H., & Renneboog, L. (2017). On the foundations of corporate social responsibility. *The Journal of Finance, 72*(2), 853–910.

Li, J., & Qian, C. (2013). Principal-principal conflicts under weak institutions: A study of corporate takeovers in China. *Strategic Management Journal, 34*(4), 498–508.

Li, X., Lou, Y., & Zhang, L. (2022). *Do commercial ties influence ESG ratings?* Evidence from Moody's and S&P. Available at https://papers.ssrn.com/sol3/papers.cfm?abstract_id=4190204

Love, E. G., Lim, J., & Bednar, M. K. (2017). The face of the firm: The influence of CEOs on corporate reputation. *Academy of Management Journal, 60*(4), 1462–1481.

Luoma, P., & Goodstein, J. (1999). Stakeholders and corporate boards: Institutional influences on board composition and structure. *Academy of Management Journal, 42*(5), 553–563.

Lu, W., & Taylor, M. E. (2016). Which factors moderate the relationship between sustainability performance and financial performance? A meta-analysis study. *Journal of International Accounting Research, 15*(1), 1–15.

Maas, K. (2018). Do corporate social performance targets in executive compensation contribute to corporate social performance? *Journal of Business Ethics, 148*, 573–585.

Marquis, C., & Qian, C. (2014). Corporate social responsibility reporting in China: Symbol or substance? *Organization Science, 25*(1), 127–148.

Marquis, C., & Tilcsik, A. (2013). Imprinting: Toward a multilevel theory. *Academy of Management Annals, 7*(1), 195–245.

Matten, D., & Moon, J. (2008). "Implicit" and "explicit" CSR: A conceptual framework for a comparative understanding of corporate social responsibility. *Academy of Management Review, 33*(2), 404–424.

Mattingly, J. E. (2017). Corporate social performance: A review of empirical research examining the corporation–society relationship using Kinder, Lydenberg, Domini social ratings data. *Business & Society, 56*(6), 796–839.

McDonald, M. L., & Westphal, J. D. (2011). My brother's keeper? CEO identification with the corporate elite, social support among CEOs, and leader effectiveness. *Academy of Management Journal, 54*(4), 661–693.

Mitchell, R. K., Agle, B. R., & Wood, D. J. (1997). Toward a theory of stakeholder identification and salience: Defining the principle of who and what really counts. *Academy of Management Review, 22*(4), 853–886.

Ocasio, W. (2011). Attention to attention. *Organization Science, 22*(5), 1286–1296.

Oh, W. Y., Chang, Y. K., & Kim, T. Y. (2018). Complementary or substitutive effects? Corporate governance mechanisms and corporate social responsibility. *Journal of Management, 44*(7), 2716–2739.

Oh, W. Y., Chang, Y. K., & Martynov, A. (2011). The effect of ownership structure on corporate social responsibility: Empirical evidence from Korea. *Journal of Business Ethics, 104*, 283–297.

Palmer, D., & Barber, B. M. (2001). Challengers, elites, and owning families: A social class theory of corporate acquisitions in the 1960s. *Administrative Science Quarterly, 46*(1), 87–120.

Palmer, D., Friedland, R., & Singh, J. V. (1986). The ties that bind: Organizational and class bases of stability in a corporate interlock network. *American Sociological Review, 51*(6), 781–796.

Parmar, B.L., Freeman, R.E., Harrison, J.S., Wicks, A.C., Purnell, L., & de Colle, (2010). Stakeholder theory: The state of the art. *Academy of Management Annals, 4*(1), 403–445.

Peters, G. F., Romi, A. M., & Sanchez, J. M. (2019). The influence of corporate sustainability officers on performance. *Journal of Business Ethics, 159*, 1065–1087.

Petrenko, O. V., Aime, F., Ridge, J., & Hill, A. (2016). Corporate social responsibility or CEO narcissism? CSR motivations and organizational performance. *Strategic Management Journal, 37*(2), 262–279.

Pfeffer, J., & Salancik, G. R. (1978). *The external control of organizations: A resource dependence perspective.* Harper & Row Publishers.

Porter, M. E., & Kramer, M. R. (2006). The link between competitive advantage and corporate social responsibility. *Harvard Business Review, 84*(12), 78–92.

Post, C., Rahman, N., & Rubow, E. (2011). Green governance: Boards of directors' composition and environmental corporate social responsibility. *Business & Society, 50*(1), 189–223.

Reed, D. (2002). Employing normative stakeholder theory in developing countries: A critical theory perspective. *Business & Society, 41*(2), 166–207.

Rees, W., & Rodionova, T. (2015). The influence of family ownership on corporate social responsibility: An international analysis of publicly listed companies. *Corporate Governance: An International Review, 23*(3), 184–202.

Reuters. (2024a, April 12). *Europe stands firms against US-driven ESG backlash.* Available at https://www.reuters.com/sustainability/sustainable-finance-reporting/europe-stands-firm-against-us-driven-esg-backlash-2024-04-12/

Reuters. (2024b, April 24). *EU parliament approves new business supply chain audit law.* Available at https://www.reuters.com/world/europe/eu-parliament-approves-new-business-supply-chain-audit-law-2024-04-24/

Saeed, A., Riaz, H., Liedong, T. A., & Rajwani, T. (2022). The impact of TMT gender diversity on corporate environmental strategy in emerging economies. *Journal of Business Research, 141*, 536–551.

Sajko, M., Boone, C., & Buyl, T. (2021a). CEO greed, corporate social responsibility, and organizational resilience to systemic shocks. *Journal of Management, 47*(4), 957–992.

Sajko, M., Boone, C., & Wernicke, G. (2021b). The multifaceted effect of CEO age on firm social and environmental practices. In *Academy of Management Proceedings* (Vol. 2021, No. 1, p. 12343). Briarcliff Manor, NY 10510: Academy of Management.

Strand, R. (2013). The chief officer of corporate social responsibility: A study of its presence in top management teams. *Journal of Business Ethics, 112*, 721–734.

Strand, R. (2014). Strategic leadership of corporate sustainability. *Journal of Business Ethics, 123*, 687–706.

Suchman, M. C. (1995). Managing legitimacy: Strategic and institutional approaches. *Academy of Management Review, 20*(3), 571–610.

Surroca, J., Tribó, J. A., & Waddock, S. (2010). Corporate responsibility and financial performance: The role of intangible resources. *Strategic Management Journal, 31*(5), 463–490.

Tang, D. Y., Yan, J., & Yao, C. Y. (2022). *The determinants of ESG ratings: Rater ownership matters.* Available at https://papers.ssrn.com/sol3/papers.cfm?abstract_id=3889395

Tang, D. Y., & Zhang, Y. (2020). Do shareholders benefit from green bonds? *Journal of Corporate Finance, 61*, Article 101427.

Tang, Y., Mack, D. Z., & Chen, G. (2018). The differential effects of CEO narcissism and hubris on corporate social responsibility. *Strategic Management Journal, 39*(5), 1370–1387.

Tang, Y., Qian, C., Chen, G., & Shen, R. (2015). How CEO hubris affects corporate social (ir) responsibility. *Strategic Management Journal, 36*(9), 1338–1357.

Velte, P., & Stawinoga, M. (2020). Do chief sustainability officers and CSR committees influence CSR-related outcomes? A structured literature review based on empirical-quantitative research findings. *Journal of Management Control, 31*(4), 333–377.

Walls, J. L., & Berrone, P. (2017). The power of one to make a difference: How informal and formal CEO power affect environmental sustainability. *Journal of Business Ethics, 145*, 293–308.

Wernicke, G., Sajko, M., & Boone, C, (2022). How much influence do CEOs have on company actions and outcomes? The example of corporate social responsibility. *Academy of Management Discoveries, 8*(1), 36–55.

Wiengarten, F., Lo, C. K., & Lam, J. Y. (2017). How does sustainability leadership affect firm performance? The choices associated with appointing a chief officer of corporate social responsibility. *Journal of Business Ethics, 140*, 477–493.

Wong, E. M., Ormiston, M. E., & Tetlock, P. E. (2011). The effects of top management team integrative complexity and decentralized decision making on corporate social performance. *Academy of Management Journal, 54*(6), 1207–1228.

Wowak, A. J., Mannor, M. J., Arrfelt, M., & McNamara, G. (2016). Earthquake or glacier? How CEO charisma manifests in firm strategy over time. *Strategic Management Journal, 37*(3), 586–603.

Yoshikawa, T., Nippa, M., & Chua, G. (2021). Global shift towards stakeholder-oriented corporate governance? Evidence from the scholarly literature and future research opportunities. *Multinational Business Review, 29*(3), 321–347.

Yoshikawa, T., Zhu, H., & Wang, P. (2014). National governance system, corporate ownership, and roles of outside directors: A corporate governance bundle perspective. *Corporate Governance: An International Review, 22*(3), 252–265.

Zajac, E. J., & Westphal, J. D. (1994). The costs and benefits of managerial incentives and monitoring in large US corporations: When is more not better? *Strategic Management Journal, 15*(S1), 121–142.

Zeitlin, M. (1974). Corporate ownership and control: The large corporation and the capitalist class. *American Journal of Sociology, 79*(5), 1073–1119.

The Effect of Green Logistics on Carbon Neutrality

12

Tomoo Noguchi

Abstract

The distribution and marketing sectors urgently need to implement green logistics in trucking to achieve carbon neutrality. This is because the transportation sector accounts for about 20% of Japan's total CO_2 emissions, with trucks contributing around 40% of these emissions. The rise of e-commerce platforms such as Amazon and Rakuten has significantly increased truck usage, which has the potential to elevate CO_2 emissions in Japanese logistics. Regarding CO_2 emissions, the logistics industry in Japan is currently facing a trade-off between extreme weather conditions and global warming and the increase in truck transportation. An effective way to address this problem is through green logistics. The author proposes a four-phase action plan based on the current level of penetration and difficulty of execution: Phase 1 (energy use reform), Phase 2 (logistics digital transformation, also known as logistics DX), Phase 3 (logistics structural reform), and Phase 4 (robotization). Each phase of the action plan involves crucial social innovations and, if implemented as envisioned, will enable a considerable reduction in CO_2 emissions and contribute toward a carbon-neutral society.

Keywords

Green logistics · Digital transformation · AI logistics · Robotization · Carbon offsetting

12.1 Introduction

This chapter examines how CO_2 emissions can be reduced in the logistics sector, which is an essential part of the goods transfer process. This type of logistics effort is called green logistics, and its practice could significantly contribute to the realization of a carbon-neutral society. CO_2 is a major greenhouse gas that contributes to abnormal weather and global warming. Japan's CO_2 emissions totaled 1,037 million tons in FY2022 (Ministry of the Environment, 2024). By sector, the industrial sector accounted for the largest share of emissions at 34.0% (352.26 million tons), followed by the transportation sector at 18.5% (191.8 million tons). Within the transportation sector, freight vehicles account for 38.0% (72.92 million tons), representing a considerable emission scale. This is why the logistics sector needs green logistics.

T. Noguchi (✉)
Waseda University, Tokyo, Japan
e-mail: tomo@waseda.jp

K. Akao and A. Washizu (eds.), *Climate Change Issues and Social Sciences*,
Sustainable Development Goals Series, https://doi.org/10.1007/978-981-96-8951-4_12

The world is currently being urged to achieve the SDGs to realize a sustainable global society, including resolving climate change issues. One of the 17 goals that the logistics industry must prioritize is Goal 7: *Affordable and Clean Energy for All*. This is because many logistics systems still run on fossil fuels, such as diesel and gasoline, which emit large amounts of CO_2 into the atmosphere. To reduce the burden on the global environment and realize a carbon-neutral society, it is necessary to either minimize the use of fossil fuels or increase their utilization efficiency.

In this chapter, the author will explore methods that seem effective in reducing CO_2 emissions (social innovations) through case studies and quantitative analysis. The author aims to examine how green logistics should be structured to achieve carbon neutrality while minimizing the impact on users' benefits as much as possible.

12.2 Background of Increasing CO_2 Emissions in Logistics

12.2.1 Growth of E-Commerce and Increase in Parcel Delivery

Currently, the total number of parcels handled in Japan is 5,058.88 million per year (of which 4,925.08 million are transported by truck and 80.8 million by air and other means) (Ministry of Land, Infrastructure, Transport and Tourism, 2023b). This substantial number is primarily due to the rapid growth of the e-commerce market driven by companies such as Amazon and Rakuten. The market size of e-commerce in the logistics-related retail sector reached 13.9997 trillion yen in 2022, representing a 5.37% increase compared to the previous year. Furthermore, the e-commerce penetration rate in this sector has steadily increased from 6.78% in 2019 to 8.08% in 2020, 8.78% in 2021, and 9.13% in 2022, indicating the widespread adoption of online purchasing (Ministry of Economy, Trade and Industry, 2023).

The high utilization of e-commerce is attributed to the convenience of online shopping. However, this increased convenience has also led to inefficiencies in transportation and a corresponding rise in CO_2 emissions. The average cargo weight per shipment has significantly decreased from 2.43 tons in 1990 to 2.13 tons in 1995, 1.73 tons in 2000, 0.95 tons in 2010, and 0.83 tons in 2021. In contrast, the number of shipments has continuously increased from 13,656 in 1990 to 15,290 in 1995, 15,964 in 2000, 24,616 in 2010, and 25,080 in 2021 (Ministry of Land, Infrastructure, Transport and Tourism, 2024).

These trends indicate a shift toward more frequent, smaller deliveries. When considered alongside the trend in load rates per truck, it becomes evident that inefficient deliveries are on the rise, unnecessarily increasing CO_2 emissions.

12.2.2 Inefficient Logistics

The author needs to clarify the loading rate of trucking to highlight the problems posed by the increasing small-lot, high-frequency nature of deliveries due to the prosperity of e-commerce. The current level of domestic freight transport volume (for commercial use) is approximately 2,557.54 million tons per year on a tonnage basis (FY2022). On a ton-kilometer basis (load volume multiplied by transport distance), it amounts to about 199.149 billion ton-kilometers (Ministry of Land, Infrastructure, Transport and Tourism, 2023c).

Dividing the number of ton-kilometers transported by the potential loading capacity, known as "capacity ton-kilometers," provides an idea of the efficiency of transport equipment. If the truck is fully loaded, this efficiency would be 100%. However, in reality, the efficiency was 50.17% in 2000, decreasing to 39.7% in FY2022. In other words, transportation efficiency is declining, and trucks running throughout Japan transport about 60% air. Incidentally, truck transportation accounts for 98.3% of the logistics used to transport freight.

12.2.3 Increase in CO_2 Emissions Due to Redelivery

Alongside the low loading rate of trucks, another inefficiency is redelivery. While left-behind delivery is common with the e-commerce giant Amazon, this practice was not common in Japan until recently due to the risk of theft and defacement, and redelivery was the norm. This issue is primarily caused by the increase in unattended homes during the daytime (owing to lifestyle changes and the increasing number of women joining the workforce). Additionally, the proliferation of high-rise apartment buildings (which increases the burden on delivery personnel to transport items to upper floors) has exacerbated this problem.

As of April 2023, the redelivery rate in Japan's logistics industry stood at 11.4% (Ministry of Land, Infrastructure, Transport and Tourism, 2023a). Although it temporarily dropped to 8.5% in April 2020, during the peak of the COVID-19 pandemic when more people were at home, this was only a temporary phenomenon, and the rate has since returned to previous levels.

Redelivery means that deliveries that could have been completed in one attempt must be made twice or thrice. Consequently, logistics companies must operate trucks more frequently, naturally leading to an increase in CO_2 emissions.

12.2.4 Increase in CO_2 Emissions Due to Returns

In e-commerce, returns inevitably occur more frequently than in brick-and-mortar stores because users do not directly see the actual product they are purchasing, and errors in order fulfillment by the business can occur. Regarding actual return rates, they vary considerably by product type. In Japan, the median return rate is generally 5–10% (ELTEX, 2021), whereas in the U.S., it is said to be around 30% (Brohan, 2013).

E-commerce companies understand that accepting customer returns is crucial for store loyalty, so they continuously seek ways to manage the inevitable returns. For example, products purchased from Amazon.co.jp and Amazon Marketplace can, in principle, be returned or exchanged within 30 days from the date of arrival. A full refund is provided if the product is defective, unused, or unopened. Returns can be made at convenience stores, Yamato sales offices, and even through delivery lockers.

Generally, such a permissive return policy is said to increase return rates (Bower & Maxham, 2012). However, companies tolerate returns because they ultimately contribute to increased sales (Janakiraman et al., 2016). Nevertheless, users should be aware that returning products incur transportation and labor costs, and trucks are inevitably used to transport returned products, leading to increased CO_2 emissions, as stated above.

12.3 Social Innovation in the Logistics Industry to Achieve Carbon Neutrality

Measures against climate change and global warming are being called for globally, and the Japanese government has announced its goal to achieve carbon neutrality by 2050. The logistics industry is also riding this wave and introducing countermeasures. For example, the All Japan Trucking Association formulated its *Environmental Vision 2030* for the Trucking Industry in March 2022. As a main goal, the association declared that the CO_2 emissions intensity of the trucking industry will be reduced by 31% by 2030, compared to the 2005 level.

To achieve this, the following *three-step* action plan has been clearly defined:

Step 1: Reduce CO_2 emissions through "transportation."

This involves implementing global warming countermeasures in the transportation business, including: (1) introduction of trucks with superior environmental performance, (2) promotion of eco-driving, and (3) promotion of more efficient transportation.

Step 2: Reduce CO_2 emissions at "offices"

This involves implementing global warming countermeasures other than those related to the transportation business, including: (1) electricity saving at offices, (2) reducing the number of business trips by using web conferencing, and (3) tree-planting projects.

Step 3: Environmental measures by "everyone"

This involves implementing environmental measures other than those in the first step, including: (1) waste reduction, (2) promotion of resource recycling, and (3) noise reduction.

However, while this action plan mentions "steps," it primarily emphasizes the first step and does not account for the varying degrees of difficulty in achieving these goals. Additionally, the perspective of social innovation is unclear. Therefore, the author proposes the following four-level action plan, which incorporates social innovations based primarily on the current level of penetration and the difficulty of execution.

12.3.1 Phase 1: Energy Use Reformation

12.3.1.1 Utilization of Renewable Energy

To reduce massive CO_2 emissions and mitigate climate change, naturally occurring "renewable energies" such as solar, wind, hydro, and geothermal have long been in the spotlight. Renewable energy accounts for approximately 25% of Japan's electricity generation (Renewable Energy Institute, 2025). In comparison with major developed countries, Canada leads with 67%, followed by Germany at 58%, Spain at 58%, Italy at 49%, and France at 29%. Japan's figure is notably lower. The U.S. has a similar rate at 24%, slightly below Japan's but nearly at the same level.

If all energy were provided by renewable sources, the occurrence of greenhouse gas emissions and extreme weather events would be curtailed, and *SDG 13: Take urgent action to combat climate change and its impacts* might no longer be necessary. However, renewable energy sources are difficult to control in terms of supply, as they are affected by weather conditions, such as those seen in solar power generation. Additionally, issues such as environmental destruction by mega solar power plants and energy storage problems make renewable energy generation challenging. Renewable energy is inherently a flow-type energy, not a stock-type energy like fossil fuels. Therefore, it is difficult to control renewable energy sources artificially in the way one might control "mining volumes."

Moreover, the biggest problem remains cost. According to estimates by the Agency for Natural Resources and Energy, as of 2020, the cost of power generation was 12.9 yen/kWh for solar (for commercial use), 30.0 yen/kWh for offshore wind, and 25.3 yen/kWh for small hydro. In comparison, the costs were 10.7 yen/kWh for natural gas-fired, 12.5 yen/kWh for coal-fired, and 11.5 yen/kWh for nuclear (Agency for Natural Resources & Energy, 2021). Clearly, the cost of power generation from traditional fossil fuels is lower.

While it is indisputable that renewable energy will continue to be emphasized in the future, currently, clean energy sources are insufficient to fully support an advanced economy and society. Fossil fuels such as natural gas, coal, and petroleum still play a significant role, producing greenhouse gas emissions in proportion to their use. Therefore, energy conservation measures targeting equipment that uses fossil fuel-derived electricity are essential and should be promoted more aggressively in the future.

12.3.1.2 Energy Conservation Measures for Distribution Centers

Logistics companies have long promoted energy conservation in warehouses and distribution centers. Examples include the installation of LED lighting, the use of energy management systems, and the application of insulation materials. These measures have not only decreased electricity costs but also contributed to the reduction of CO_2 emissions.

A relatively recent success story is Tokyo Tatemono Co.'s giant logistics facility, T-LOGI Akiruno, completed in May 2023. The facility has implemented solar panels to generate power in excess of its electricity consumption and has established a *self-commissioning scheme*, a system whereby excess power is directly transferred to other properties owned by the company. This method is the first of its kind among Japanese real estate companies, with 2,210 MWh/year, or about 58% of the total power generation (approximately 3,820 MWh/year), being transferred to other facilities as surplus power (Tokyo Tatemono Co., Ltd., 2023). Such efforts will undoubtedly contribute to *SDG 7: Affordable and Clean Energy for All*.

12.3.1.3 Electric and Fuel Cell Trucks

Modern logistics providers are transitioning to a more *environmentally friendly* society by introducing more fuel-efficient vehicles such as plug-in hybrid vehicles (PHVs), electric vehicles (EVs), and fuel cell vehicles (FCVs). However, the current general penetration rate of EVs in Japan is unusually low. Although EV sales had been increasing since 2022, as of February 2024, EVs accounted for only 3.4% of new car sales (ENECHANGE, 2024). Scandinavian countries such as Norway (88%) and Sweden (51.9%) have very high EV penetration rates, while Western European countries such as Germany (24.7%), the United Kingdom (23%), and France (20.2%) also have significantly higher levels compared to Japan. The United States, which has the world's largest EV sales share with Tesla, has a penetration rate of only 6%, similar to Japan. However, China, which aims for global dominance in EVs, has many manufacturers, including BYD, and a relatively high penetration rate of 11%. In Japan, there are challenges related to recharging stations and the high cost of vehicles. However, the government has launched a subsidy program to promote the introduction of clean energy vehicles, allocating approximately 129.1 billion yen for fiscal 2023. The program aims to achieve 100% EV sales for new cars by 2035. Therefore, the penetration rate is expected to increase significantly in the future.

Regarding electric trucks in the logistics industry, models such as the eCanter from Mitsubishi Fuso Truck and Bus Corporation and the ELF-EV from Isuzu Motors Limited have been introduced and are contributing to achieving carbon neutrality in transportation. Seven-Eleven Japan, the largest company in the convenience store industry, is a leading ESG (Environment, Social, Governance) investor in this area. With the aim of reducing CO_2 emissions, the company purchased three eCanters from Mitsubishi Fuso Truck and Bus Corporation in March 2018 and two electric trucks from Isuzu in July 2019. Additionally, the company has introduced clean diesel engine vehicles, natural gas vehicles, hybrid vehicles, and fuel cell trucks fueled by hydrogen, actively practicing green logistics. As of February 2022, 4,424 (71.3%) of the company's 6,208 delivery vehicles were *eco-friendly*.

12.3.2 Phase 2: Logistics DX (Digital Transformation)

The Japanese government defines logistics DX as an activity that goes beyond digitalization and mechanization. It will revolutionize traditional logistics by driving operational improvements and work style reforms, ultimately transforming the business model of the logistics industry itself (Ministry of Land, Infrastructure, Transport and Tourism, 2021b). In other words, the objective of logistics DX is to fundamentally transform old logistics activities and metamorphose them into a new business model through digital innovation. This will result in more efficient and accurate logistics activities, while simultaneously reducing CO_2 emissions. Here, the author would like to clarify the main logistics DX measures that will be important in the future.

12.3.2.1 Data Sharing and Visualization Among Trading Companies

Sharing and visualizing data in cyberspace among trading companies is an important issue. If the flow of goods and logistics equipment can be shared in real-time, truck loading efficiency can be improved, and lead times can be shortened. For example, Hakobu's dynamic management application "MOVO" can track the location of trucks and other vehicles every 5 s using GPS, allowing for visualization of both cargo arrival and stops, as well as CO_2 emissions. By sharing these data with business partners, a sense of security and trust can be fostered while also reducing work hours. Kyocera Corporation, which implemented MOVO Vista, has reduced the number of person-hours required to arrange truck transportation by about 60% (Hakobu, 2023). Data sharing and visualization among trading companies will make logistics systems more efficient, resulting in less wasteful deliveries and reduced CO_2 emissions.

12.3.2.2 Improving Transportation Efficiency with AI

The social innovation of AI logistics has the potential to significantly reduce CO_2 emissions. E-commerce giant Amazon offers an AI-powered app called Amazon AI for last-mile delivery contractors. This system, which began operating around 2021, presents optimal routes and cargo volumes. This AI can be combined with a geographic information system to calculate optimal transportation routes, considering factors such as traffic conditions, road restrictions, and delivery destination priority. In some cases, data can be analyzed not only in advance but also in real-time to suggest alternative routes in case of changes or delays during transportation.

Major convenience stores, such as Seven-Eleven and Lawson, are working to optimize delivery routes with AI at the center. Consequently, they have successfully reduced the number of deliveries of perishable items with short expiration dates from four to three times a day and from three to two times a day, respectively. Additionally, Family Mart is using its proprietary AI system for deliveries, which has enabled it to reduce the number of conventional delivery routes by approximately 10%. The company estimates that if its unique AI logistics system were expanded to all its stores, it could cut transportation costs by more than 1 billion yen per year and reduce CO_2 emissions by 1,300 tons (AI Research Institute, 2024).

Furthermore, combining AI with tracking systems can improve transportation efficiency. For example, Maven Machines, a U.S.-based company that aims to automate trucking operations, uses a cloud-based platform to track truck locations in real-time, provide alerts for estimated arrival times, and manage the dispatch of trucks that do not meet their carrying capacities. Modern tracking systems not only track truck operations in real-time but also utilize sensors to monitor temperature, humidity, vibration, and other factors to ensure that goods are being delivered in proper condition. For instance, Easy Tracking, a cold chain support solution provided by Allion Corporation, uses a cloud-based management platform to perform *proper temperature and humidity control and tracking and reporting during transportation* in real-time. More specifically, this system synchronizes monitored data with the operation manager to sequentially check the status of the site. It automatically sends alert messages in the event of abnormalities and visualizes temperature information and transportation route schedules. As described above, a logistics system with AI at its core can conduct efficient management sequentially and, as a result, significantly contribute to the realization of a carbon-neutral society.

12.3.3 Phase 3: Logistics Structural Reform

There have been attempts to improve transportation efficiency and reduce CO_2 emissions by fundamentally reforming logistics structures and methods that have been traditionally constructed. In the following, the author will describe what the author thinks to be a social innovation to achieve carbon neutrality.

12.3.3.1 Modal Shift

Modal shift refers to the distribution of goods over long distances in a relay fashion, with coordination among partners in the supply chain. By integrating trucks, ships, and railroads into the transportation process, it is possible to achieve efficient delivery of goods while significantly reducing

CO_2 emissions. Modal shift is particularly significant because CO_2 emissions from freight vehicles account for 38% of the entire transportation sector's emissions, while coastal shipping accounts for only 5.3% and rail transport is even lower at 3.8% (Ministry of Land, Infrastructure, Transport and Tourism, 2021a). In other words, by combining ferries and railroads, which emit much less CO_2 than trucks, the causes of climate change can be significantly reduced, contributing to *SDG 13: Take urgent action to combat climate change and its impacts.* A specific example of this is the collaboration between Lion, Kewpie, Japan Pallet Rental (JPR), and Kanko Kisen, which has been implementing a modal shift using ferries since 2018. This has resulted in significant reductions in CO_2 emissions and truck driving time.

12.3.3.2 The Physical Internet

The growth of e-commerce due to the COVID-19 pandemic has brought into the limelight a term that has traditionally not been discussed much in the logistics industry: the "physical Internet." This term was coined by Paul Markillie in *The Economist* (2006), stating that "like information on the internet, goods are moving around the world with ever greater efficiency." Applying this term to logistics, it can be interpreted as an effort to increase truck loading rates through collaboration among multiple companies. This is an extremely important initiative in the Japanese logistics industry, where demand for small-lot, high-frequency deliveries has increased dramatically owing to the rise of e-commerce. Despite a shortage of drivers and massive CO_2 emissions, only less than 40% of the cargo space on trucks in operation is currently being utilized.

The term "Internet" is included because the concept is analogous to the Internet's communication method, where multiple data packets are sent and received over a single line through packet switching. Similarly, when drivers operate trucks, there are always costs associated with labor and diesel fuel. Therefore, the fundamental idea is that it is more rational to operate trucks with as many loads as possible at the same time.

An example of the physical Internet in Japan is the joint delivery initiative by Asahi Beverages, Nissin Food Products, and Nippon Express, which began in September 2020. Beverages are high-weight products and could not be stacked in two layers on the truck bed, resulting in poor loading efficiency. Asahi and Nissin unified their different pallets, placing beverages on the first tier and lightweight instant noodles on the second tier, achieving efficient mixed loading that reduces the amount of available space. Nippon Express managed the logistics of the mixed cargo, transporting it from the Kanto region to both companies' distribution centers in Kyushu.

12.3.3.3 Joint Delivery Among Rival Companies

The above example of joint delivery by Asahi Beverages, Nissin Food Products, and Nippon Express involved companies that were not direct rivals. Since these companies are not competing for the same market, there is little need for secrecy, making collaboration easier. However, the situation is completely different when it involves rival companies, as they must disclose diverse information to each other to collaborate. Nevertheless, exemplary cases have emerged where such resistance and barriers have been overcome to achieve joint logistics for the purposes of *improving transportation efficiency and carbon neutrality.*

One such example is the *F-LINE* food logistics platform, established in 2015 by six food manufacturers (Ajinomoto Co. Inc., Kagome Co. Ltd., Nisshin Oillio Group Ltd., Nisshin Seifun Welna Inc., House Foods Group Inc., and Mizkan Holdings Co. Ltd.). In practice, this organization integrates the logistics operations of Ajinomoto Logistics Co., Ltd., Kagome Logistics Service Co., Ltd., and House Logistics Service Co., Ltd. (part of their business), achieving groundbreaking progress in modern logistics challenges. These efforts include improving logistics efficiency through joint delivery, alleviating heavy driver labor by establishing relay points, utilizing modal shifts by employing rail and ship transportation, standardizing and streamlining operations with advanced material handling equipment, computerizing shipping slips, and standardizing outer packaging sizes.

Under the motto "Compete with products, collaborate in logistics," a joint delivery program was implemented in the Hokkaido area in April 2016. From October 2023, the two storage and delivery sites were consolidated into a single location to improve loading efficiency and reduce the number of deliveries. Estimates suggest that this initiative will reduce CO_2 emissions by 16% (F-LINE, 2023). The simple idea is that instead of individual companies maintaining similar functions separately, as in the case of F-LINE, they should pursue efficiency and economies of scale by promoting joint and integrated functions wherever possible. However, many difficulties arise along the way, making it challenging to create a seamless, integrated organization.

Standardization is essential as a prerequisite for collaboration that transcends the boundaries among rival companies. Moreover, *data sharing* is crucial to put such collaboration into practice. However, a company's internal data is often considered "treasured" and "highly confidential." Requiring them to disclose this information to competitors can result in strong resistance. Therefore, while the need for "collaboration" has long been advocated in many industries, practical examples have been few.

Although not as drastic a case of structural reform as F-LINE, there have been efforts to avoid the risk of cargo not being carried and to reduce CO_2 emissions by sharing logistics in the last-mile. One such initiative is the joint delivery executed by the Japan Business Machine and Information System Industries Association (JBMIA), involving major Japanese MFP and printer manufacturers, including Canon and Ricoh. The methodology involves consolidating the companies' products in the warehouses of logistics companies and delivering them to customers in the target area in the form of "consolidation." While the association has a long track record (more than 20 years) in joint collection and recycling of unwanted goods, it has been implementing joint forward logistics, rather than reverse logistics, for the first time since November 2021. This joint operation was driven by the increased risk of not being able to transport cargo due to a shortage of drivers, and the urgent need to reduce CO_2 emissions as highlighted by the SDGs. The results of the joint operation in Hokkaido indicate that CO_2 emissions and the number of trucks used could be reduced by as much as 50% (JBMIA, 2022).

12.3.3.4 Relay Point Delivery

##When we think of a distribution warehouse, we typically envision a facility for storing inventory to be picked and shipped as orders are received. These are known as distribution centers. However, there are also transfer centers (TCs), which do not store inventory but instead ship goods immediately after sorting at relay points. These facilities serve purely as transit points, where goods are cross-docked (passed like a baton) to the next stage in the transportation process. Essentially, these facilities do not hold inventory; therefore, they incur no storage costs.

Of course, not all distribution centers will be converted into TCs due to their different functions; however, the importance of such centers is certain to increase in terms of improving transportation efficiency and achieving carbon neutrality. A notable case of successful decarbonization through the skillful use of relay points is Next Logistics Japan (funded by 20 companies including Hino Motors, Asahi Group, Nissin Foods, Bridgestone, and Konoike Transport). They have cross-dock centers in Sagamihara City and Nishinomiya City, where drivers change trucks at intermediate points (Toyota City and Hamamatsu City) and then return to their original locations. This method allows cargo to move toward its destination while enabling drivers to return to their starting point. As it avoids one-way long-distance deliveries, it eliminates the waste of empty return trips. In other words, by utilizing high-load-factor trucks in a switchback relay at intermediate points, transportation efficiency is increased, and CO_2 emissions are reduced.

12.3.4 Phase 4: Robotization

Under the logistics DX trend, robotization (automation) in the logistics industry is being promoted quite aggressively. This is because Japan is suffering from a chronic shortage of drivers, and in particular, the enforcement of the Workplace Reform Law in April 2024 limits drivers' overtime hours and requires the deployment of multiple drivers for long-distance transportation. The initial idea was to use robots to fill the roles of drivers if there were not enough human drivers available. Regarding robotization in the logistics industry, unmanned forklifts and robotic arms have been introduced in logistics centers to save considerable manpower. However, their primary role is to replace human labor, and they do not directly contribute to carbon neutrality.

Various unmanned initiatives are also being seen in transportation operations, including automated vehicles, automated trucks in convoys, automated delivery robots, and drone delivery. However, similar to the situation in logistics centers, it is generally believed that these initiatives do not directly contribute to CO_2 reductions. Nevertheless, it is not accurate to say that these technologies do not contribute to decarbonization or climate change mitigation at all. The following sections will discuss the main areas where they are involved.

12.3.4.1 Self-Driving Vehicles

Automated driving is considered primarily an innovation to address the driver shortage and may not directly contribute to carbon neutrality. However, control of truck driving by machines, which are tireless and more accurate than humans, can provide smooth and optimal operation. Consequently, the energy consumption efficiency of trucks would increase, potentially decreasing CO_2 emissions to a minimum. In addition, if electric and hydrogen trucks, which will be the mainstay of future automated vehicles, become widely used, CO_2 emissions from truck operations will be reduced to almost zero. This is precisely the kind of green logistics that can be realized through the effective use of clean energy.

The current achievements in automated driving can be summarized as follows. Presently, automated trucks are mainly limited to long-distance transportation or transportation within specific areas such as mines and ports. In 2014, Germany's Daimler developed a semi-automated vehicle equipped with sensors, stereo cameras, radar, and other sensing devices for long-haul trucks targeted at highway operations. Similarly, in 2016, Volvo of Sweden developed a self-driving truck capable of navigating underground mines. This truck was remarkable for its ability to travel automatically over 7 km on rough and rugged terrain 1,320 m underground. Naturally, radio waves cannot reach such deep underground locations, making remote control impossible. The truck used LiDAR (Light Detection and Ranging) to shine a laser beam into the mine shaft and relied on the reflected light and sensors to measure distances, allowing it to navigate. This was the world's first development of an automated truck in such a harsh environment.

Chinese manufacturers are also actively pursuing long-distance transportation using automated trucks. In 2019, Plus (Chika Technology) partnered with S.F. Holding (Shunfeng Holding) to traverse the United States, covering about 4,500 km. Additionally, in 2021, the company conducted a test run of over 3,000 km in China (Suzhou to Dunhuang). In Japan, results are already being achieved in the limited areas mentioned above. T2, a developer of automated driving systems, succeeded in autonomous driving of a truck on an expressway (between Yatsu Funabashi I.C. and Wangan Narashino IC on the Higashi-Kanto Expressway) in April 2023.

Beyond experimentation, Wal-Mart, the king of the retail industry, has achieved groundbreaking results in practical applications. The company completed delivery activities with self-driving trucks in its home state of Arkansas in 2019 and Louisiana starting in 2020. This effort, accomplished in partnership with Gatic, a developer of box trucks, is notable for its *entirely unmanned operation* on public roads. Naturally, a driver was on board during the verification phase, but since 2021, the trucks have been delivering entirely on their own over distance of approximately 11 km. Middle-mile deliveries are likely to be challenging in Japan's intricate city centers, with their narrow streets; therefore, the main application for self-driving trucks will likely be on highways for the time being.

However, in the future, it is envisioned that a mobility pool will be installed in the *next-generation core logistics facility* (Joyo City, Kyoto Prefecture) that Mitsubishi Estate plans to develop. From there, Japan's major arteries will be connected by Level 4 self-driving trucks (Mitsubishi Estate, 2023), resulting in an optimal transportation system with high energy consumption efficiency.

12.3.4.2 Automated Convoy Trucks

In contrast to the above-mentioned single unit-based automated driving, there are also efforts to automate the operation of multiple trucks, allowing them to move together as if they were a single, enormous trailer. This is known as *convoy driving or platooning*. Before discussing this, it is worth mentioning the rise of *double-coupled trucks* in recent years. This approach attempts to curb greenhouse gas emissions through the traditional method of "upsizing." The method is quite simple: by adding an additional trailer behind the bed of a regular large truck, one driver can haul about twice as much cargo.

The January 2019 revision of the Special Vehicle Permit Standards allowed double-coupled trucks to extend from the previous length of 21 m to 25 m. In March of the same year, the Ministry of Land, Infrastructure, Transport, and Tourism approved a *Comprehensive Efficiency Plan* for Yamato Transport, Nippon Express, Seino Transport, and Japan Post for home-delivery cargo using this double-coupled truck system for joint transportation. Under the plan, larger trucks would reduce CO_2 emissions by 216.5 tons per year and driver operating time by 9,157 h (Ministry of Land, Infrastructure, Transport, and Tourism, 2019). While this approach can be seen as the pursuit of economies of scale typical of the last century, new initiatives in the twenty-first century involving automated driving have made it possible to achieve both economies of scale and CO_2 reductions. This is known as *automated convoy trucks*, where trucks communicate directly with each other via V2V (Vehicle-to-Vehicle) technology and are controlled by ADAS (Advanced Driver-Assistance Systems), allowing them to drive in unison while maintaining a certain distance between vehicles. In this system, a human driver is required in the lead vehicle, but the trailing vehicles are operated by ADAS, eliminating the need for additional drivers. Consequently, labor costs can be reduced relative to the amount of cargo transported, and energy can be saved by reducing the air resistance of the trailing vehicles.

Convoy driving has been implemented by companies such as Daimler and MAN in Germany and Volvo in Sweden since 2016, using highways for cross-border operations. Daimler connected three trucks via Wi-Fi and successfully traveled in formation from Stuttgart to Rotterdam. In Japan, Toyota Tsusho Corporation has been actively researching convoy driving since 2016 and has been conducting demonstration tests on the Shin-Tomei Expressway with unmanned trailing vehicles since January 2019. Three trucks travel at 70–80 km/h while maintaining a 10-m distance between vehicles, utilizing the Coordinated Adaptive Cruise Control (CACC) system, GPS, and LiDAR tracking.

Telecommunications carrier SoftBank is leveraging its expertise by utilizing 5G technology to support convoy driving. The *5G News Radio (5G-NR)* technology allows trucks in a convoy to communicate directly with each other. Prior to this, communication had to go through a base station, causing delays and potential connectivity issues in enclosed spaces like tunnels. SoftBank's 5G-NR has successfully achieved low-latency communication by directly connecting multiple trucks, reducing the wireless segment delay to less than 1 ms (1/1000th of a second). In terms of specific outcomes for convoy trucks, Peloton Technology in the U.S. has reported average fuel savings of more than 7%, with annual fuel savings per truck amounting to $7,000–$10,000 (Peloton, 2019).

It is worth mentioning that there was a remarkable achievement related to convoy driving that can be called a Japanese accomplishment. As a result of proactive efforts, including the first demonstration tests on public roads on the Shin-Tomei Expressway, Japan proposed an international standard for convoy driving. This proposal was issued by ISO (International Organization for Standardization)/TC204 (ITS Intelligent Transport Systems)/WG14 (Driving Control) in September 2022. This means that the standardization that originated in Japan has become an international standard, allowing mixed convoy driving with trucks from other countries.

12.3.4.3 Automatic Delivery Robots

Automatic delivery robots are systems that enable unmanned logistics for the last-mile. Users can order products using a smartphone or PC, and an unmanned, box-shaped vehicle with four or six wheels, measuring about one meter in length, width, and height, stores the products inside and delivers them to the user's desired location. The vehicle operates autonomously at a speed of about 5–6 km/h, can avoid obstacles, and provides a contactless delivery method, which was particularly valued during the COVID-19 pandemic.

However, these robots cannot travel long distances like unmanned trucks and are unable to climb steps of even 10 cm, making complete outdoor delivery challenging. Additionally, despite their contribution to unmanned deliveries, their delivery range is limited, typically to a one-kilometer radius, due to their low speed on sidewalks. A more critical problem lies in the impediment to user convenience. Even if an automated delivery robot delivers the package to the house as ordered, it only travels to the front of the house and notifies the resident of its arrival, leaving the rest of the process to the user. If the user is not at home or unable to answer the door, the package must be taken back.

Of course, redelivery is possible, and if the delivery can be left in a drop-off location or delivery locker, the need for redelivery is eliminated. However, box-type automated delivery robots do not have arms and legs; thus, they cannot handle such tasks or even perform a simple drop-off. If the user is unavailable to receive the package on time, redelivery becomes necessary, resulting in inefficiency. Therefore, as far as logistic systems go, these robots do not significantly contribute to achieving carbon neutrality.

However, while it is currently unclear whether these issues can be fundamentally resolved, there is one interesting initiative. Ford, the well-known American automaker, is working on a bipedal delivery robot. Simply put, this is an attempt to have a humanoid robot with arms and legs perform automated deliveries, and it has been undergoing testing since 2019. If this robot can operate as effectively as a human, it could carry packages to the front door, ring the intercom, and deliver the packages. Even if the user cannot receive the package, it could place it in a delivery locker or leave it at the door. In other words, if such social innovations are realized, they could simultaneously reduce the need for human labor and the hassle of redelivery, making significant progress toward carbon neutrality.

12.3.4.4 Drone Delivery

In the United States, there is a trend toward utilizing drones for door-to-door delivery. Walmart is currently deploying a drone-based product delivery service in several states and plans to expand it further in the future. The company is also actively partnering with drone companies, one of which (DroneUp), estimates a potential demand of over 1 million deliveries per year.

As Japan's population ages, an increasing number of older adults are gradually becoming unable to visit physical stores. Additionally, in depopulated areas, the number of so-called "shopping refugees"—those who no longer can easily access shops and supermarkets to purchase their daily necessities—has risen to 9.04 million (Ministry of Agriculture, Forestry and Fisheries, 2024), and this number continues to grow. To address this vulnerable segment, drone logistics could become a significant addition in the future.

The advantage of drone delivery lies in its ability to quickly airlift goods to remote, mountainous areas with poor road conditions, without relying on labor-intensive drivers or trucks powered by diesel or gasoline. Notably, the environmental benefits of drones are substantial. According to a study by Smithsonian (2018), greenhouse gas emissions from drones were reduced by more than 54% compared to trucks. Although flight performance and economics are not without their problems at present (Noguchi, 2024), a partial shift from trucks to drones as a means of transportation in the last-mile may be inevitable in the future, given environmental problems and the growing depopulation of Japan.

12.4 Limitations of Zero Emissions and Future Marketing

12.4.1 An Empirical Study on Carbon Offsetting

In the logistics industry, as described above, there are various social innovations to realize a carbon-neutral society. Figure 12.1 organizes these innovations into a conceptual diagram. This shows that the degree of difficulty of execution is roughly inversely proportional to the current level of penetration. As long as there is a geographical gap between production and consumption, the transfer of goods will require trucks running on diesel or gasoline for some time, making it impossible to achieve zero emissions.

However, social innovations have already been developed to mitigate this situation. One such innovation is *carbon offsetting*, a mechanism whereby a company compensates for the CO_2 emitted in the distribution process of its products by investing in CO_2 reduction activities (such as energy conservation, use of renewable energy, and tree planting) conducted by another organization. The government certifies the amount of decarbonization as "J-credits," which can be traded financially. This is a realistic and significant effort to achieve carbon neutrality, and some online retailers, such as Canada's Shopify, are beginning to implement it.

The author expects carbon offsetting to expand in the logistics industry in the future. To identify effective marketing strategies for promoting carbon offsetting, an empirical study was conducted based on an original survey in February 2024. Details of the factors, the analytical model constructed, and the results are included in the Appendix (Fig. 12.2). The essence of the findings is highlighted here.

12.4.2 Eco-Engagement Marketing

In recent years, *engagement*, which refers to the collaborative relationship between companies and consumers, has become an important research topic in the field of marketing. Marketing efforts aimed at increasing this level of engagement are termed "engagement marketing" (Harmeling et al., 2017). The motivations that increase engagement include intrinsic (Ryan & Deci, 2002) and extrinsic factors (Barari et al., 2021). For instance, in eco-activities, intrinsic factors might involve the pleasure derived from the activity or the desire for recognition, while extrinsic factors could include eco-points and information from mass media and social networking sites. Ultimately, increased engagement is likely to enhance business performance (Pansari & Kumar, 2017).

Analyzing these causal relationships using Structural Equation Modeling (SEM) revealed that to effectively increase eco-engagement and subsequently improve business outcomes, companies should focus on enhancing the cathartic nature of intrinsic factors and boosting peer pressure through information elicitation from others (Fig. 12.2).

Specific marketing implications include identifying individuals who find pleasure in eco-activities and fostering their influence as key strategies. Additionally, the study found that the most effective way to increase eco-conscious pressure is through public relations (PR) activities by companies to promote environmental measures. Therefore, logistics companies should also actively engage in PR activities related to carbon offsetting. In contrast, the impact of pragmatic factors, such as eco-points, on eco-engagement was relatively low, indicating that the driving factors for eco-activity transcend mere economic gain.

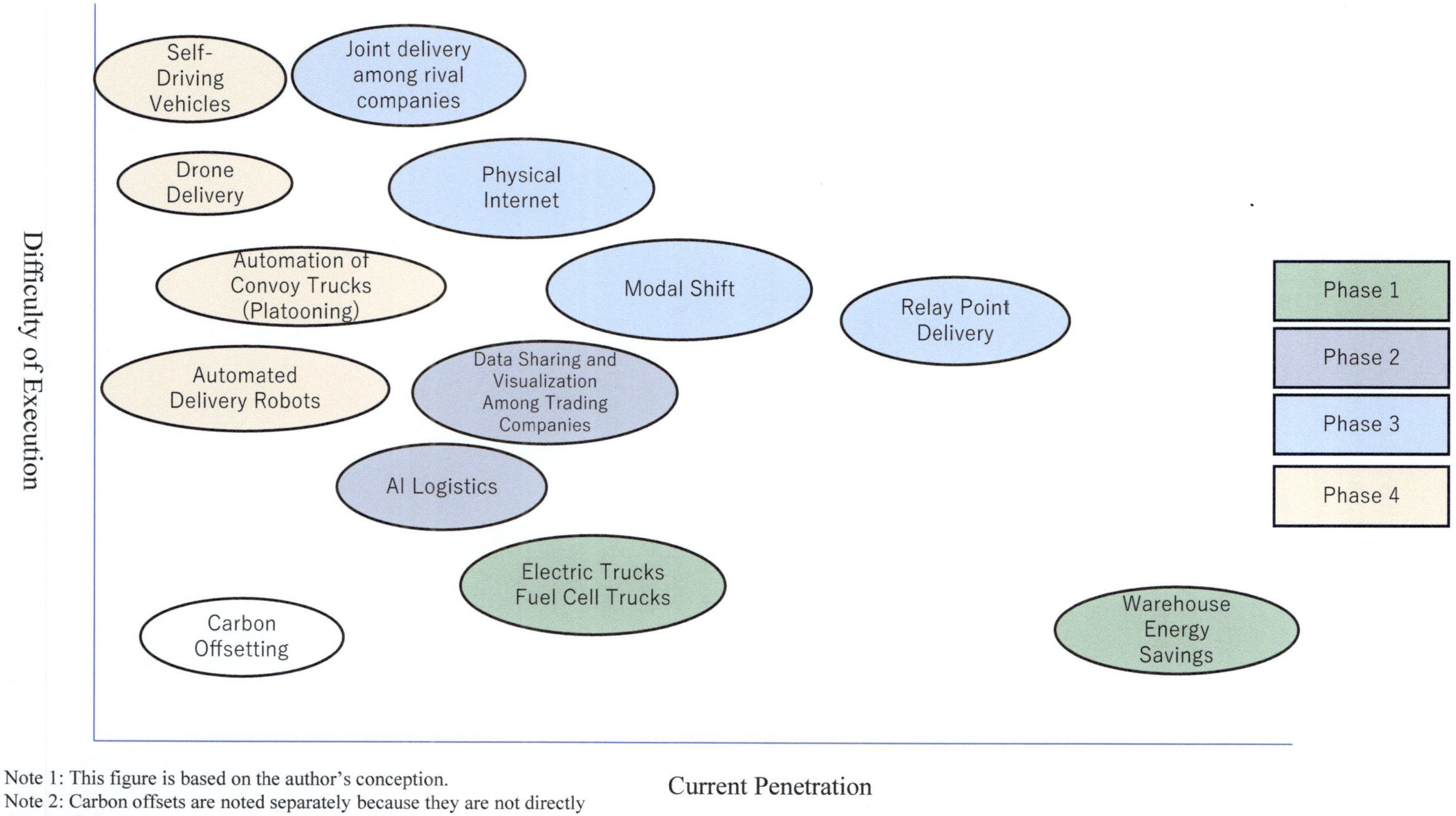

Note 1: This figure is based on the author's conception.
Note 2: Carbon offsets are noted separately because they are not directly decarbonizing.

Fig. 12.1 Green logistics' steps toward a carbon-neutral society

12.5 Conclusion

In this chapter, the author has examined the methods and examples of green logistics that have been implemented in the logistics sector to date and how they should contribute to the realization of a carbon-neutral society. In terms of energy, while efforts to incorporate renewable energy and improve energy efficiency are underway at distribution centers, the future requires further development and popularization of electric trucks and FCVs for transportation.

In terms of logistics DX, in Japan, where more than 99% of companies are small, medium, or micro enterprises, analog methods such as handwritten documents, telephone calls, and faxes are still prevalent. Astonishingly, even major nationwide convenience store chains sometimes send handwritten fax orders to warehouses. Digital data sharing and visualization are essential, as is the utilization of AI to improve transportation efficiency.

Of particular importance in contributing to a carbon-neutral society is the reform of logistics structures. Modal shifts, the physical Internet, joint delivery, relay point delivery, and other innovative combinations that break down existing vertical competitive structures and transcend the boundaries between rivals will dramatically increase transportation efficiency and realize environmentally friendly logistics.

Looking to the future, it is also crucial for the logistics industry, which is facing a serious labor shortage, to replace its operations with robots. For example, automated driving robots are tireless, never complain, and provide optimal driving. Consequently, driver shortages will be eliminated, energy consumption efficiency will be increased, and CO_2 emissions will be reduced to a minimum. There is no doubt that the green logistics initiatives described above will contribute significantly to the realization of a carbon-neutral society. However, as long as the transfer of goods is necessary, achieving complete zero emissions is impossible. Therefore, carbon offsetting, where a company offsets its own CO_2 emissions by funding reduction activities undertaken by other organizations, will become increasingly important. Empirical studies have shown that PR activities promoting environmental measures by companies enhance eco-engagement with consumers and improve corporate outcomes.

Appendix

See Fig. 12.2.

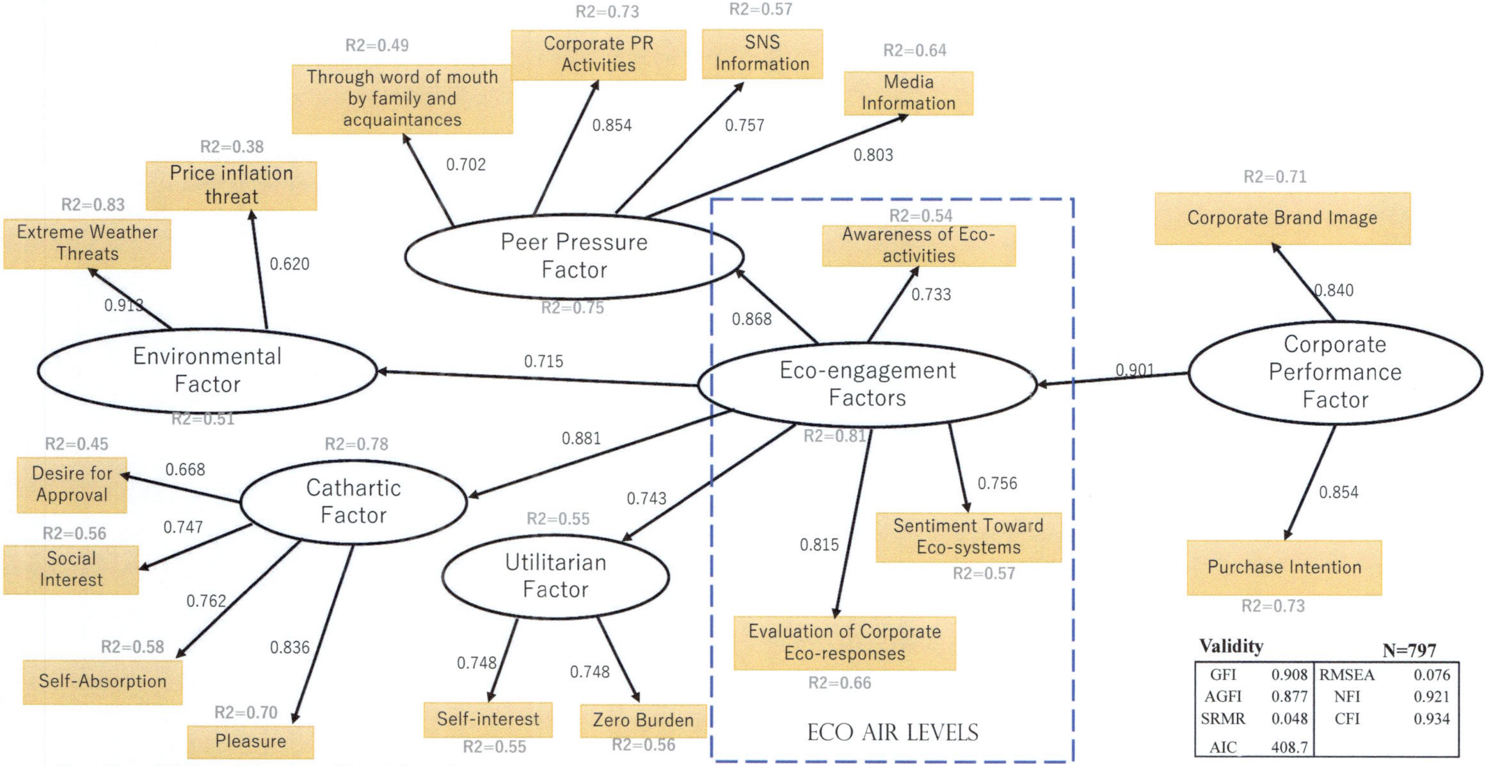

Note: The reliability of all path coefficients is $p < 0.001$.

Fig. 12.2 SEM analysis: Model and results

References

AI Research Institute. (2024). *14 AI applications in the logistics industry*. Retrieved from https://metaversesouken.com/ai/ai/logistics-applications/

Agency for Natural Resources and Energy. (2021). *What does it cost to produce electricity?* Retrieved from https://www.enecho.meti.go.jp/about/special/johoteikyo/denki_cost.html

Barari, M., Ross, M., Thaichon, S., & Surachartkumtonkun, J. (2021). A meta-analysis of customer engagement behaviour. *International Journal of Consumer Studies, 45*(4), 457–477.

Brohan, M. (2013). *Reducing the rate of returns*. Retrieved from https://www.internetretailer.com/2013/05/29/reducing-rate-returns

Bower, A. B., & Maxham, J. G. (2012). Return shipping policies of online retailers: Normative assumptions and the long-term consequences of fees and free returns. *Journal of Marketing, 76*(5), 110–124.

Eltex. (2021). *Survey of mail order business involved persons 2021*. Retrieved from https://dc.eltex.co.jp/research_column/research/20210824/

ENECHANGE. (2024). *What is the E.V. penetration rate in Japan?* Retrieved from https://ev-charge-enechange.jp/articles/033/

F-LINE. (2023). *Environmental initiatives*. Retrieved from https://www.f-line.tokyo.jp/esg/environment/

Hakobu. (2023). *At the logistics department, Kyocera Kagoshima Kawauchi Plant*. Retrieved from https://hacobu.jp/case_study/6406/

Harmeling, C. M., Moffett, J. W., Arnold, M. J., & Carlson, B. D. (2017). Toward a theory of customer engagement marketing. *Journal of the Academy of Marketing Science, 45*(3), 312–335.

Janakiraman, N., Syrdal, H. A., & Freling, R. (2016). The effect of return policy leniency on consumer purchase and return decisions: A meta-analytic review. *Journal of Retailing, 92*(2), 226–235.

Japan Business Machine and Information System Industries Association (JBMIA). (2022). *MFPs and other arterial logistics joint delivery starts in selected areas!* Retrieved from https://www.jbmia.or.jp/whatsnew/detail.php?id=1496

Ministry of Agriculture, Forestry and Fisheries. (2024). *Estimated food inaccessible population in 2020*. Retrieved from https://www.maff.go.jp/j/press/kanbo/kihyo01/240227.html

Ministry of Economy, Trade and Industry. (2023). *FY2022 digital transaction environment improvement project* (Market research on electronic commerce). Retrieved from https://www.meti.go.jp/press/2023/08/20230831002/20230831002-1.pdf

Ministry of Environment. (2024). *Greenhouse gas emissions and absorption* in FY2022. Retrieved from https://www.env.go.jp/content/000215754.pdf

Ministry of Land, Infrastructure, Transport and Tourism. (2019). *Joint transportation by double-connected trucks begins*. Retrieved from https://www.mlit.go.jp/report/press/20190328_SF25.html

Ministry of Land, Infrastructure, Transport and Tourism. (2021a). *Carbon dioxide emissions in the transportation sector*. Retrieved from https://www.mlit.go.jp/sogoseisaku/environment/sosei_environment_tk_000007.html

Ministry of Land, Infrastructure, Transport and Tourism. (2021b). *Comprehensive logistics policy outline* (FY2021-FY2025). Retrieved from https://www.mlit.go.jp/seisakutokatsu/freight/content/001409564.pdf

Ministry of Land, Infrastructure, Transport and Tourism. (2023a). *Redelivery rate of parcel delivery service decreased to about 11.4% in April 2023*. Retrieved from https://www.mlit.go.jp/report/press/content/001616051.pdf

Ministry of Land, Infrastructure, Transport and Tourism. (2023b). *Survey and methods of aggregating the number of parcels handled in FY2022*. Retrieved from https://www.mlit.go.jp/report/press/content/001625915.pdf

Ministry of Land, Infrastructure, Transport and Tourism. (2023c). *Automobile transportation statistical survey (2022)*. Retrieved from https://www.e-stat.go.jp/dbview?sid=0003442538

Ministry of Land, Infrastructure, Transport and Tourism. (2024). *Current situations and issues surrounding logistics*. Retrieved from https://www.mlit.go.jp/road/ir/ir-council/buturyu_douro/pdf01/03.pdf

Mitsubishi Estate. (2023). *Building a logistics network of automated trucks*. Retrieved from https://www.mec.co.jp/news/detail/2023/06/30_mec230630_t2

Noguchi, T. (2024). *Nippon no buturyu mondai* [Logistics problems in Japan]. Tokyo: Chikuma Shobou.

Pansari, A., & Kumar, V. (2017). Customer engagement: The construct, antecedents, and consequences. *Journal of the Academy of Marketing Science, 45*, 294–311.

Peloton. (2019). *Peloton announces its vision for the trucking industry: Drivers lead, and technology follows*. Retrieved from https://peloton-tech.com/peloton-announces-its-vision-for-the-trucking-industry-drivers-lead-and-technology-follows/

Renewable Energy Institute. (2025). *Statistics international energy*. Retrieved from https://www.renewable-ei.org/statistics/international/

Ryan, R. M., & Deci, E. L. (2002). Overview of self-determination theory: An organismic dialectical perspective. In E. L. Deci & R. M. Ryan (Eds.), *Handbook of self-determination research* (pp. 3–33). University of Rochester Press.

Smithsonian. (2018). *Is drone delivery good for the environment?* Retrieved from https://www.smithsonianmag.com/innovation/drone-delivery-good-for-environment-180968157/

Tokyo Tatemono Co. (2023). *T-LOGI Akiruno, the 10th in the T-LOGI series, completed*. Retrieved from https://prtimes.jp/main/html/rd/p/000000219.000052843.html

Impact of Policies on Traditional Resource Management: Case Studies on Subak in Bali, Indonesia

13

Naori Miyazawa

Abstract

This study examines policies that have been applied to traditional water resource management using case studies in Bali, Indonesia. The paddy fields of Bali have long been managed by the traditional irrigation organization, "Subak." The paddy fields such as in Tabanan Regency, Bali, were registered as a World Cultural Heritage in 2012 by UNESCO as "cultural landscape," which have been created as a basis of harmony between humans and nature. However, with the recent surge in global tourism, Bali has experienced a considerable reduction in paddy field areas, posing challenges to agricultural land conservation. In response, the government adopted Subak-based policies. This study employed field surveys to assess the impact of the Subak government program, examining socioeconomic conditions of local farmers, paddy field reduction, income changes, and the effectiveness of subsidies and policies supporting the Subak system. The results revealed a complex relationship between tourism-driven economic pressures and the sustainability of traditional farming. The key findings comprise decreased paddy field areas, alterations in farmer incomes and livelihoods, and challenges in preserving the cultural and environmental integrity of the Subak system. This study offers broad insights into sustainability, cultural preservation, and balancing modernization and traditional practices.

Keywords

Traditional resource management · Subak · Bali · Tourism

13.1 Introduction

This chapter offers insights into the ways in which traditional systems such as Subak can contribute to sustainable land and water management practices. By examining the balance between cultural heritage, local governance, and environmental change, it contributes to ongoing discussions on pathways toward a more sustainable future.

Community-based natural resource management has gained increasing attention in global discussions on carbon-neutral societies and environmental conservation. It has been clarified and reassessed that traditional systems manage and conserve natural resources in several regions. The major features of the traditional system are its high compatibility with the natural conditions of the region and its small-scale sustainability.

N. Miyazawa (✉)
Waseda University, Tokyo, Japan
e-mail: naori.miyazawa@gmail.com

K. Akao and A. Washizu (eds.), *Climate Change Issues and Social Sciences*, Sustainable Development Goals Series, https://doi.org/10.1007/978-981-96-8951-4_13

Furthermore, it has been indicated that centralized resource management has faced some challenges. In particular, several developing countries tend to lack environmental laws and implementation systems necessary for managing natural resources (JICA, 2001). Although laws and regulations exist, they are occasionally ineffectively implemented.

The wave of decentralization has modified this top-down approach. Resultingly, a policy shift has been observed in local and regional focus, particularly in development and resource management, based on community participation.

Therefore, traditional systems were reassessed for resource management purposes. For example, the Paris Agreement, adopted at the 21st Conference of the Parties to the United Nations Framework Convention on Climate Change in 2015 emphasizes the importance of traditional systems as follows: "*Traditional knowledge, knowledge of indigenous peoples and local knowledge systems, with a view to integrating adaptation into relevant socioeconomic and environmental policies and actions*" (UNFCC, 2015).

13.2 Water Irrigation System in Bali: Subak

The paddy fields have been managed by the Subak in Bali, and its practice began in the ninth century (UNESCO). A Subak is built on community-based organizations that manage irrigation and its facilities and is jointly managed by the members of the organization.

The purpose of the Subak was to provide a stable supply of water for agriculture, primarily for rice cultivation.[1] Its activities cover agricultural water supply, as well as social, cultural, and environmental functions. The Subak governs water distribution for irrigation and serves as a social organization that regulates agricultural practices based on customary law, known as "Awig-awig." Roth and Sedana (2015) and Sedana et al. (2025) have identified Subak as a system that embodies the Balinese philosophy of "Tri Hita Karana," which emphasizes the harmonious relationship between humans, nature, and the divine. Recently, it has been observed that the growing impact of political and economic pressures on Subak institutions, particularly due to factors such as urbanization, tourism, and land conversion.

As Bali has undergone rapid development in recent decades and Subak systems have been confronted with considerable challenges, especially with the growth of global tourism. The reduction in paddy field areas, particularly in tourism areas, is a clear indication of the impact of these pressures. The Subak is currently facing challenges for sustainability, and several researchers have highlighted the necessity of formal intervention to conserve Subak as a valuable cultural and environmental resource (Lorenzen & Lorenzen, 2011; Windia et al., 2017).

13.2.1 Impact of Development on the Subak

The development-driven decline in Subak areas has been a central theme in the literature. As Davidson and Henley (2007) describe, the post-colonial era has seen a reconfiguration of traditional systems such as Subak in the context of modern state governance and economic priorities. The rapid decline of paddy fields in Denpasar, as evidenced by 36.2% in the past 13 years (Fig. 13.1).

13.2.2 Role of Customary Law (Awig-Awig) in the Subak

"Awig-awig," or customary law, are the Customary Rules of Subak and forms the backbone of the Subak governance structure, regulating water distribution, land use, and other agricultural practices. Study by Roth (2014) have emphasized the adaptability of "Awig-awig" to changing social and environmental conditions; however, they also warn of the fragility of these systems when external interventions, such as rapid development or market forces, come into play. The literature highlights the correlation between compliance

[1] For the dry farming, such as for vegetables, there are different systems, called *Subak abian*.

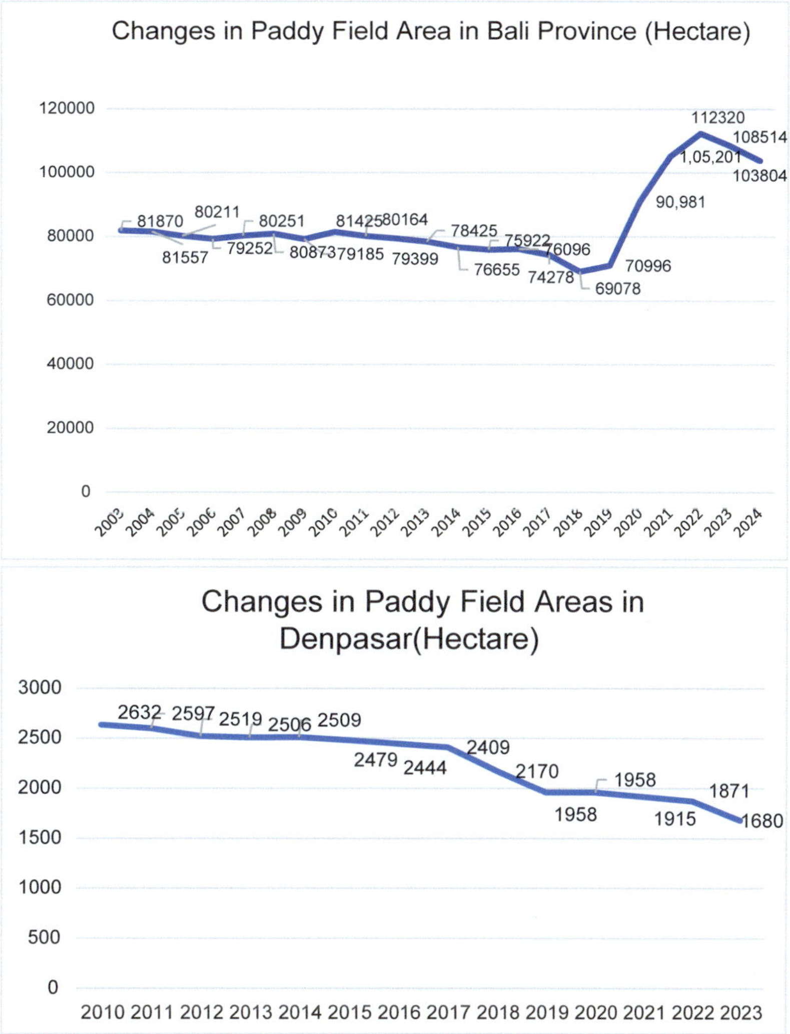

Fig. 13.1 Changes in paddy field areas in Bali province and Denpasar (hectare)[2]

with "Awig-awig" and strong community engagement, as well as the ongoing reinforcement of traditional norms through regular meetings and collective decision-making processes to regulate Awig-awig (Lansing, 2006).

The main regulations in Awig-awig include the following aspects: membership provisions, organizational responsibilities, ritual ceremonies, water distribution, paddy field management, arbitration of the dispute, and installation of *sunari* (bamboo flute) for mouse control. Violators of these rules are usually required to pay fines and hold ritual ceremonites.

[2]Source: Department of Agriculture, Government of Indonesia.

13.2.3 Governance and Community Participation in the Subak

The governance structure of Subak which is based on collective decision-making and community participation has long been admired as a model for sustainable resource management. Ostrom et al. (1999) used the Subak system as an example of how local communities can manage common-pool resources effectively without top-down regulation. The frequency of meetings within the Subak plays a crucial role in this governance model by allowing for a regular review of laws, conflict resolution, and collective planning.

13.2.4 Formalizing the Subak: Historical Background

Legal plurality is a characteristic of Indonesia: state law, *Adat*, and customary law. During the Dutch Colonial era, taxes were collected through the Subak (Roth & Sedana, 2015). The Subak began to form in 1972 with the enactment of the Provincial Irrigation Regulation (ibid.). Through the rapid spread of the Green Revolution and related irrigation development programs in the late 1970s, state intervention in the Subak increased (Horst, 1996). However, introducing blueprint forms of water division technology based on "modernized" irrigation increasingly created conflicts about water allocation and distribution (ibid.). From the 1980s, the Subak was rediscovered as a successful "community-based farmer-managed irrigation system" (FMIS), to be studied for irrigation development policies. Thus, the focus has shifted not only from "hardware" (physical infrastructure such as canals, etc.) but also to the institution consisting of rules and practices of Subaks.

Currently, the government recognizes Subaks as an integral part of Balinese culture, as follows:

> Regional Regulation Of Bali Province has an article "About *Subak*". (Regulation Number 9, Year 2012)
>
> "*As a part of Balinese culture, Subak Institution is a social organization having the potential to improve people's welfare through its main activities of regulating water use for irrigation of paddy fields, so that its existence and traditional rights need to be recognized and respected.*" (Regulation Number 9, Year 2012)

13.3 Challenges Faced by Subaks

Although the Subak system has proven resilient over centuries, it currently encounters several challenges under the rapid global tourism development in Bali. One major challenge is the decline of paddy fields. Several areas of agricultural land have been converted for non-agricultural uses, such as the construction of hotels and villas. Lorenzen and Lorenzen (2011) indicated that the expansion of non-agricultural land uses, such as housing and hotels, poses a direct threat to the Subak system.

The reasons for the conversion of agricultural land are complex. One of the major issues is the inadequate income generated from agricultural products. Farmers in Subaks earn limited income from paddy fields earning approximately Rp.1–2 million/month on average. This is insufficient for farmers to support their families, particularly considering the costs of educating their children. The minimum income level in Bali, as regulated by the Bali Provincial Decree, was Rp. 2,813,672 as of 2024.[3]

Farmers also face high tax burdens. Increasing land prices near tourist areas have resulted in higher taxes, consequently pressurizing farmers to sell their land. The conversion of paddy fields to tourist facilities undermines traditional agricultural practices, as well has broader implications for environmental sustainability and local culture. This transformation, driven by market forces and rapid tourism development, exerts financial pressure on farmers who face limited incomes from agriculture and increased tax burdens due to rising land prices. Such pressures are compounded by broader economic and social dynamics, including changes in generational attitudes toward farming and the allure of alternative employment in tourism.

In response to these challenges, various sustainability initiatives have been introduced to maintain Subak systems. Government projects have aimed to preserve these systems by improving the infrastructure and codifying customary laws, as seen in the Subak Sustainability Project. Windia (2010) underscores the importance of these projects in preventing further degradation of paddy fields, particularly in urban areas where development pressures are the strongest.

---

[3] https://gajimu.com/garment/Salaries-in-Garment/minimum-wage-garment/provincial-and-regencies-minimum-wage-in-bali#:~:text=The%20Bali%20Provincial%20Government%20announced,of%20the%20Province%20of%20Bali.

13.3.1 Government Policy on Subaks

The government has adopted new policies to support the Subak system. First, financial support of Rp. fifty million/year was provided to each Subak by the Bali Provincial Government. After the COVID-19, the amount of the financial support has been decreased. Some farmers expressed concerns that the total amount provided was the same, despite the size of the Subak. In reality, certain Subaks cover larger areas with a greater number of members.

In addition, the provincial government facilitates promotional activities called "subak contests" (lomba subak). Each Subak competes for aspects such as regulations, culture, and management.

13.3.2 Government Program for Subaks in Denpasar: "Sustainable Subak Program: Subak Lestari"

The Government of Denpasar, the capital of Bali, initiated a program to support Subaks in 2015. This governmental program is called "Sustainable Subak program: *Subak Lestari*."[4]

This program was intended to support Subaks in terms of facility improvement, subsidies, and tax exemptions for farmers' land. For instance, to improve the facilities in Subaks, the Government supports building small roads in Subak areas and irrigation canals.

The Denpasar government provides a subsidy of 75% when farmers buy seeds and the central government of Indonesia provides a subsidy of 35% for buying fertilizers.

Furthermore, the government purchases rice through subsidies for farmers. When farmers sell rice, the government purchases it in addition to the selling price of Rp. 200/kg under the Subak sustainability program.

[4] Interview with Director, Department of Agriculture. Bali provincial Government, 2018.3.29, 2018.8.23. Denpasar.

The government also provides tax exemptions for farmers. The tax on farmland is determined by the location and not by the amount of agricultural production. Land prices closer to tourist areas are higher, thereby resulting in higher taxes and land prices. Land taxes are becoming a heavy burden for farmers, putting pressure on them to sell their land. The government of Denpasar has provided full tax exemption for farmers' land in Denpasar City since 2014. Other regencies have similar tax-exemption policies, for example, the Tabanan Regency has provided a 40% tax exemption for farmland since 2012.

Interestingly, Bank Negara Indonesia (BNI) provides private sector support to Subak farmers. Using a BNI card, farmers can defer payment for agricultural inputs until after the harvest in some areas.

13.4 Research Gap, Objectives, and Methodological Approach

Although previous studies on the Subak system have extensively explored its cultural, environmental, and socio-political significance, a considerable portion of the literature has focused on the historical resilience of Subak and its role in traditional water management. However, there is a clear gap in the empirical examination of direct impacts of modern government interventions on the sustainability of Subak systems, particularly in rapidly urbanizing regions such as Denpasar. The majority of studies have either discussed the decline of Subak in general or focused on the philosophical foundations of Subak governance, but few have conducted a detailed analysis of the impact of specific sustainability projects on the socio-economic situation of farmers, compliance with customary law, community participation, and conflict resolution within Subak communities.

The aim of this research was to fill that gap by examining the effect of targeted government sustainability projects on the socio-legal and governance structures of Subak. In particular, it examines the differences between project-targeted Subak and non-project Subak in terms of economic condition, meeting frequency, compliance with customary law "Awig-awig," and conflict resolution processes.

13.4.1 Key Objectives of the Study

This study seeks to examine the multifaceted implications of the Subak Sustainability Project on local communities in Bali. First, it aims to assess the project's impact on the socioeconomic conditions of farmers by comparing project-targeted Subak with non-project Subak, with particular attention to farmer income, the frequency of meetings, and the degree of community engagement. Second, the research analyzes the extent of compliance with Awig-awig (customary law) across both project and non-project Subak, thereby evaluating whether the project reinforces or alters adherence to traditional governance norms. Finally, the study investigates the functioning and effectiveness of conflict resolution mechanisms in both project and non-project Subak, highlighting the role of the project in shaping institutional responses to disputes.

13.4.2 Methodology

This study adopted a mixed-methods approach that combined quantitative and qualitative data to provide a comprehensive analysis. A field survey was conducted with 200 farmers across both project-targeted and non-project Subak to collect quantitative data on socioeconomic condition of farmers, meeting frequency, compliance with "Awig-awig," and conflict resolution practices. Additionally, interviews with key stakeholders, such as "Pekaseh" (Subak leaders), government officials, and professors from a local university, were conducted to obtain enhanced understanding of the government involvement and the perceived assessment of the sustainability project. The quantitative data were analyzed using statistical methods to identify significant differences between project and non-project Subak, whereas qualitative data was thematically analyzed to explore the broader socio-cultural implications of the findings.

By addressing this gap in the literature, the aim of this study was to provide an evidence-based understanding of the evolving role of government interventions in sustaining traditional systems such as the Subak. It also seeks to offer practical recommendations for improving the design and implementation of future sustainability projects in Bali's agricultural and cultural landscapes.

13.4.3 Field Survey in Subaks, Denpasar

A field survey of Subaks has been conducted in Denpasar since 2018 intensively and afterwards in collaboration with professors and students from local universities in Indonesia. The objective of the field survey was to assess the impact of the Subak Sustainable Program supported by the government and to understand the situation in Subaks and the conditions of farmers. To assess the impact of the Subak Sustainable Project, the field survey compared the situation of Subaks under the Subak Sustainable Project and those not under the project, as shown in Table 13.1. Two hundred farmers were interviewed using a questionnaire.

Three Subaks were selected from the government projects: Anggabaya, Umadesa, and Umalayu. Another Subak, Kedua, which was not targeted by a government project in adjacent areas, was chosen for comparison. One hundred farmers were randomly selected from each Subak category (Subak under project and Subak Non-project) and interviewed using the same questionnaire for both groups.

In addition, interviews with policymakers, namely, the Department of Agriculture, Denpasar Government, university professors, and Subak leaders, were also conducted.

Table 13.1 Target areas of field survey

	Subak under Project			Subak Non-Project
Name of Subak	Anggabaya	Umadesa	Umalayu	Kedua
Number of Respondents (n)	41	28	31	100
Area (ha)	28	11	27	91

13.5 Results

13.5.1 Changes in Paddy Field Areas

The paddy field area managed by Subaks in Denpasar has decreased by 36.2% in the past 13 years, with 956 ha lost in the area (see Fig. 13.1). In some areas, paddy fields were completely lost, thereby resulting in Subak temples being managed by village members and not by Subak members.[5] The paddy fields in Denpasar are decreasing at a faster rate than those on Bali Island. The area of paddy fields in Bali decreased until 2018. The data show, however, a drastic increase since 2020. This correlates with the timing of outbreak of COVID-19 in 2020. This recent trend has some relationship between farming and COVID-19. Bali, widely recognized as a major international tourist destination, relies heavily on tourism, which employs nearly 70% of its workforce. The outbreak of COVID-19 and the implementation of social restrictions severely disrupted the tourism sector, resulting in economic stagnation and a sharp decline in revenues. Consequently, household incomes on the island decreased by approximately 66% (Kopernik, 2020). Under these circumstances, agriculture has re-emerged as a critical sector for safeguarding food security (Miyazawa, 2024). Importantly, new agricultural initiatives have been undertaken, with participation extending to groups previously disengaged from farming, including the younger generation (Miyazawa, 2024).

Figure 13.2 shows the changes in the area targeted by the field survey. The areas of Subaks targeted by government projects have not changed. However, the area of Subak Kedua, which had not yet been targeted by the project, was slightly reduced (from 93 in 2013 to 91 in 2017).

13.5.2 Comparison of Income

In the field survey, questions were asked regarding the farmers' incomes to compare their incomes before and after the project. The project started in the targeted Subaks, Anggabaya, Umadesa, and Umalayu, in 2015. Therefore, income level data for 2014 and 2018 were compared. One hundred farmers in the project area were surveyed regarding their incomes before and after the project (n = 100). The interview results showed that the monthly income increased, costs decreased, and monthly profits increased (Fig. 13.3). Statistically significant results were obtained for income and profit (t-value > double, p-value <5% level) (Table 13.2). However, the monthly costs showed no statistically significant changes. In more detail, the monthly income has increased by 20% when comparing the period before the project with the period after its completion. In addition, the monthly profit has increased by 95% when comparing the period before and after the project.

The survey compared 100 farmers in the project target area (n = 100) and 100 farmers in the non-project area (n = 100). First, the monthly income, costs, and profits per hectare were compared (Fig. 13.4). The project-targeted area had lower monthly income and cost. In contrast, monthly profits were higher in the project-targeted area. Only differences in monthly costs were statistically significant.

However, these aspects must be closely examined. Comparing rice productivity, the non-project areas had higher productivity than the project target areas. (non-project area: 5779 kg/ha; project

[5] Interview with Prof.Windia, 2018.8 Denpasar. 2022.7 Denpasar

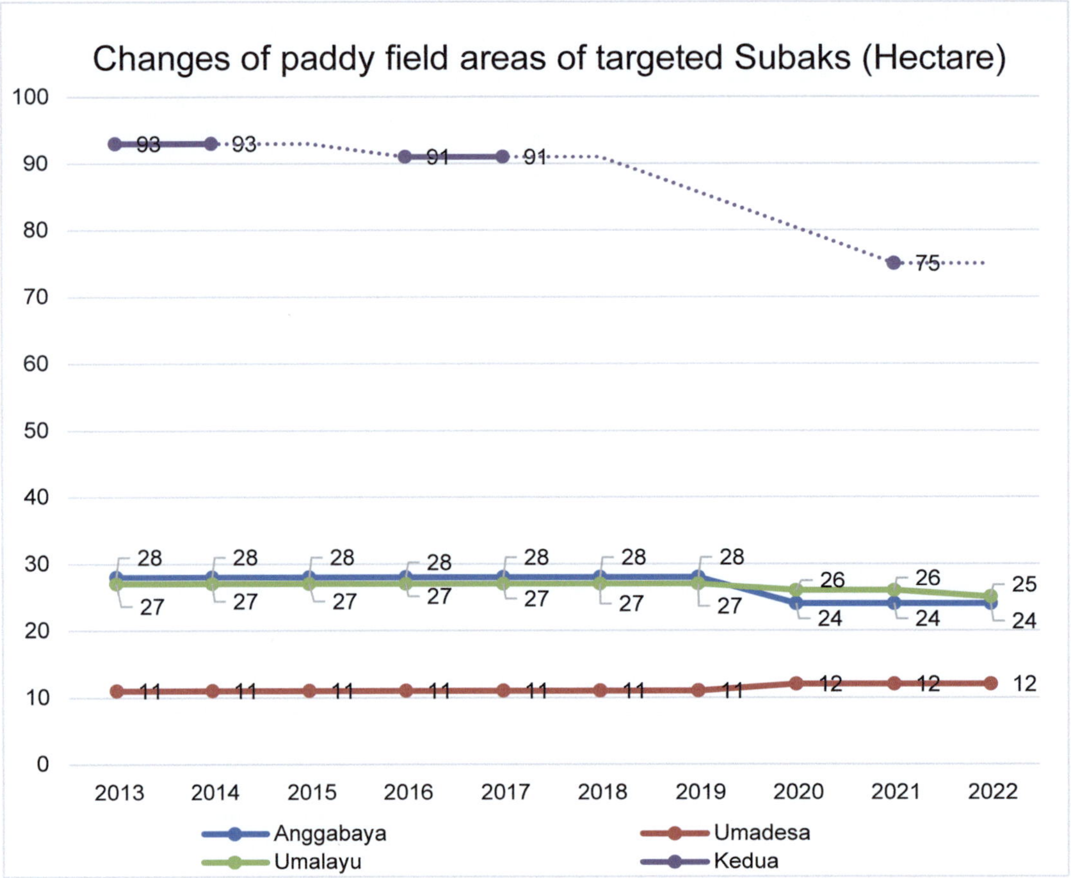

Fig. 13.2 Changes of paddy field areas of targeted Subaks (hectare)[6]

target area: 5771 kg/ha). In addition, a larger amount of rice was sold during one harvest season in the non-project areas (Fig. 13.5), with the consumption of rice by family households being lower in these areas. These factors may have affected the differences in income from rice sales.

13.5.3 Impact of Subsidies

The impacts of the subsidies were also compared. Farmers were asked questions about the subsidies provided when selling rice, and purchasing seeds and fertilizers in the project target areas and non-project areas. Higher subsidies were obtained in the project area when selling rice and purchasing fertilizer, and statistically significant results were obtained (Fig. 13.6). However, fertilizers are subsidized in non-project areas at a higher price, and are also subsidized by national programs. This aspect needs to be investigated in greater detail.

13.6 Result of Statistical Analysis

The table provides detailed statistical comparisons of farmers' income, costs, and profits before and after the Subak Sustainable Program as well as between project and non-project areas. The following discussion is based on the key findings from the table.

[6]Ibid.

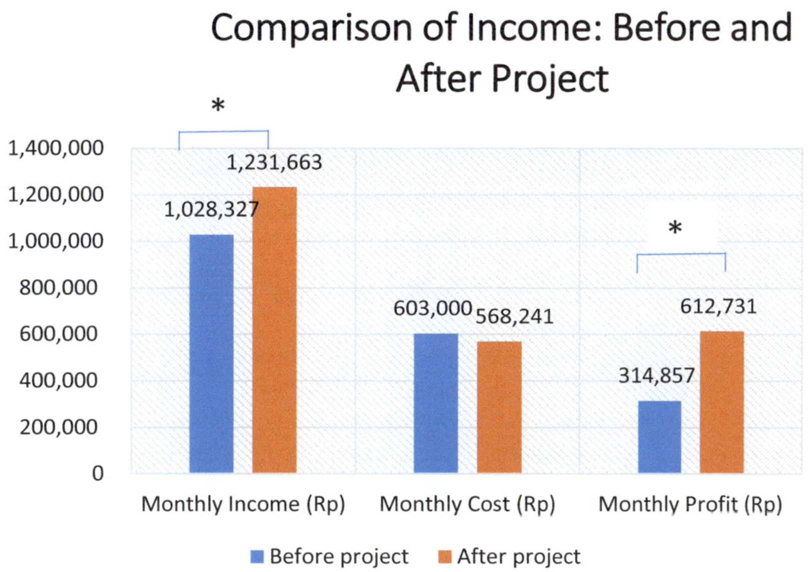

Fig. 13.3 Comparison of income: Before and after the project

13.6.1 Income Before and After the Project

The research results indicate that the mean monthly income increased significantly after the project, rising from 1,028,327 IDR to 1,231,663 IDR, with a mean difference of 203,336 IDR ($p < 0.001$). Monthly production costs decreased slightly from 603,000 IDR to 568,241 IDR, but the difference of 34,759 IDR was not statistically significant ($p = 0.424$). Consequently, the monthly profit increased significantly from 314,857 IDR before the project to 612,731 IDR after the project, representing a mean difference of 297,875 IDR ($p < 0.001$). These results suggest that the project had a positive and statistically significant impact on both monthly income and profitability, primarily due to increased revenues rather than reduced production costs.

13.6.2 Comparison of Project and Non-Project Areas

The results show that the mean monthly income per hectare was higher in the non-project area (4,509,866 IDR) compared to the project area (3,693,445 IDR), with a mean difference of 816,421 IDR. However, this difference was not statistically significant ($p = 0.058$). In contrast, the monthly production cost per hectare was significantly lower in the project area (1,861,430 IDR) than in the non-project area (2,586,144 IDR), with a mean difference of –724,714 IDR ($p = 0.009$). As a result, the project area recorded a higher mean monthly profit per hectare (2,048,663 IDR) compared to the non-project area (1,668,517 IDR), although this difference of 380,146 IDR was not statistically significant ($p = 0.542$). These findings suggest that while gross income tends to be higher in the non-project area, the project area benefits from reduced production costs, leading to relatively higher profitability.

13.6.3 Rice Production and Sales

The results show that the mean value of rice sold per hectare in one season was slightly lower in the project area (18,740,664 IDR) compared to the non-project area (20,067,009 IDR), with a mean difference of –1,326,345 IDR. However, this difference was not statistically significant ($p = 0.449$). In terms of quantity, the mean amount of rice sold per hectare per season was significantly

Table 13.2 Result of questionnaire survey with statistical analysis

Comparison of Income: Before and After Project (sample=100)						
	Before project	After project	Mean Difference	t-value	p-value (two-sided test)	Conclusion
Monthly Income (Rp)	1,028,327	1,231,663	203,336	4.43	p<0.001	Statistically significant
Monthly Cost (Rp)	603,000	568,241	34,759	0.80	0.424	
Monthly Profit (Rp)	314,857	612,731	297,875	3.80	p<0.001	Statistically significant
Comparison of Income: Project Area (n=100) and Non-project area (n=100)						
	Project Area	*Non-project area*	*Mean Difference*	*t-value*	*p-value(two-sided test)*	*Conclusion*
Monthly income / ha	3,693,445	4,509,866	**−816,421**	−1.90	0.058	
Monthly cost/ha	1,861,430	2,586,144	**−724,714**	2.65	0.009	Statistically significant
Monthly Profit/ha	2,048,663	1,668,517	380,146	0.61	0.542	
	Project Area	*Non-project area*	*Mean Difference*	*t-value*	*p-value(two-sided test)*	*Conclusion*
Rice Sold in 1 season/ha (Rp.)	18,740,664	20,067,009	**−1,326,345**	−0.75	0.449	
Rice Amount Sold in 1 season/ha (Kg)	3,391	4,345	**−954**	−2.61	0.010	Statistically significant
Rice production 1 season Kg/ha	5,771	5,779	−8	−0.04	0.960	
Rce consumption by family (kg/month)	48.9	44.5	4.4	1.70	0.090	
Impact of Subsidy						
	Project Area	*Non-project area*	*Mean Difference*	*t-value*	*p-value(two-sided test)*	*Conclusion*
Seed subsidy (Rp/kg)	6,064	4,729	1,335	2.84	0.005	Statistically significant
Fertilizer subsidies (Rp/kg)	2,496	3,506	**−1,010**	−4.69	p<0.001	Statistically significant, but oppposite result
Price of unhulled rice (Rp/kg)	4,591	4,380	211	2.29	0.023	Statistically significant
Rice Sale subsidies(Rp/kg)	457	0				

"Statistically significant" shows t-value > double, p-value <5% level

lower in the project area (3,391 kg) than in the non-project area (4,345 kg), with a mean difference of –954 kg ($p = 0.010$). By contrast, the mean total rice production per hectare was almost identical between the two areas (5,771 kg vs. 5,779 kg), and the difference was not statistically significant

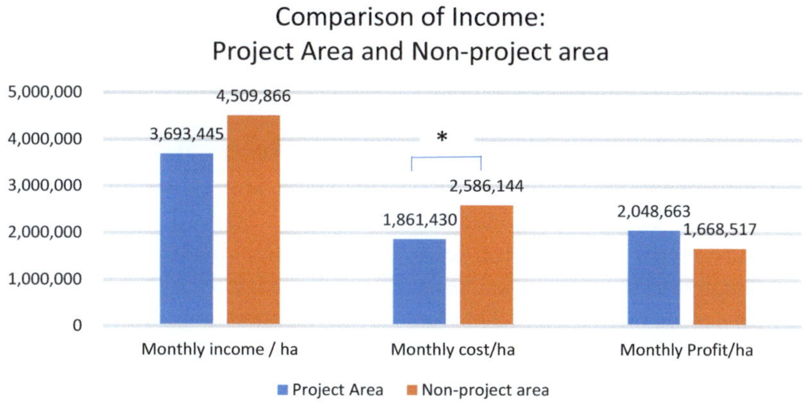

Fig. 13.4 Comparison of income: Project area and non-project area

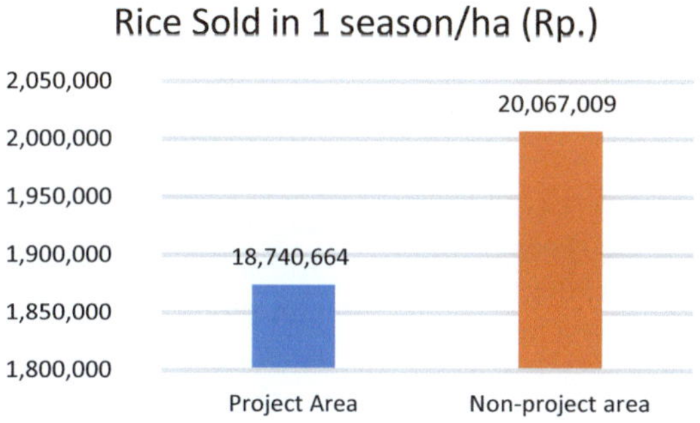

Fig. 13.5 Quantity of rice sold in 1 season/ha (Rp.)

Fig. 13.6 Amount of subsidy farmers received for each item (Rp/kg)

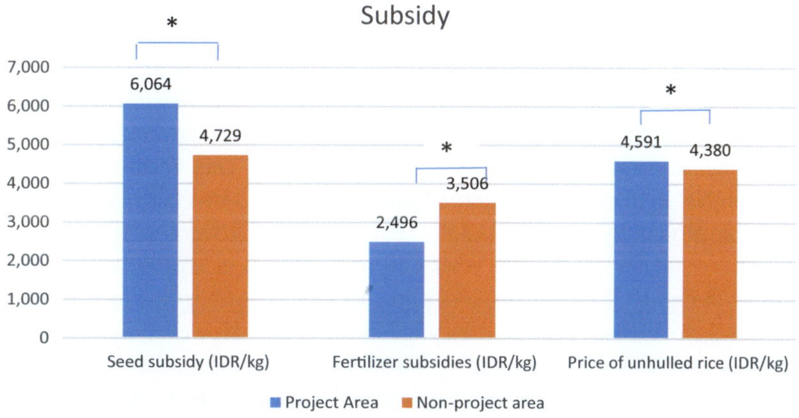

(p = 0.960). Regarding household consumption, families in the project area reported slightly higher rice consumption (48.9 kg/month) compared to those in the non-project area (44.5 kg/month), but this difference of 4.44 kg was not statistically significant (p = 0.090).

13.6.4 Impact of Subsidies

The findings show that farmers in the project area received significantly higher seed subsidies (6,064 Rp/kg) compared to those in the non-project area (4,729 Rp/kg), with a mean difference of 1,335

Rp/kg ($p = 0.005$). By contrast, fertilizer subsidies were significantly lower in the project area (2,496 Rp/kg) than in the non-project area (3,506 Rp/kg), with a mean difference of –1,010 Rp/kg ($p < 0.001$). This indicates a statistically significant but opposite result, where project farmers less from fertilizer subsidies. The price of unhulled rice was significantly higher in the project area (4,591 Rp/kg) compared to the non-project area (4,380 Rp/kg), with a mean difference of 211 Rp/kg ($p = 0.023$). In addition, only project farmers benefited from rice sale subsidies, averaging 457 Rp/kg, while no such support was reported in the non-project area.

13.6.5 Improvement in Facility After Subak Sustainability Project

Farmers were inquired about the types of infrastructural enhancements observed after the project was implemented in Subaks. The majority of farmers (99 of 100) responded that Subak roads were improved, followed by canals, water dams, and Subak temples (Fig. 13.7).

13.6.6 Frequency of Meetings in Subaks

Questions were asked about the frequency of meetings in Subaks. The frequency of meetings was higher in the project-targeted Subaks than in the non-project Subaks (Fig. 13.8).

The frequency of meetings is shown in different colors in Fig. 13.8. In Subaks under project 98 out of 100 farmers responded that they held meetings at least once every 2 or 3 months. Meanwhile, in Subaks where the project was not implemented, only 60 farmers replied that they held meetings at least once every 2 or 3 months. It was observed that Subaks participating in the project held meetings more frequently than those not included in the project.

The reason for this difference seems to be that there are several opportunities to hold meetings with the government on the project, such as creating rules and planning the project in Subaks.

13.6.7 Compliance of Farmers Toward Customary Law by Each Subak

Each Subak has its own regulations based on customary law, called *Awig-awig*, as previously mentioned. Some Subaks have already codified the regulation, but other Subaks also have regulations passed down by word of mouth from generation to generation. *Awig-awig* usually regulates water distribution, usage, and other aspects based on a consensus among Subak members.

The compliance level of farmers who followed customary law, *Awig-awig*, was interviewed, and the compliance rate is shown in Fig. 13.9. It was found that project-targeted Subaks had a higher compliance rate with customary law than the non-project Subaks.

Seventy out of 100 farmers in the project targeting Subaks responded that farmers comply 100% with the regulation of *Awig-awig*. In the non-project Subaks, only 12 out of 100 farmers responded that they complied 100%.

One of the reasons for this difference is that the Subak members reviewed the customary law and agreed on the regulations targeting Subaks when the project was implemented.[7]

Furthermore, when the Subak members reviewed customary law, the government held meetings with them about new provisions on what the government intended to reflect in its policy. This means that government regulations were reflected in the customary law. For example, the government intends to include a regulation that paddy fields should not be changed to land for use other than agriculture, such as housing and hotels. However, the regulations in customary law recommended by outsiders, including the government, seem to be less effective than local regulations based on Subak culture.[8]

---

[7] Interview with Prof.Windia,2018.11 Denpasar.

[8] Interview with Director, Department of Agriculture. Bali provincial Government, 2018.3.29, 2018.8.23. Denpasar.

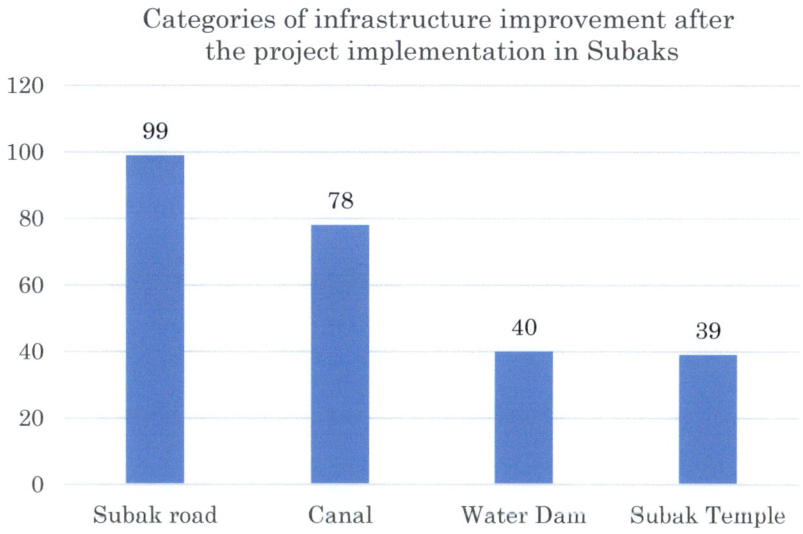

Fig. 13.7 Improvement in facilities after the Subak Sustainability Project

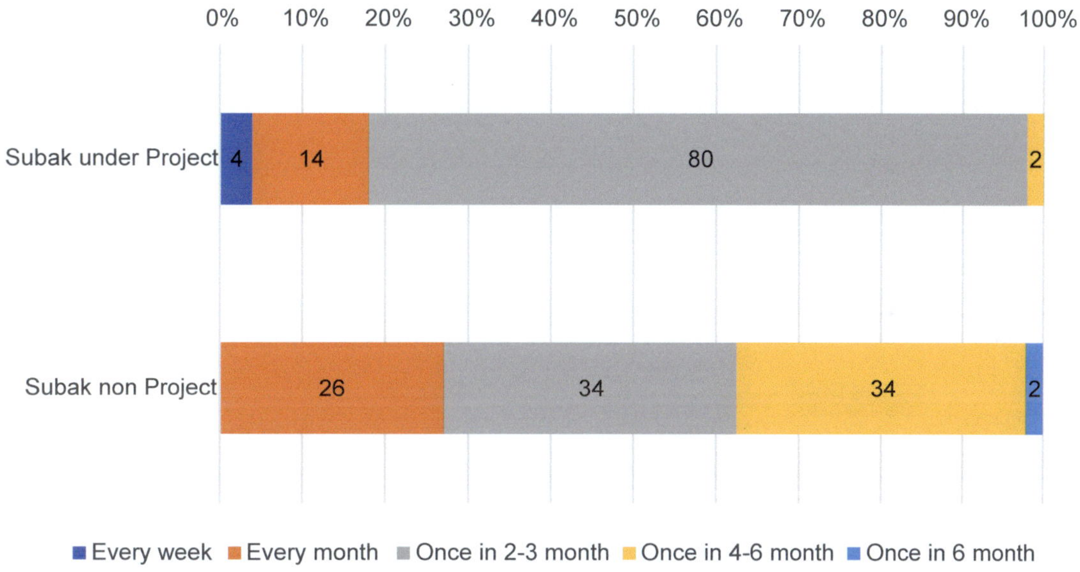

Fig. 13.8 Frequency of meetings in Subaks

13.6.8 Types of Sanctions Imposed in Cases of Violations of Customary Law

A question was asked, "When farmers violate the rule of customary law, what type of sanction the violators have to obey" (Fig. 13.10). In the majority of cases, the violators must pay monetary sanctions. If the violator does not comply with this sanction, the water canal for the violator is closed. In addition, ceremonies must be held occasionally. The ceremony is important because the violator must apologize for the mistake; the violator must bear the cost of the ceremony.

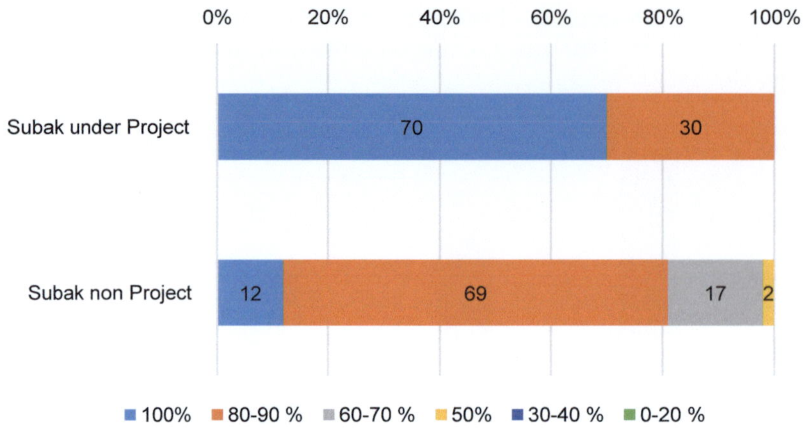

Fig. 13.9 Compliance level of farmers toward customary law within each Subak

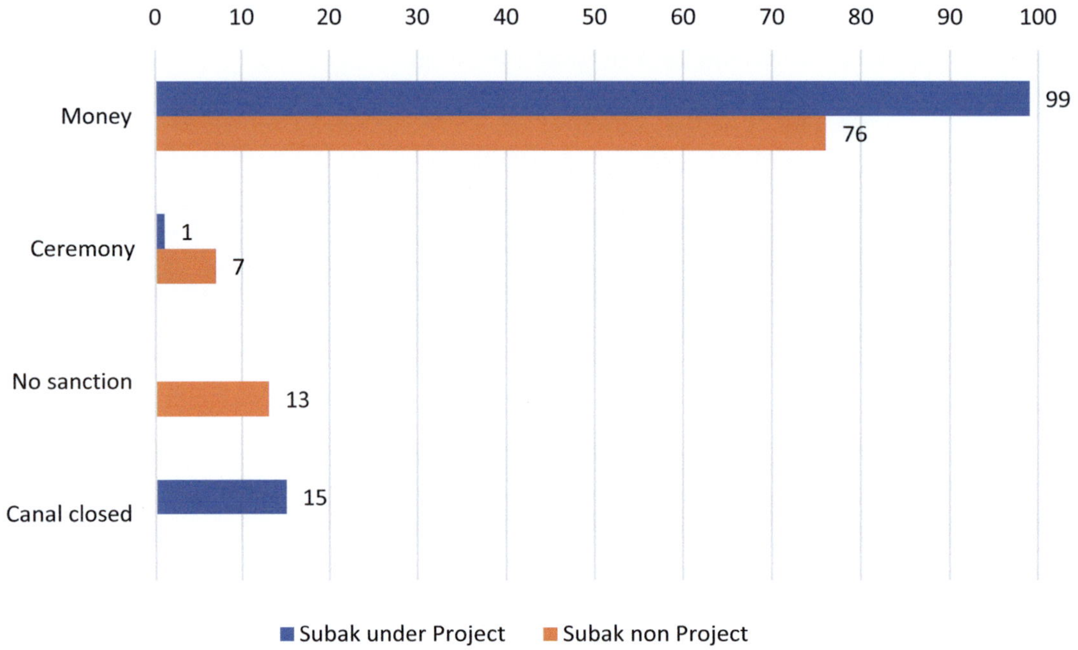

Fig. 13.10 Types of sanctions imposed in cases of violations of customary law

13.6.9 Form of Conflict Resolution

Some questions were asked regarding conflict resolutions. First, the form of conflict resolution, in the project-targeted Subaks, 53 farmers replied that they resolve conflict by dialogue in groups and by public dialogue (Fig. 13.11).

In the non-project Subaks, more than fifty-seven percent of farmers solved conflicts by dialogue between the two parties. The farmers in Subaks under the project tended to solve conflicts in a more public way than farmers in the non-project Subaks.

13.6.10 Person to Consult for Conflict Resolution

The next question concerning the conflict was, "With whom do farmers consult in case of a

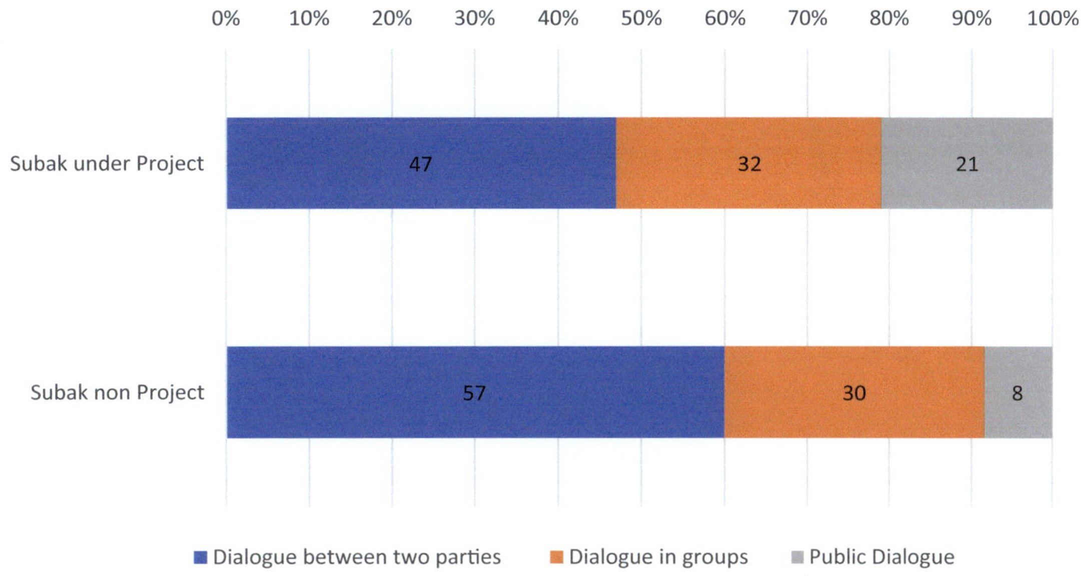

Fig. 13.11 Form of conflict resolution

Fig. 13.12 Types of entities to consult for conflict resolution

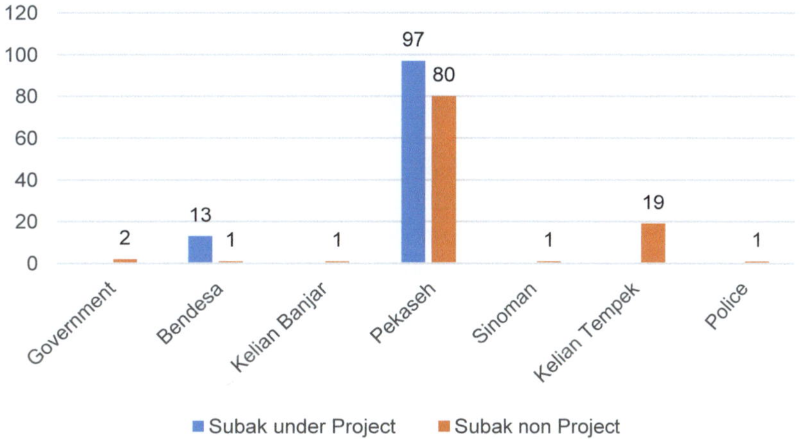

dispute?" In both areas, the leader of Subaks called *Pekaseh* was the most prominent conflict mediator (Fig. 13.12).

What is characteristic is that in the Subak project area, there is no person other than Pekaseh, a leader of Subak, and the village organization leader, named *Bendesa* for consulting. In contrast, in the non-project Subaks, the person for consultation is listed with more variety: government officials or the police. Other leaders of village organizations (Kelian Banjar) and sub-leaders of Subaks called Kelian Tempeks are also people who are consulted for conflicts.

13.6.11 Financial Institutions for Agriculture Activity

The farmers were asked a question about financial institutions that they borrow money from for agriculture activity (Fig. 13.13).

13.6.12 Types of Support Needed in Subaks

A question was asked regarding the types of support that were needed in Subaks. Different

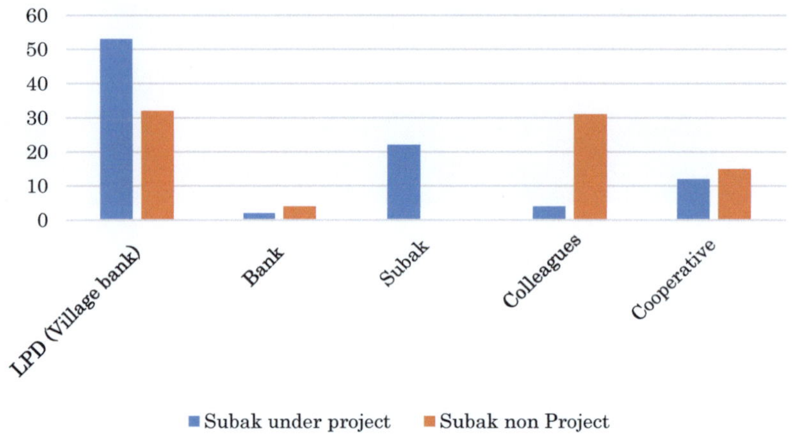

Fig. 13.13 Financial institutions from which farmers borrow money for agriculture activity

opinions were expressed by farmers in the project and non-project Subaks (Fig. 13.14). In the Subak project, farmers mentioned that more support for the *soft* component was needed, such as (1) marketing, (2) training, (3) establishing cooperatives, (4) subsidy, (5) irrigation facilities, and (6) Subak Temple (most orders). Meanwhile, farmers in the non-project Subaks mentioned that the following support is needed: (1) subsidy, (2) training, (3) irrigation facilities; and (4) marketing, (5) tax exemption. Farmers in the non-project Subaks have not yet received substantial government support compared to the project targetted Subaks. The support on "training" is expressed in both areas commonly with more than 30%. This implies that training is an area in which additional budget and support should be allocated to farmers. The majority of the farmers in Subaks under project also replied that the project was very helpful or helpful (44 and 56 farmers, respectively, out of 100).

13.7 Discussion

13.7.1 Change in Paddy Field Areas

The data presented indicates a significant decline in the paddy field areas managed by Subak systems in Denpasar over the past 12 years, showing an 36.2% reduction, which equates to a loss of approximately 952 hectares (Fig. 13.1). This decline is concerning, as it affects agricultural production, as well as the traditional, cultural and social roles of Subaks in Bali. The reduction in Subak-managed land has led to situations in which Subak temples, traditionally managed by Subak members, are currently being supervised by village members because of the diminished presence of farmers.

The more rapid reduction observed in Denpasar than in Bali highlights the heightened pressure on agricultural land in urban areas. Urbanization, land conversion for development, and other socioeconomic factors are likely to have contributed to this accelerated decline, particularly in the capital city of Bali, where land values and competing demands for space are higher.

Figure 13.2 shows the contrast between areas targeted by government projects and those that were not. The targeted Subak areas appear to have remained stable, suggesting that government intervention may help prevent further losses. However, "Subak Kedua," which was not part of the government project, saw a slight reduction, from 93 ha in 2013 to 91 ha in 2017. Although this decline is less dramatic than the overall reduction in Denpasar, it indicates that Subak systems outside the targeted conservation or revitalization projects are vulnerable to further shrinkage.

This trend underscores the critical need for comprehensive policies that protect the Subak

Fig. 13.14 Types of support needed in Subaks

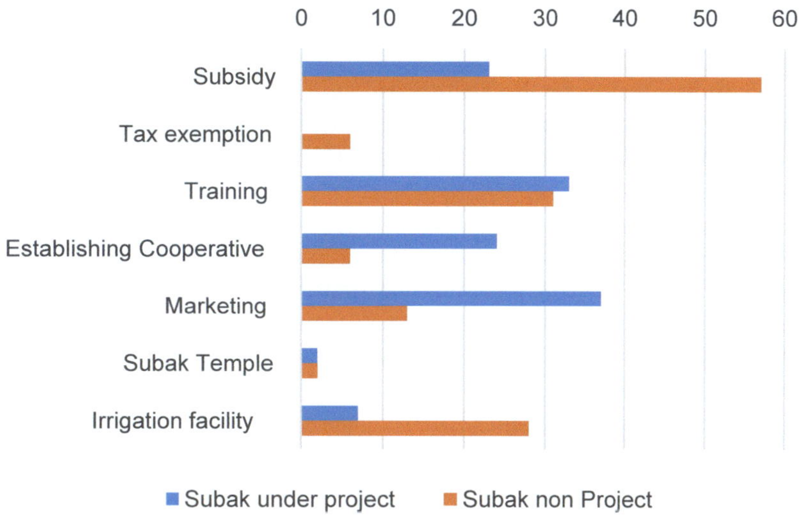

systems targeted by government projects and those not yet included. Without broader intervention, the continued decline of these agricultural systems could result in significant cultural, economic, and environmental impacts, particularly as Subak systems play an essential role in Bali's water management, agriculture, and heritage conservation. In addition, the loss of Subak areas could undermine local food security and disrupt the balance between traditional practices and modern urban development.

13.7.2 Impact on Farmers' Income and Profit After the Project

The data show that the project had a statistically significant positive impact on farmers' income and profits. Despite the increase in costs, the overall increase in profits was driven by gains in income and subsidies received by farmers, thereby leading to a net increase in profits. Additionally, although the project areas experienced some disadvantages in terms of rice production and fertilizer subsidies, the overall financial outcomes for farmers in these areas improved, particularly through income and seed subsidies. This suggests that the project, although successful in increasing income, may need to be adjusted to address inefficiencies in production and cost management to maximize the benefits for farmers.

13.7.3 Impact on Social Aspect After Project

The Subak Sustainability Project has positively impacted infrastructure development, community engagement, and adherence to traditional regulations. Farmers in project-targeted Subaks benefit from improved facilities, more frequent meetings, higher compliance with customary laws, and more organized conflict resolution mechanisms. However, the project also introduced challenges in balancing external government regulations with traditional Subak practices. The expressed need for further support in areas such as marketing, training, and cooperative development suggests that, although the project has made significant strides, continued assistance and refinement are necessary to ensure long-term sustainability and market access.

13.7.4 Awig-Awig

This research's findings that project-targeted Subak hold more frequent meetings than non-project areas align with the literature. Lorenzen and Lorenzen (2011) argue that these meetings are essential for maintaining the social fabric of Subak and ensuring that water distribution and agricultural activities are conducted fairly and efficiently. In contrast, in non-project areas where meetings are less frequent.

As seen in the Subak Sustainability Project, the government's attempts to regulate land conversion reflect broader concerns regarding the future of traditional farming in Bali. The literature suggests that for Subak to survive in the long term, a balance between traditional practices and modern interventions is needed. Windia (2010) argues that government support, although necessary, must be sensitive to the cultural and social foundations of Subak. Projects need to respect these traditions, as the community's participation and adherence to "Awig-awig" are crucial for the system's functioning.

13.7.5 Support and Capacity Building for Farmers

The need for additional support for farmers, particularly in terms of training, marketing, and cooperative development, is a common theme in literature (Roth & Sedana, 2015). The findings that farmers in both project and non-project areas prioritize training and marketing support are consistent with these studies. Training is crucial in assisting farmers adapt to new agricultural techniques and market demands, and improved marketing strategies are required to ensure the economic viability of small-scale farming in Bali.

13.7.6 Conflict Resolution and Sanctions in Subaks

Conflict resolution in Subak is traditionally handled through dialogue and consensus-building, with the "Pekaseh" (Subak leader) playing a central role in mediating disputes. The research findings show that conflict in project-targeted Subak is more likely to be resolved through public dialogue, whereas in non-project areas, private discussions between two parties are more common. This is consistent with Suamba et al. (2013), who suggested that formal projects may encourage more transparent and inclusive conflict resolution processes.

The use of sanctions in Subak, particularly financial penalties and ritual ceremonies, is another aspect that has been well documented in the literature. Roth (2014) highlighted how these sanctions, when combined with community-based conflict resolution, aid in maintaining harmony and ensuring compliance with customary laws. These sanctions serve as punitive measures, as well as mechanisms for reinforcing social norms and restoring spiritual balance within the community.

13.8 Conclusion

This study highlighted some findings to illustrate the impact of the Subak Sustainability Project. After the project was implemented, the income and profit of farmers from agriculture increased compared with those in 2014 (before project implementation). Farmers considered that facilities, such as Subak roads, canals, water dams, and temples, in Subaks had been improved. The Subak areas have not significantly decreased since the project started, whereas the non-project Subak areas have decreased largely. There was higher compliance with customary law (awig-awig) in Subak project areas; 70 out of 100 farmers in project-targeted Subak reported 100% compliance with awig-awig regulations. More frequent meetings were held in Subaks under the project than in non-project Subak areas.

Based on the findings of this survey, it can be concluded that the Subak Sustainable Project has had a positive impact, to a certain extent, on the income and profit of farmers, facilities in Subaks, and socialization and solidarity among farmers. In the future, farmers wish to have more training opportunities for agriculture and establish cooperatives. It would be effective to provide capacity-building opportunities for farmers to increase their marketing abilities and strengthen Subaks' economic function. Adding value to rice and marketing products would also increase farmers' income.

By promoting localized and environmentally conscious farming methods and reducing the pressure to convert Subak areas into urban land, this program aligns with the broader goals of environmental conservation. This sustainable development model can serve as a foundation for future initiatives that balance economic growth with ecological responsibility and can aid in achieving environmental sustainability.

References

Davidson, J. S., & Henley, D. (Eds.). (2007). *The revival of tradition in Indonesian politics: The deployment of adat from colonialism to indigenism*. Routledge.

Horst, L. (1996). Intervention in irrigation water division in Bali, Indonesia: A case of farmers' circumvention of modern technology. In G. Diemer & F. Huibers (Eds.), *Crops, people and irrigation: Water allocation practices of farmers and engineers* (pp. 34–52). Intermediate Technology Publications.

JICA. (2001). *Second Environmental Aid Study Group report*. https://openjicareport.jica.go.jp/pdf/11657384.pdf. Accessed on 25 June 2024.

Lorenzen, R., & Lorenzen, S. (2011). Changing realities—Perspectives on Balinese rice cultivation. *Human Ecology, 39*(1), 29–42.

Kopernik. (2020). *The impact of COVID-19 on livelihoods in Bali*. https://v1.kopernik.info/documents/publication/1591847503909_4236.pdf. Accessed on 1 March 2025.

Miyazawa, N. (2024). Proyek Penelitian Aksi Subak: Mendukung Akses Petani Lokal Bali ke Pasar Pertanian yang Lebih Luas. In Widianingsih et al. (Eds.), *Social Design: Studi Fenomena Sosial dalam Perspektif Mahasiswa Indonesia dan Jepang* (pp. 61–78). UNPAD press. English Translation: "The action research project to support the access to market by local farmers: Case study in Bali, Indonesia" UNESCO Website. https://whc.unesco.org/en/list/1194/?utm_source=chatgpt.com. Accessed on 25 August 2025.

Ostrom, E., Burger, J., Field, C. B., Norgaard, R. B., & Policansky, D. (1999). Revisiting the commons: Local lessons, global challenges. *Science, 284*(5412), 278–282.

Roth, D. (2014). Environmental sustainability and legal plurality in irrigation: The Balinese subak. *Current Opinion in Environmental Sustainability, 11*, 1–9.

Roth, D., & Sedana, G. (2015). Reframing Tri Hita Karana: From "Balinese culture" to politics. *The Asia Pacific Journal of Anthropology, 16*(2), 157–175.

Sedana, G., Wirawan, K., & Miyazawa, N. (2025, March). Implementation of organic rice farming based on local wisdom in Bali Province, Indonesia. *The Philippine Agriculture Scientist, 108*(1), 72–78. https://doi.org/10.62550/EN18053023

Sriartha, I. P., Suratman, & Giyarsih, S. R. (2015). The effect of regional development on the sustainability of local irrigation system. *Forum Geografi, 29*(1), 31–40.

Suamba, I. K., Mustadjab, M. D., Koestiono, D., & Windia, W. (2013). The advantage of agrobusiness for farmers on the subak system of Bali. *Journal of Basic Applied Science Research, 3*(3), 237–245.

Sutawan, N. (1987). Farmer-managed irrigation systems and the impact of government assistance: A note from Bali, Indonesia. In *Public intervention in farmer-managed irrigation systems* (pp. 49–69). International Irrigation Management Institute.

UNFCC. (2015). https://unfccc.int/sites/default/files/english_paris_agreement.pdf. Accessed on 25 June 2024.

Windia, W. (2010, October 13). *Sustainability of subak irrigation system in Bali*. Paper presented at the Seminar on the History of Irrigation in Eastern Asia, organized by ICID-IID, Yogyakarta.

Windia, W., Sumiyati, I Kt. Suamba, I Wy. Tika, I DG. A. Diasana P., Md. Mudra. (2016). *Rencana Aksi Pembentukan Subak Lestari Made Ayu Intan Di Subak Anggabaya, Subak Umalayu, Subak Umadesa, Subak Intaran Timur, Dan Subak Intaran Barat, Dinas Pertanian Tanaman Pangan danHortikultura Kota Denpasar, Denpasar*. https://simdos.unud.ac.id/uploads/file_penelitian_1_dir/9beb3261c1a55804f114d58a21ee362f.pdf

Windia, W., Sumiyati, Suamba, I. K., & Tika, I. W. (2017). Water management technique on subak irrigation system in Bali. *Journal of Agriculture and Rural Development in the Tropics and Subtropics, 118*(2), 1–10.

14 The Impact of Climate Change on Vulnerable Families in Southeast Asia: A Case Study in Cambodia

Yuko Shimazaki

Abstract

Climate change is a matter of human rights. To achieve a carbon-neutral society, understanding the challenges faced by the vulnerable groups is necessary. This chapter adopts a case study approach, using interviews, observation, and documentary analysis to examine the impact of climate change on economically disadvantaged and socially vulnerable populations in the Global South. The case of a Cambodian village is examined to define the concept of vulnerability within this context and gain a deeper understanding of the impact of climate change on vulnerable groups. Understanding the reality of people's lives makes it possible to develop a sustainable society that enables social inclusion.

Keywords

Climate change · Global South · Vulnerability · Cambodia

Y. Shimazaki (✉)
Tokai University, Tokyo, Japan
e-mail: yshimazaki@tokai.ac.jp

14.1 Introduction

The consequences of climate change have been observed in all geographical regions worldwide. However, there are notable differences in social and economic impacts on individuals in the Global North and South, with climate change having a markedly detrimental impact on impoverished communities in the Global South. For example, the impact of climate change on child poverty and vulnerable families is considerably more significant in the Global South than in the Global North, with serious implications for their well-being. Furthermore, climate change has been linked to increased risks of human trafficking, displacement, and forced migration, which are causes of concern at the international level (IIED, 2022; UNODC, 2022). These problems in the Global South illustrate that climate change is not merely an environmental concern but a human rights issue as well (UNHRC, 2021).

The influences of climate change on human life are also of disparate severities. Those in the Global North are better equipped to protect themselves from the effects of climate change and have access to a greater level of support in the event of adverse events. By contrast, those in the Global South, particularly those living in poverty, lack the resources and resilience to cope with the challenges posed by climate change. This implies

K. Akao and A. Washizu (eds.), *Climate Change Issues and Social Sciences*, Sustainable Development Goals Series, https://doi.org/10.1007/978-981-96-8951-4_14

that the impact of climate change on vulnerable populations is contingent on factors such as poverty, structural inequality, and the deprivation of historical and developmental backgrounds. Furthermore, the effects of climate change on the human body are life threatening.

Thus, this chapter addresses the question of how climate change affects the poor, with a particular focus on the necessity of a more profound understanding of the circumstances prevailing in the Global South that can facilitate sustainable development. Cambodia's case studies show the reality of vulnerable people's situation. The Mekong region of Southeast Asia is widely recognised as being significantly affected by climate change. However, the specific circumstances within individual countries remain insufficiently understood, and there is a notable lack of attention to the ways in which the poor are affected. This chapter adopts Cambodia as a case study to examine the effects of climate change on the impoverished.

The current consideration of issues is being addressed by governments and international society, and measures are being taken to address climate change worldwide. One such focus is the growing expectation for carbon-neutral societies. In considering the realisation of a carbon-neutral society, it is essential to begin by focusing on the phenomenon of climate change and its impact on the poor in the Global South, and to grasp the reality of the world in which we live.

14.2 The Impact of Climate Change on the Global South

14.2.1 Asia and Climate Change

Asia exhibited the highest frequency of natural disasters among all regions between 2003 and 2022, at 77.8% (CRED, 2024, p. 6). In terms of the direct impact on people in 2023, including drought, earthquakes, extreme temperatures, floods, mass movement, storms, volcanic activity, and wildfires, 71.4% of the total affected population was located in Asia (CRED, 2024, p. 6). The Intergovernmental Panel on Climate Change (IPCC) has identified several risks associated with climate change in Asia, including those related to access to water, the cryosphere, health, food, energy, urbanisation, and habitats. Southeast Asian countries have voiced particular concern regarding the potential impacts of flooding, sea level rise, rising temperatures, and extreme rainfall (IPCC, 2022).

The potential consequences of these changes include an anticipated rise in migrant labour and an increased likelihood of difficulty for people to maintain their current homes. In Southeast Asia, Myanmar and Thailand were identified as the countries most affected by climate change between 2000 and 2019, resulting in a significant number of people being forced to leave their homes and seek employment (IOM, 2022). These migrant workers are among the most impoverished people. Reports indicate that they are frequently involved in unsafe migration and work with inadequate access to safe mobility and information.

In the Asian region, approximately 66.4 million individuals have been impacted by natural disasters such as floods, typhoons, and cyclones in 2023. This has resulted in injuries, displacement, and other forms of hardship for some communities (CRED, 2024, p. 6). UNICEF (2021) determined that the global regions identified as particularly vulnerable to climate-related hazards are those where the health conditions and risks faced by the population are the most severe (UNICEF, 2021, p. 12).

The region affected is the Mekong Basin (which is spread across Myanmar, Vietnam, Laos, Cambodia, Thailand, and China). There are the highest risks where climate and environmental risks are considered significant (UNICEF, 2021, p. 79).[1] The regions exhibiting a correlation between vulnerability and climate change are primarily located in Southeast Asia. The Mekong Basin region has also been identified as

[1] Of the 10 worst cases, China's score is 9.0, Vietnam's 8.8, Myanmar's 8.3, Thailand's 8.4, Laos's 7.5, and Cambodia's 7.2.

a priority area for climate change-related initiatives at the United Nations and other international levels (IPCC, 2022, pp. 1468–1510).One illustrative example is the impact of floods on low-income households with high levels of vulnerability, particularly those residing in rural areas and engaged in agricultural activities such as rice cultivation (IPCC, 2022). For instance, in Vietnam, it is estimated that 39% of the households in the Mekong River Delta will be affected by persistent flooding in the future (IPCC, 2022, p. 1495). Moreover, the projected rise in temperature and alteration in precipitation patterns in the Mekong region will result in a transformation of the local population's agricultural practices, which will, in turn, affect yield (IPCC, 2022, p. 1495).

Furthermore, in recent years, international organisations such as UNICEF have raised concerns about the impact of rising temperatures on vulnerable households. Despite the Mekong River region having abundant water resources, the existing infrastructure is fragile and there is insufficient water for agricultural, industrial, and domestic use. Consequently, during the dry season, residents of rural and marginalised areas with inadequate infrastructure are forced to live in difficult conditions. This situation has led to people being compelled to purchase water, which is a further strain on their already limited financial resources. Households with limited financial resources for purchasing water face challenges in meeting their daily water demands. This predicament has far-reaching consequences, including the deterioration of sanitation infrastructure such as toilets, resulting in a pernicious cycle.[2] The issue of rising temperatures is emerging as a reason for households experiencing difficulties in securing water.

14.2.2 Key Agreements, Protocols, and the UN

This section provides an overview of the changes in the initiatives taken by the UN in relation to climate change. The term 'common but differentiated responsibilities' (CBDR)[3] is a frequently encountered phrase in the climate change debate. However, the concept of CBDR did not emerge independently. It was preceded by the 1972 United Nations Conference on the Human Environment in Stockholm, which constituted a significant step forward in this regard.This inaugural global conference prioritising environmental concerns constituted a seminal multilateral environmental agreement. Subsequently, the Montreal Protocol on Substances that Deplete the Ozone Layer (1987) was adopted, and in the following year, the Intergovernmental Panel on Climate Change (IPCC) was formed.[4] The IPCC was established by the World Meteorological Organization and the United Nations Environment Programme with the objective of assessing all aspects of climate change and its impacts based on available scientific information.

It has been a long road to greater recognition of the relationship between the environment and human rights. It has taken the form of the following processes in UN agencies. Once these fundamental principles were established, several core agreements emerged, providing the basis for further development. Key events included the UN Earth Summit in Rio de Janeiro (1992), the United Nations Framework Convention on Climate Change (UNFCCC) (1992/coming into force in 1994), the Millennium Development Goals (2000), and the United Nations Conference on Sustainable Development ('Rio+20', 2012). Additionally, the Rio+20 conference and the Paris Agreement (2015) represent significant milestones in this regard. The 2015 Paris Agreement, designated as COP21, comprised 195

[2] Interviews with villagers and village leaders (M.S, C.K, H.K), NGO members by author, September 2022 in Svey Rieng, Kandal, and Phom Penh.

[3] The principle (Principal 7) was established in 1992 at the first Earth Summit in Rio de Janeiro.

[4] It followed approval from the United Nations General Assembly in UN General Assembly Resolution 43/53 of 6 December 1988.

countries that collectively adopted the inaugural universal and legally binding global climate agreement, leading to the establishment of the Sustainable Development Goals (SDGs) and the UN Climate Change Conference (COP26). In 2021, the UNHRC made a notable decision, recognising the right to a healthy and sustainable environment as a 'human right' for the first time.[5]

The 2030 Agenda for Sustainable Development employs a thematic approach to evaluate the efficacy of programmes and initiatives implemented by UN agencies. This method of analysis divides the subject matter into four distinct categories: environmental, economic, social, and project themes. Arguably, environmental issues extend beyond the domain of natural disasters, encompassing a multitude of factors that directly affect human well-being. Nevertheless, despite the implementation of measures at the national and international levels, several issues related to climate change require immediate attention. Those residing in the Global South are confronted with challenges associated with climate change, and the situation of these individuals must be addressed in ways that will affect their daily lives.

To address climate change, it is important to consider the factors that have led to the expansion of the Millennium Development Goals (MDGs) to 17 SDGs. The SDGs are designed to facilitate dialogue between developed and developing countries globally. They address fundamental issues related to the creation of a just and sustainable world, including poverty, hunger, food security, and human rights. Without understanding the underlying causes of global issues, identifying a clear path towards a sustainable society is challenging. The fundamental question guiding this debate is why the SDGs were launched as an initiative involving the Global North rather than focusing on measures for the Global South alone. The origins of global challenges, including environmental problems, are formulated within the context of the relationships between the Global North and South. Furthermore, a shared understanding is necessary among the international community regarding the entities responsible for addressing these issues and the manner in which the global community should progress towards implementing solutions.

Moreover, and this may be regarded as a critical perspective, an examination of contemporary global social trends reveals a transition from the term 'SDGs' to terms such as 'climate change' and 'green investment' (World Economic Forum, 2024). It is imperative to acknowledge and address environmental concerns and climate change. However, it is important to note that the measures being considered for global issues under the SDGs do not all involve a shift towards environmental issues. It is imperative for the international community to scrutinise the circumstances confronting those in the most precarious positions, and comprehend the true essence of the SDGs, climate change and vulnerability.

14.3 What Is the Meaning of Vulnerability and Poverty Related to Climate Change?

International organisations are broadening their approach to tackle climate change. They are beginning to consider the impact on people and society, as well as the natural environment. When examining vulnerability in the context of climate change, its interconnection with poverty and marginalisation in the Global South must be acknowledged. This section attempts to determine the meaning of vulnerability by concentrating on the interconnections between exclusion, vulnerability, climate change issues, and the negative vicious circle relating them.

From the perspective of climate change, the UNFCCC has identified eight key indicators of vulnerability to climate change, grouped under the following categories: (1) education level by age group, (2) primary occupation type, (3) household assets and facilities, (4) remoteness, (5) source of drinking water, (6) sanitation facilities, (7) dependency ratio, and (8) frequency of occurrence of extreme climate events (The Kingdom of Cambodia, 2022, p. 103). The analysis of these indicators reveals a picture of people in the

[5] This resolution was named 48/13.

Global South such as Cambodia who have become socially vulnerable because of the multifaceted nature of poverty within the country. These individuals are not solely those residing in impoverished rural communities; they are also those situated on the periphery of urban development, such as those who have been involuntarily displaced from urban areas and are confronted with the challenges of climate change.

Bhalla and Lapeyre (1994) argue that those who are subject to social exclusion in the context of globalisation are more likely to experience isolation and difficulties in obtaining positive social recognition. They are at risk of exclusion from multiple perspectives. In addition to economic, social, and geographic perspectives, they are also disadvantaged in symbolic contexts, a conceptualisation of social exclusion and poverty that encompasses a number of interrelated components (Bhalla & Lapeyre, 1994). The human rights of a target social group are violated in several ways, resulting in the individual or group being isolated from their social networks and protections. This phenomenon is typified by discriminatory behaviour and utterances of disregard for human rights. This exclusion is subsequently carried out by dominant groups that occupy or monopolise resources for their own benefit. It is therefore evident that the term 'poverty' in the developing countries is not synonymous with the extreme poverty defined by the World Bank.

To gain an accurate understanding of poverty and vulnerability in the Global South, a wide range of frameworks must be considered. These include economic poverty,[6] social poverty resulting from social and cultural factors such as education, health, gender norms, and political situations, as well as human deprivation situations based on the UNDP Human Development Index. Moreover, the analysis can be conducted in accordance with the social exclusion theory proposed by Percy-Smith (2000). In an analysis, the elements of social exclusion that render individuals vulnerable are facilitated by a combination of 'compounded disadvantages'. The aforementioned aspects can be classified into the following categories: economic, social, political, situational, personal, spatial, and collective (Percy-Smith, 2000). These structural elements are responsible for driving the poor into more vulnerable positions in modern times, thereby creating a poverty trap and a vulnerable class of people in the Global South. Closer examination of this issue reveals that climate change is not merely a matter of responding to the immediate consequences of natural disasters but encompasses a broad range of concerns. In light of the aforementioned considerations, an examination of the situation of the poor and vulnerable, who are particularly susceptible to the adverse effects of climate change, and the associated negative vicious circle is undertaken using Cambodia as a case study.

14.4 Cambodia

14.4.1 Cambodia and Climate Change

The aim of this case study is to examine impacts of rising temperatures on vulnerable households in one rural Cambodian village. Despite the recognition of the problem (IPCC, 2022; UNICEF, 2021; World Bank, 2024a) in the Mekong region, there is a lack of qualitative studies in Cambodia cases. Cambodia is situated in Southeast Asia and has a population of 16.5 million, 76% of whom reside in rural areas (World Bank, 2024b). However, the country is undergoing rapid urbanisation (World Bank, 2024b). This process has continued unabated, with the movement of people from rural to urban areas becoming a routine aspect of the social landscape. However, the Cambodian population remains heavily reliant on agriculture and fishing, which collectively account for 25% of the country's gross domestic product (GDP) (World Bank, 2024b). In 2011, the GDP experienced a growth rate of 7.1%, with estimates indicating a stabilisation of around 7%–8% in 2013 (Royal Government of Cambodia, 2013). This indicates a rapid growth trajectory over the past decade. Cambodia's primary economic

[6] According to the World Bank, the extreme poverty line is 3.0 USD per person per day (June 2025).

sectors are garments, tourism, construction, and agriculture. Notably, agriculture represents the dominant employment sector for the rural population, comprising 49% of the total labour force (World Bank, 2024b).

The consequences of climate change on the agricultural sector and other areas are undoubtedly substantial. Cambodia has been reported to be significantly affected by climate-related disasters, including floods, droughts, and heat waves (World Bank, 2023, p. 7). The impact of heat stress on urban environments is of particular concern in the context of climate change. Cambodia is among the countries most affected by climate change, with the highest temperatures recorded in Poipet, Siem Reap, Phnom Penh, and Battambang in 2020 (World Bank, 2023, p. 7).

To gain insight into the reality of climate change in Cambodia, poverty must be considered. This is because the areas identified by UNICEF and the IPCC are the most vulnerable to the effects of climate change, coinciding with locations where the poorest populations are concentrated. In this context, the term 'poverty' is not merely an indicator of economic hardship; rather, it denotes a state of vulnerability within a complex societal structure shaped by a multitude of interconnected economic, social, and political factors.

The poverty rate reached its highest point in 2007, followed by the implementation of national poverty-reduction strategies that exceeded 50% in rural areas. Since then, poverty has declined notably. According to Cambodian statistics for 2018, the poverty rate was 13.3% in the capital city, Phnom Penh, 11.5% in other cities, and 17% in rural areas. Despite the reduction in Cambodia's overall poverty rate, the prevalence of undernourishment remains a significant concern, with an estimated 15% of the population affected (The Kingdom of Cambodia, 2022; World Bank, 2023, p. 6). Moreover, the UNDP Human Development Index, which assesses longevity, health, access to knowledge, and adequate standard of living, ranks Cambodia 148th among 193 countries in 2022 (UNDP, 2024). This suggests that there are still unaccounted-for individuals left behind in poverty because of this lack of development.

According to the Ministry of Environment and National Council for Sustainable Development report, a study on vulnerability to climate change indicated that at least 33.5% of the population is highly vulnerable (The Kingdom of Cambodia, 2022, p. 103). In other words, to gain a full understanding of the reality of climate change in Cambodia, the demographic groups affected, their economic and social conditions, and the reasons for their vulnerability must be considered.

In Cambodia, where the effects of climate change are likely to become severe, a national plan has been established to address this development. The Cambodian Policy for Climate Change 2014–2023 (2013) has eight strategic objectives: (1) The promotion of climate resilience through the improvement of food, water, and energy security; (2) the reduction of sectoral, regional, gender vulnerability, and health risks of climate change impacts; (3) the assurance of climate resilience of critical ecosystems, biodiversity, protected areas, and cultural heritage sites; (4) the promotion of low-carbon planning and technologies to support sustainable development; (5) the enhancement of the capabilities, understanding, and awareness of climate change responses; (6) the advocation for adaptive social protection and participatory approaches in the mitigation of losses and damages resulting from climate change; (7) the reinforcement of institutions and coordination frameworks for national climate change responses; and (8) the reinforcement of collaboration and active participation in regional and global climate change processes (Royal Government of Cambodia, 2013, p. 17). In the future, it would be beneficial to make these policies more concrete and practical to facilitate their implementation. Furthermore, it is essential to find the most effective means of operationalising national policies in such a way that they have a substantial and practical impact on the lives of individuals at the community level.

14.4.2 Case Studies: Social Environment and Vulnerability

The case studies in this section show that rising temperatures can negatively affect the economic well-being of vulnerable households. For example, poor people in Cambodia face many challenges, including arguments caused by foreign investment over land, economic hardships, limited access to healthcare and information, poor education, and social isolation. These challenges are similar to those faced by people with climate change experience, as highlighted by UNICEF and UNFCCC. It is important to understand that events related to climate change have a greater effect on the daily lives of people living in poverty. This case study explores how rising temperatures and their associated factors affect vulnerable households.

The survey method employed was semi-structured interviews. Interviews were conducted with the village leader and other key people in the village, as well as a random sample of the village people. The interviews are conducted individually. Qualitative research was employed. The main survey period is from 2023 to 2024. The content of the questionnaires included an analysis of economic and social conditions, experiences of migrant work and family structure, changes in daily life and rising temperatures, and the physical effects (direct and indirect) of these. In addition, questions on living conditions included the presence or absence of chronic diseases such as food shortages, health, malnutrition, and diarrhoea, and the prevalence of dengue fever and malaria.

This case study examines the realities of poverty and climate change among residents of a forced relocation site in Kandal Province, situated approximately 40 km from Phnom Penh.[7] The analysis of those affected by climate change is often limited to natural disasters such as floods, rising water temperatures, and landslides. However, this approach does not fully capture the diverse impacts of climate change on the Global South. The effects of climate change are not limited to people directly affected by natural disasters, but also affect everyday life.

14.4.2.1 Conditions of Life in This Village

The poverty level in this village is extremely high, with some households engaging in activities such as begging, collecting plastic items for sale, and collecting other materials found on the streets for cash.[8] In addition to economic deprivation, the social environment is characterised by a lack of access to basic necessities, such as healthcare, food, and a stable living environment. By 2023, the village had 140 registered households. However, by February 2024, the number of households in the village had decreased to approximately 90,[9] because individuals had migrated from the village in pursuit of seasonal employment opportunities and daily labour, which provide them with a source of income.

The housing conditions observed in these settlements are characterised by inadequate ventilation and the absence of sanitation facilities in some households.[10] In the absence of an adequate water supply, the provision of sanitation is slow and deficient. The lack of a sewage system and inadequate water supply impede access to sanitation facilities, even when they are available.[11]

The village is in an area that has not been developed for a decade, and the working environment is not conducive to productivity. The residents do not own their agricultural land. A

[7] This is when people were forced to leave their homes in the Borei Keila district of Phnom Penh in 2012. In January 2012, the army and police, following orders from the government, violently confronted residents who were against the evictions (Cambodian Human Rights Centre/CCHR, 2012). The author did more research in June 2013, February 2015, February–March 2023, and February 2024. Land concessions are defined as contracts between the government and other entities that set out how a particular area of land will be managed.

[8] Interview conducted in 2016–2024 by the author.

[9] Interview conducted by the author on February 2024 from a survey of interviews with the leaders of this community.

[10] Despite the presence of sanitation facilities, using them is precluded by the lack of sewage systems and an inadequate water supply.

[11] Interviews conducted by the author on August 2023 and February 2024.

considerable proportion of the local population is engaged in unstable employment, resulting in prolonged periods of unemployment. Most inhabitants lack access to land for agricultural activities, making them dependent on external sources for sustenance. Moreover, the remoteness of the area, when considered alongside the lack of transportation options, presents a substantial challenge to securing employment in other regions. This situation compels a considerable number of residents, including children, to seek income opportunities outside the area, predominantly through migrant labour. Many residents in these communities receive financial assistance in the form of remittances from family members who have migrated for work, as well as from other sources. I identified households in which grandparents or other caregivers were responsible for the care of their grandchildren, while parents were engaged in migrant work. In one, the eldest daughter, who was a young teenager, ceased her education to assume responsibility for her younger sister and brother, while her parents were absent because of their employment as migrants.[12]

Many of those engaged in low-paid work in Phnom Penh return to their communities only on weekends or once every few months. Furthermore, the interviews indicated that children who remain in the community and in certain instances, men, traverse the area on foot to nearby tourist attractions, where they solicit alms from tourists or proceed to nearby temples to participate in food sharing. In the case of boys, some children remain near the temples to receive a portion of their food from the temple's contributions.

14.4.2.2 Relationship Between Water Access and Poverty

Based on interviews with village people, this case study explored the relationship among local communities, poverty, access to water, and heatstroke, especially during the dry season.[13] This area is located near the aforementioned region in Cambodia where a temperature increase has been recorded (World Bank, 2023, p. 7). For individuals who lack basic access to water, rising temperatures are perceived as a grave threat to life. The results of the interviews indicated an increase in the frequency of heatstroke among infants and young children in this settlement in recent years.[14] In this area, it is not unusual for mothers to be away for a long time because they are migrant workers. In these cases, infants are fed powdered milk, which requires water. However, the answers from the interviews indicated that not much water was available.

The following interview responses also provided insights into this issue. A 32-year-old mother (see Fig. 14.1) who has been experiencing rising temperatures year after year and who has been looking after her child's health reported that '*When it is hot, babies and young children have fever and are sometimes sick. There is not enough water or cooling, so they have to sleep outside*' (A,[15] female, age 32).[16]

The voices of older women caring for their grandchildren on behalf of the migrant parents expressed their concerns as follows: '*Grandchildren often have a fever due to heat stroke. We have to buy drinking water, so we cannot give them much when we are short of money*' (B, female, age 75).[17]

Another older woman said that the following:

> '*In recent years, fever in children and older people has become more common on high-temperature days. In addition, water consumption is increasing,*[18] *and one tank (1,000 L) is purchased.*[19] *This water is boiled before use. However, this water is for domestic use and cannot*

[12] Interview conducted by the author on February 2024.

[13] The dry season is from April to November.

[14] August 2023 and February 2024.

[15] To maintain confidentiality, the names of the individuals in question have been replaced with letters that do not correspond to the actual names.

[16] February 2024, in Kandal Province in Cambodia, interviewed by the author.

[17] February 2024, in Kandal Province in Cambodia, interviewed by the author.

[18] The village has no infrastructure and the only access to water is through a well. The water is muddy and cannot be used as domestic or drinking water.

[19] The price was 24,000 riel (approximately 6 USD) in February 2024. The water cannot be used as drinking water.

Fig. 14.1 A, Female, age 32, answering an interview, Photographed taken by author on Feb. 2024

be used for drinking. During the rainy season, the price changes, and 1000 L costs 20,000 Riel.[20] *The temperature is rising, it is hot, and people are drinking more water than before. We are in need of water. However, we cannot buy water without money. There is not enough water for drinking'.*

(C, female, age unknown, who takes care of grandchildren instead of the parents).[21]

According to data provided by the World Bank (2024a), there has been a marked increase in the average minimum temperature throughout Cambodia. Between 1971 and 2020, this average temperature rose by 0.29 °C per decade, while the average maximum temperature increased by 0.32 °C per decade over the same period (World Bank, 2024a, pp. 11–12). Cambodia is experiencing higher minimum and maximum temperatures and warmer apparent conditions owing to high atmospheric humidity. Consequently, the country is categorised as an extreme heat risk area, and there are concerns regarding the implications of future temperature increases on human health (World Bank, 2024a, pp. 14–15).

Rising temperatures have clearly affected the lives of the poor. Moreover, houses in the affected areas of Cambodia are made from cheap tin and scrap wood, which increases the temperature inside the houses even more. These homes are poorly ventilated, which exacerbates the situation. Recently, temperatures have increased, meaning that households without electricity are now trapped in hot houses, as the following interviewee notes:

'The temperature is obviously higher than before. It is hot. We get a fever. When the fever is high, it does not go down easily. In particular, children are getting fevers more often. Our home has a fan, but the power frequently goes out, so there is not much electricity available. Even when the fan is on, the outside air only rotates, and the warm air makes it get hotter. Water is necessary. The amount of water purchased is increasing every year. We pay 5 USD for a bottle of water every 10 days. The cost of water is contingent upon its quality. Prices fluctuate depending on whether the water in the purchase is intended for consumption or domestic (daily) use. Drinking water from wells is not potable, so I boil it and use it. There is no water. The soil is dried up. In the past, when there was no food, they used to take clams and fish for food from the neighbour's pond, but now there are no clams or fish because the soil has dried up. And the well is experiencing a depletion of its water supply, resulting in a cessation of flow. The area that previously served as a water source has been reclaimed, rendering it unfit for use as a water source. Electricity costs 40,000 riel

[20] This is approximately 5 USD (February 2024).

[21] February 2024, in Kandal Province in Cambodia, interviewed by the author.

[approximately 10 USD][22] *a month, and because I cannot pay 40,000 Riel, we have to borrow money from people in the same village to pay the bill. We don't have any land of our own. There is no land for farming. There is no water. .... My two grand-children recently had to drop out of school to earn money'.* (D, female, age 75, see Fig 14.2)[23]

As stated previously, mothers living in the community reported that their infants and young children frequently experienced fever and nausea. Furthermore, adult residents reported a persistent occurrence of headache on days with elevated temperatures. These symptoms were consistent with those associated with heatstroke. In other words, the adverse physical effects of daily high temperatures affect not only the children but also the adults who are economically disadvantaged and particularly vulnerable to the adverse effects of heat stress and its associated effects on their health and productivity. This is because they are often engaged in daily work that requires them to be outdoors, and they reside in overcrowded housing that lacks adequate ventilation and cooling facilities.[24] UNICEF has determined that Cambodia is significantly affected and susceptible to the adverse effects of climate change on children (UNICEF, 2021, p. 14).

In rural areas of the Global South, areas with inadequate infrastructure, or areas of poverty, saving lives is challenging, and there is significant inequality in life expectancy. In other words, climate change represents the disparity between the lives of haves and have-nots. People who have access to essential resources are able to cool their bodies and protect themselves. However, individuals who cannot afford to obtain water may not have the means to do so, putting their lives at risk.

IPCC data indicate a scarcity of water in developing countries (IPCC, 2022). The financial burden of water expenditure results in a reduction in income, which exacerbates poverty. Furthermore, it is crucial to acknowledge the stark contrast in infrastructure accessibility between urban and rural Cambodia, particularly regarding water resources. In Phnom Penh, 81.7% of the population uses a pipe-dwelling system for drinking water, whereas only 10.1% of the rural population employs tubed/piped wells or boreholes owing to inadequate infrastructure. Furthermore, 11.5% of the population in rural areas rely on water supernatants (e.g. ponds, rivers, and streams) (National Institute of Statistics, Ministry of Planning Cambodia Socio-Economic Survey 2019/20, 2020, p. 30). Approximately only 51% of the population residing in urban areas have access to safe drinking water (The Kingdom of Cambodia, 2022, p. 54).

Access to water is unequal in Cambodia, which can cause problems. As temperatures rise, people require water and other methods to keep themselves safe. Human health worsens without access to water. People must buy more water, which worsens the situation of the poor. Therefore, it is important to examine how poverty and water insufficiency are related to rising temperatures, which worsens poverty for the already poor (the poverty cycle).

In accordance with Percy-Smith's (2000) conceptualisation of the constituent elements of social exclusion, an examination of the community as a whole revealed that residents exhibited the characteristics of a negative social cycle of social vulnerability. The people in this study are clearly experiencing the physical effects of rising temperatures because of climate change in addition to the physical vulnerabilities that have arisen owing to difficulties in accessing water.

The vulnerable groups within this community are characterised by the same factors identified by the UNFCCC. Certain demographic groups are particularly susceptible. The author observed that poverty in Cambodia has a multifaceted and complex impact on society, affecting different social groups in diverse ways. The research questions addressed in this chapter demonstrate a clear link between climate change and vulnerability among the most marginalised and vulnerable groups with the physical effects of rising temperatures and water shortages being the most common causes. Notably, this issue is not solely a consequence of climate change but indicates

[22] February 2024.

[23] February 2024, in Kandal Province in Cambodia, interviewed by the author.

[24] February 2024, in Kandal Province in Cambodia, interviewed by the author.

Fig. 14.2 D, Female, age 75. Taking care of grandchildren instead of their parents. Photographed taken by author on Feb. 2024

a correlative problem, namely, the inability to purchase water due to economic poverty.

Consequently, focusing solely on climate change is insufficient in the context of climate change and the measures implemented to address it in developing countries. A multidisciplinary approach is required, encompassing not only environmental concerns but also the social and economic contexts, particularly infrastructure and the economic situation of the poor. Such multifaceted measures are required to ensure that vulnerable groups are not disadvantaged by climate change.

14.5 Conclusion

This chapter demonstrates that climate change can have a significant impact on the lives of people in Southeast Asia. In areas with inadequate infrastructure or high poverty levels, implementing effective life-saving measures is challenging, and leads to disproportionate impacts on the most vulnerable populations. Thus, the impacts of climate change represent a disparity between the lives of economically privileged and disadvantaged people. In this context, it is important to consider how developing countries can safeguard their resources and infrastructure effectively. This is because financial resources are required to procure water, and until such resources are accessible, the ability to protect lives is limited. One possible approach to achieving this is for future studies to consider the environment in terms of carbon neutrality and how it can be achieved.

Sustainable development related to climate change must be implemented to consider the lived realities of the local population in a specific local context. For policymakers, this study points to the need for socially inclusive climate change measures to be implemented, including those designed to benefit impoverished communities. In order to consider initiatives that address the current situation of vulnerable people, an analysis of the case studies of heat waves and access to water indicates the necessity of providing economic and social support. Such initiatives include access to water, health care, food security, housing, and transportation, which are essential for the well-being of the most vulnerable members of society.

Such an approach could directly protect the lives of the most vulnerable members of local communities who are often the poorest. Additionally, the implementation of socially inclusive climate change measures could reduce the increasing economic disparity in communities. No one should be denied a life of dignity in the face of climate change.

Note While this chapter was being written, an unprecedentedly large typhoon Yagi struck Sapa, Lao Cai, in northern Vietnam, one of the author's research sites. The storm caused damage to many people in impoverished ethnic minority communities. This chapter is dedicated to those who were affected by this disaster.

References

Bhalla, A. S., & Lapeyre, F. (1994). *Poverty and exclusion in a global world* (2nd ed.). Palgrave Macmillan.

The Cambodian Center for Human Rights (CCHR). (2012). CCHR Case Study Series, Volume 5. *The Continuing Borei Keila Tragedy - May 2012*. Case Study Fact Sheet. CCHR.

The Centre for Research on the Epidemiology of Disasters (CRED). (2024). *Disasters in Numbers, 2023*. CRED. https://files.emdat.be/reports/2023_EMDAT_report.pdf. Accessed March 4, 2024.

Human Rights Now (HRN). (2012). Fact finding report for Cambodia: In Cambodia, people are deprived of land. *Human Rights Now*. http://hrn.or.jp/eng/activity/HRN%20Cambodia%20Report%20on%20Land%20Rights%202012.pdf. Accessed August 8, 2024.

Intergovernmental Panel on Climate Change (IPCC). (2022). *Climate change 2022: Impacts, adaptation and vulnerability IPCC*. Intergovernmental Panel on Climate Change. IPCC.

International Institute for Environment and Development (IIED). (2022). *Climate change, migration and vulnerability to trafficking*. IIED.

International Organization for Migration (IOM). (2016). *The climate change-human trafficking nexus*. IOM.

International Organization for Migration (IOM). (2022). *Asia-Pacific migration data report 2021*. IOM.

The Kingdom of Cambodia, The National Council for Sustainable Development, The Ministry of Environment. (2022). *The Cambodia's Third National Communication, Submitted under the United Nations Framework Convention on Climate Change*. The General Directorate of Policy and Strategy, the Ministry of Environment/the National Council for Sustainable Development. https://unfccc.int/sites/default/files/resource/20220921_Third%20National%20Communication_Cambodia.pdf. Accessed June 13, 2024.

National Institute of Statistics, Ministry of Planning. (2020). *Report of Cambodia Socio-Economic Survey 2019/20, Ministry of Planning*. Accessed July 13, 2024.

Percy-Smith, J. (2000). Introduction: The contours of social exclusion. In J. Percy-Smith (Ed.), *Policy responses to social exclusion: Towards inclusion?* (pp. 1–21). Open University Press.

Royal Government of Cambodia. (2013). *Cambodia Climate Change Strategic Plan 2014–2023*. National Climate Change Committee. https://ncsd.moe.gov.kh/resources/document/cambodia-climate-change-strategic-plan-2014-2023-cccsp-2014-2023en. Accessed June 15, 2024.

Shimazaki, Y. (2021). *Human trafficking and the feminization of poverty*. Lexington Books.

The UN Refugee Agency (UNHCR). (2021). *The human right to a clean, healthy and sustainable environment*. A/HRC/RES/48/13. Accessed June 15, 2024.

United Nations Children's Fund (UNICEF). (2021). *The climate crisis is a child rights crisis: Introducing the children's climate risk index*. UNICEF. https://www.unicef.org/media/105376/file/UNICEF-climate-crisis-child-rights-crisis.pdf. Accessed May 6, 2024.

United Nations Development Programme (UNDP). (2024). *Human Development Reports. Cambodia*. https://hdr.undp.org/data-center/specific-country-data#/countries/KHM. Accessed August 15, 2024.

United Nations Framework Convention on Climate Change (UNFCCC). (2025). *Cambodia's Initial Biennial Transparency Report (BTR1)*. UNFCCC. https://unfccc.int/sites/default/files/resource/Cambodia%20BTR1%E2%80%8B.pdf. Accessed March 25, 2025.

United Nations Office on Drugs and Crime (UNODC). (2022, October). *Policy brief: Climate, crime and exploitation: The gendered links between climate-related risk, trafficking in persons and smuggling of migrants*. UNODC.

World Bank. (2023). *Country climate and development report*. The World Bank Group. documents1.worldbank.org. Accessed May 6, 2024.

World Bank. (2024a). *Climate risk country profile, Cambodia*. https://climateknowledgeportal.worldbank.org/sites/default/files/country-profiles/16814-WB_Cambodia%20Country%20Profile-WEB.pdf. Accessed June 15, 2024.

World Bank. (2024b). *Climate change overview: Country summary—Cambodia*. https://climateknowledgeportal.worldbank.org/country/Cambodia. World Bank. Accessed June 15, 2024.

World Economic Forum. (2024). *Why investing in nature makes ecological and economic sense*. https://www.weforum.org/agenda/2024/06/investing-in-nature-world-environment-day/. Accessed July 13, 2024.

15 Transition to a Carbon-Neutral and Sustainable Society

Ayu Washizu, Ken-Ichi Akao, and Yudi Liu

Abstract

Creating a carbon-neutral, sustainable society requires revolutionary social innovations. The previous chapters of this book have attempted to present what kind of social innovation is needed and what methodologies are necessary to create a carbon-neutral and sustainable society based on the specialized perspectives of different humanities and social science disciplines. The final chapter shows the social innovations necessary to realize a carbon-neutral society, which were concluded from discussions in diverse social science fields, and the transition stages (first to third steps) to achieve a carbon-neutral society. However, research into the various synergies and trade-off effects that occur during a simultaneous resolution of multiple issues is a key future task.

Keywords

Carbon-neutral society · Social innovation · Sustainable society · Transition stages

A. Washizu (✉) · K.-I. Akao · Y. Liu
Waseda University, Tokyo, Japan
e-mail: washizu@waseda.jp

K.-I. Akao
e-mail: akao@waseda.jp

Y. Liu
e-mail: yudiliu@aoni.waseda.jp

The ideal image of a carbon-neutral and sustainable society is actively discussed in various fields. However, it is insufficient to merely envision an ideal image, and discussions on how to realize it are indispensable. Therefore, a methodology for approaching the realization of the ideal image is needed. Central to this methodology is how revolutionary social innovations can be triggered.

Figure 15.1 illustrates the process by which social innovation changes society (Geels, 2019; Loorbach, 2010). Social innovations are initially implemented in demonstration fields but gradually spread throughout society by involving and expanding stakeholders. In addition, it is usually difficult for social innovations to be accepted by existing regimes (policies, institutions, market environment, corporate and citizen mindsets, scientific and technological trends, etc.) when they are first created. Therefore, it is essential to build a new regime suitable for new innovations.

The previous chapters of this book have attempted to present methodologies for the fruition of such social innovations in the formation of a carbon-neutral and sustainable society based on the specialized perspectives of different humanities and social science disciplines. In this chapter, we summarize the most important social innovations for achieving a carbon-neutral society and key points during this transition.

K. Akao and A. Washizu (eds.), *Climate Change Issues and Social Sciences*,
Sustainable Development Goals Series, https://doi.org/10.1007/978-981-96-8951-4_15

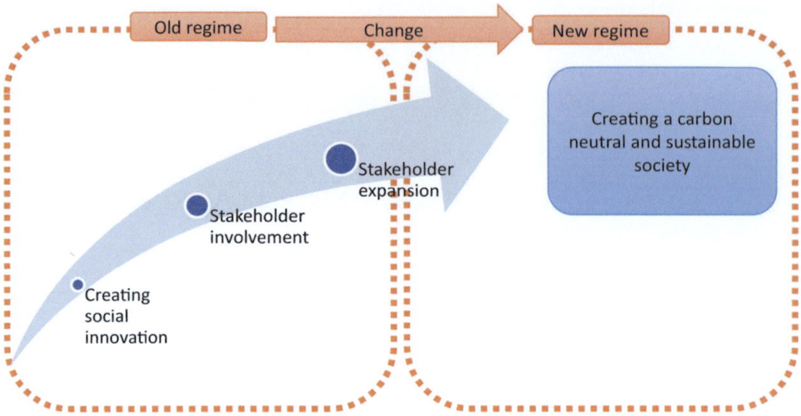

Fig. 15.1 Process diagram of how social innovation changes society

The innovation presented in Chap. 1 focuses on the power of stories and narratives. It is about clarifying the cognitive history of the environment; that is, how Japanese people in the past imagined the environment surrounding human beings. The cultural codes of such imagination are found in the accumulation of historical stories passed down in minimal communities (local communities). It is necessary to explore these to continue to educate about the restoration and power of imagination to remake and change the environment. From a humanistic perspective, the transition to a carbon-neutral society will be accomplished as a long-term, sustained change over decades or centuries. The unique narratives of minimal communities, which were forged through mutual influence and the incorporation of myths, maxims, and omens since ancient and medieval times, gradually lost their efficacy with the birth of modern society and the modern state and were replaced by the illusion of a standardized nation-state. The restoration of the minimalist local imagination that is being lost will not be solved by a simple regression. We must incrementally and gradually foster a culture that can share the imagination for nature that minimal communities once had by relativizing anthropocentrism and going beyond the framework of humans.

The innovation outlined in Chap. 2 involves creating a mechanism that bridges education and the policymaking process to strengthen people's intergenerational altruism and establish an ethic of intergenerational equity supported by the majority. As a step toward a carbon-neutral society, the strengthening of intergenerational altruism will proceed in three stages: (1) many people come to understand the long-term consequences of climate change, including the circumstances faced by distant future generations; (2) on the premise that the expected circumstances of future generations are tragic, people's intergenerational altruism will be enhanced based on the empathy-altruism hypothesis; and (3) strong decarbonization policies, grounded in intergenerational altruism, will be supported and implemented. Regarding an ethic of intergenerational equity, its establishment will proceed as follows: (1) it will become widely recognized that economic policies can and should be proposed and implemented not based solely on economic principles (e.g., efficiency) or politically expedient solutions to equity, but on moral principles supported by the majority; (2) strong sustainability, as the ethical foundation behind the 1.5-degree Celsius target, will be justified through logic widely accepted by the public, such as a Rawlsian social contract; and (3) strong decarbonization policies, rooted in strong sustainability, will be supported and implemented. Through these processes, it is expected that the following feedback effects will occur: altruism will strengthen support for moral principles, and the spread of morality will generate a sense of fulfillment or pleasure in adhering to it, thereby reinforcing support for policies based on utilitarian reasoning. However, such social innovation will require time since it engages individuals' inner lives. A more practical and faster-acting transition scenario toward a carbon-neutral society, already becoming a reality, includes the following steps: (1) each country implements decarbonization policies to achieve the goals agreed upon under the United Nations Framework Convention on Climate Change and the Paris Agreement, with developing countries adopting these measures as the most viable means of securing funding and technology transfers from developed countries; (2) R&D investments related to decarbonization become attractive business opportunities, leading to a concentration of resources in this area; and (3) consequently, the cost of carbon dioxide removal decreases significantly, strong decarbonization policies gain widespread support, and countries are able to fulfill their commitments under the Paris Agreement.

The innovation derived from Chap. 3 involves fundamentally redefining the concept of moral responsibility attributed to individuals. Specifically, it means shifting from the "agency-based causal responsibility" model—which evaluates individual actions on an individual basis—to a "non-agential shared responsibility" model, which locates individual actions within a social system, recognizing that the number and scale of these actions affect their moral evaluation. Under this new system, our lifestyles and consumption behaviors, such as meat consumption and driving, must be modified in a more sustainable direction. The transition toward a carbon-neutral society can be outlined in five steps: (1) strengthening education and public awareness. To encourage changes in individual behavior, create opportunities for people to reflect on the ethical implications of their everyday consumption and transportation choices, while promoting understanding of the "non-agential shared responsibility" model and raising awareness of its necessity; (2) research and dissemination of ethical frameworks. Academic institutions, policymakers, and civil society should work together to advance research and practical application of ethical frameworks that support the "non-agential shared responsibility" model; (3) transforming cultural and social values. Recognizing that our interdependence and systemic interconnectedness extend beyond the individual level to the global scale fosters global citizenship and moves away from conventional anthropocentric values and social models. This shift aims to heighten awareness and effect change in people's mindsets; (4) policy support and infrastructure development. To make sustainable lifestyles more accessible, strengthen social and economic infrastructures through policy measures—for example, by expanding the supply of plant-based foods and lowering their costs—so that consumers can more easily make environmentally friendly choices; and (5) introducing legal and regulatory measures. Implement legal and regulatory measures that reflect the shared responsibility model. This may involve legislation that holds companies and industries collectively responsible for damages, as well as policies ensuring that responsibility is fairly distributed among individuals and groups.

The innovations outlined in Chap. 4 highlight transportation policy. The history of the coupling of urban transportation and land use are examined, and the need to re-examine a policy legacy that was not always intentionally designed is highlighted. In the past, Japan's urban transportation systems and policies were implemented haphazardly, with coexisting policies encouraging low-carbon railways and carbon-emitting roads for cars following market dynamics. As path dependence was formed, it is important to shift paths to redesign both policies from a carbon-neutral perspective. The steps for transitioning to a carbon-neutral society are as follows. The first step is to recognize path dependence. This will reaffirm the history of the system and promote a structural understanding of the historical power relations and contingencies in which policies were formed, rather than conducting conventional descriptive research. The second step is to reevaluate the process of path-dependent policy practice. For example, one will discuss whether the public investment framework created in the past, which rushed to build roads, is appropriate in the new era of carbon neutrality. The third step is to explore reforms that can address its shortcomings. Support is given to public transportation in remote suburbs and regional cities to overcome the substantial carbon emissions from automobiles and associated low-density urban areas.

The innovation outlined in Chap. 5 is to build a smart society using renewable energy. Here, a smart society is a society in which decarbonization and improvement in people's well-being are compatible by implementing detailed management in each field through the use of advanced information and communication technology. In a large-scale centralized society in the past, efficiency could only be achieved by pursuing economies of scale. However, by using smart society technology, fine optimization through detailed management in each field becomes possible. By accumulating each of these optimizations, it becomes possible to transition to a society in which efficiency can be pursued even in a small-scale decentralized manner. The first step to transitioning to a carbon-neutral society is to develop niche innovations in technologies that utilize renewable energy and smart society technologies for meticulous management in various fields. We will make maximum use of advanced information and telecommunications technology to create not only the latest technological innovations but also to improve existing technologies, combine several existing technologies in new ways to create new value, and develop aging technologies to carefully maintain the accumulation of past achievements. The second step is to reform the social regime so that the developed technologies for renewable energy utilization and smart society technologies can be fully utilized. In addition to reforming social systems in line with new technologies or new ways of using existing technologies, it is desirable to monetize and visualize the environmental and social value created by new technologies so that they will be actively adopted based on economic principles. Here, voluntary credits are a concrete system for monetizing environmental and social values, and it is desirable to develop such a system in the future. In the third step of the transition, the scope of fine-grained optimization by smart society technologies will be extended from decarbonization issues to various other social issues to complete the transition to a small-scale decentralized society that is both decarbonized and efficient in improving people's well-being.

Chapter 6 discusses the spread of renewable energy. The goal is to realize the local production and consumption of energy and build a society that uses local renewable energy and satisfies energy justice. Under such a society, the development of local industries can be expected, and zero emission household energy will be possible. Here, energy justice considerations are important. For example, subsidies for rooftop PV can be available to the rich who have their detached houses, but not to the poor who rent an apartment, which may violate energy justice. The steps for transitioning to a carbon-neutral society are as follows. The first step is for local governments to designate decarbonization promotion areas and build large-scale renewable energy power plants. The second step is to install small-scale renewable energy power generation facilities, such as rooftop solar power generation facilities, and enable local energy management companies to control them. Additionally, the third step is to improve the capabilities of local energy management companies and build local microgrids.

Nuclear power is an option to realize a carbon-neutral society. In this case, the innovation attributed to Chap. 7 is to "prepare a mechanism for those who have the same risk to pay each other" before the accident in the nuclear damage compensation mechanism. The first step in the transition to such a system is to "consider how to reduce high risks before they occur", and the second step is to "create a system to reduce risks limited to those with the same risk". The third step requires "establishing a risk reduction mechanism and reviewing it periodically".

The innovation resulting from Chap. 8 is to shape public opinion in favor of policy changes to achieve a carbon-neutral society and to develop methodologies to do so. To consider how to increase the environmental orientation of public opinion, it is important to understand the factors that shape and structure current environmental orientation. Research results showed that, even after controlling age and education level, people who obtain information from online media are less environmentally conscious than those who obtain information from traditional media (newspapers and television). Based on this, the transition procedure to a carbon-neutral society should take the following two steps. The first step will focus on more traditional media coverage of climate change, with a greater amount of alert and investigative reporting. To this end, it is necessary to consider designing incentives for news organizations to promote such reporting. In the second step, a media system will be developed to increase the amount of climate change coverage circulating online media and to promote incidental information contact for voters who are mainly interested in online media. This is because voters who mainly encounter online media will preferentially access information that interests them; therefore, in a situation where the amount of information on climate change is limited, it is assumed that they will have fewer opportunities to access it.

The social innovation suggested by Chap. 9 is the adoption of disclosure legislation for sustainability information, including climate change. In Japan, the Cabinet Office Ordinance on Disclosure of Corporate Information was revised in 2023, and the disclosure of sustainability information, including climate change, in securities reports was legalized. Thus, this innovation can be said to have been achieved by these developments. However, in the future, it is hoped that disclosure content will be strengthened stepwise and the scope of application will be expanded. Regarding greenhouse gas (GHG) emissions, which are currently a symbolic disclosure item, the process of strengthening disclosure content will be as follows. In the first step, Scope 3 disclosure is enforced (already envisaged) by law for large, listed companies. In the second step, this is extended to all listed companies. Finally, in the third step, this is extended to large unlisted companies.

For multinational corporations (MNCs) to formulate actions to achieve a low-carbon society, consideration should be given to the trade-offs/synergies between low-carbon challenges and other important sustainability issues in the countries/regions in which they operate. To achieve this, MNCs need to innovate their business models. This is the social innovation suggested by Chap. 10. New business models must incorporate issues such as choosing the best mix of management resources and country/region-specific resources, implementing mechanisms to prevent negative spillovers to other countries/regions, and reorganizing location portfolios and supply chains. The first step in transitioning to such a business model is for MNCs to visualize GHG emissions in the destination country and develop reduction plans, such as improving energy efficiency and introducing renewable energy. The second step is to visualize trade-offs/synergies with other SDGs and international spillovers and to support the selection of means to implement the above mentioned reduction plans. The third step is to introduce a circular economy model at the local level, build partnerships with local communities to efficiently utilize local resources, and maximize synergies between SDGs.

Company-level efforts alone are not likely to realize significant change to mitigate sustainability issues. Therefore, the innovation resulting from Chap. 11 is to strengthen corporate sustainability efforts through capital market pressures and mandatory corporate disclosure requirements. The process of strengthening companies' commitment to sustainability issues is as follows. In the first step, regulators implement policies that require companies in their jurisdiction to include sustainability performance in their disclosure practices. In the second step, public investment funds consider companies' sustainability performance when selecting companies to invest in. Doing so will put pressure on companies to improve their sustainability performance in the third step, as some funds are already doing.

The social innovation for a carbon-neutral society that Chap. 12 suggests is the realization of green logistics in the transportation sector. To achieve this, it is necessary to combine truck transportation, which accounts for most long-distance transportation, with ferry and rail transportation, which have a lower CO_2 impact. The first step of such change is the substitution of transport energy (effective use of renewable energy, utilization of energy-saving energy management systems, EVs, etc.). The second step is logistics digital transformation (data sharing and visualization between companies, AI logistics, etc.). The third step is logistics structural reform (modal shift, physical internet, joint delivery, relay point delivery, etc.). The fourth step is robotization (autonomous driving, truck platooning, automated delivery robots, drone delivery, etc.). In addition, the use of carbon offsets will be important.

The key innovation presented in Chap. 13 is the revitalization of community-based natural resource management. Traditional systems have proven effective in managing and conserving natural resources in many countries. These systems are characterized by their strong alignment with local environmental conditions and their sustainability. The first step toward the revitalization of traditional systems is to promote sustainable agricultural practices and maintain green spaces as carbon sinks. The second step is to promote community-based natural resource management, environmentally friendly agricultural practices, and reduce pressure to urbanize agricultural land. Additionally, the third step is to use this sustainable development model as the basis for future efforts to balance economic growth with ecological responsibility.

Since the direct impacts of climate change are strongly felt by the most socially vulnerable people, the social innovation that Chap. 14 focuses on is to elucidate the reality of the impacts on the poor in developing countries and to build a sustainable society based on social inclusion. The first step in the transition to such a society in developing countries is to understand the specific impacts faced by vulnerable groups who are strongly affected by climate change. The second step is to understand the nature and causes of their "vulnerability" in relation to climate change. The third step is to explore locally appropriate methods and means from the perspective of social inclusion based on the above.

The chapters above show that social innovations are necessary to realize a carbon-neutral society, which were concluded from discussions in diverse social science fields (e.g., literature, environmental economics, environmental ethics, urban transportation planning, input–output theory, administrative law, election theory, corporate law, international management, distribution theory, development studies). These discussions explain their respective the transition stages (first to third steps) to ensure a carbon-neutral society. Naturally, there are various transition methodologies for various objectives. We can choose or combine appropriate methods from a rich and diverse array of methodologies to address the challenges we face in achieving a carbon-neutral society. However, in many cases, a simultaneous resolution of multiple issues is required, where various synergies and trade-off effects may occur among objectives or methods. Synergies should preferably be maximized, while trade-off effects should be minimized. We would like to discuss this measure in a future research project.

References

Geels, F. W. (2019). Socio-technical transitions to sustainability: A review of criticisms and elaborations of the Multi-Level Perspective. *Current Opinion in Environmental Sustainability, 39*, 187–201.

Loorbach, D. (2010). Transition management for sustainable development: A prescriptive complexity-based governance framework. *Governance: An International Journal of Policy, Administration, and Institutions, 23*(1), 161–183.

MIX
Papier aus verantwortungsvollen Quellen
Paper from responsible sources
FSC® C105338

If you have any concerns about our products,
you can contact us on
ProductSafety@springernature.com

In case Publisher is established outside the EU,
the EU authorized representative is:
Springer Nature Customer Service Center GmbH
Europaplatz 3, 69115 Heidelberg, Germany

Printed by Libri Plureos GmbH
in Hamburg, Germany